Cytoskeleton

Alexander D. Bershadsky
and
Juri M. Vasiliev

Cancer Research Center
Moscow State University
Moscow, USSR

Plenum Press · New York and London

Library of Congress Cataloging in Publication Data

Bershadsky, Alexander D.
  Cytoskeleton.

  (Cellular organelles)
  Bibliography: p.
  Includes index.
  1. Cytoskeleton. I. Vasil'ev, IU. M. (IUrii Markovich) II. Series. [DNLM: 1.
Cytoskeleton. QH 603.C96 B535c]
  QH603.C96B47   1988   574.87′2   87-38489
  ISBN 0-306-42508-4

This volume is published under an agreement with the
Copyright Agency of the USSR (VAAP)

© 1988 Plenum Press, New York
A Division of Plenum Publishing Corporation
233 Spring Street, New York, N.Y. 10013

Printed in the United States of America

Preface

This book, like other monographs of the *Cellular Organelles* series, is not a comprehensive review, but an introduction to the study of cytoskeleton. Accordingly, we describe only the main facts and concepts related to cytoskeleton. Needless to say, selection and interpretation was influenced by the personal interests and opinions of the authors, although we attempted to be as fair as possible. We wished to familiarize the reader not only with well-established facts, but with current unsolved problems. Therefore, the words "possibly," "maybe," "not known," and "not clear" are much more frequent in this text than in many others.

In accordance with the style of the series, relatively short lists of additional readings are given at the end of each chapter; these lists contain mostly the recent reviews and a few original papers describing certain phenomena in detail. Few references are cited in the text; these citations are given to help the reader find the source of certain new data and theories, which are not discussed at length in the reviews. In contrast, many well-established facts and widely known theories are not cited.

The manuscript was read in whole by A. A. Neyfakh, Jr., and in part by V. I. Gelfand, D. Murphy, E. Nadezhdina, and S. M. Troyanovsky; we thank them for numerous helpful suggestions. We appreciate the many friendly discussions we had on cytoskeleton with D. Murphy during the preparation of this book. For many years the authors have worked in close cooperation and friendship with I. M. Gelfand; his profound influence on our scientific life is deeply appreciated. We are very grateful to I. S. Tint and to T. M. Svitkina for their help in printing the numerous photographs. M. L. Anfimova and V. I. Guelstein generously helped prepare the lists of references. We owe a special debt of gratitude to E. N. Vasilieva and S. A. Greenberg, who corrected the typescript and typed numerous letters. We are deeply grateful to the many scientists who kindly provided us with figures; their names are given in the appropriate legends.

Alexander D. Bershadsky
Juri M. Vasiliev

Moscow

Cytoskeleton

CELLULAR ORGANELLES

Series Editor: PHILIP SIEKEVITZ
Rockefeller University, New York, New York

CHLOROPLASTS
J. Kenneth Hoober

CYTOSKELETON
Alexander D. Bershadsky and Juri M. Vasiliev

MITOCHONDRIA
Alexander Tzagoloff

Contents

Chapter 2

Systems of Microtubules

Chapter 3

Systems of Intermediate Filaments

Chapter 4

Unconventional Fibrillar Structures in the Cytoplasm

II. General Organization and Function of Cytoskeleton

Chapter 7

Reorganization of Cytoskeleton: Morphogenesis and Locomotion of Pseudopod-Forming Cells

Chapter 8

Reorganization of Cytoskeleton: Cell Division

Chapter 9

Neoplastic Transformations: Possible Unexplored Functions of Cytoskeleton

Introduction

Cytoskeleton is the system of fibrillar structures in the cytoplasm of eukaryotic cells (Fig. I.1). The three main types of fibrils forming these structures are microtubules, actin filaments, and intermediate filaments (Fig. I.2).

Cytoskeleton undoubtedly deserves its name: it forms the supporting internal framework of the cells. This, however, is only one of its many functions. Cytoskeletal fibers carry out all the types of movement performed by and taking place within eukaryotic cells, e.g., muscle contraction, cilia and flagella beating, mitotic movements of chromosomes, and many others. Therefore, the system of these fibers could as reasonably be termed cytomusculature. However, based on the comparison of these intracellular structures with the organ systems of the whole animal, the terms are somewhat misleading. The cytoskeletal structures are unique because they can be highly dynamic. Many of these structures, like mitotic spindle or the actin cortex in the extended pseudopod, can be fully formed *de novo* in a short time and then disappear again after performing their function. Imagine an animal forming a new limb with complex skeleton and musculature in several minutes and then dissolving it at a similar rate! Polymerization of fibers from protein subunits and the reverse process, depolymerization, are the main molecular reactions responsible for rapid reorganization of cytoskeletal structures.

Cytoskeleton is also unique because it is the only structure in the cell that directly contacts all the other cellular organelles from the nucleus to the external membrane. Cytoskeletal fibers can be reversibly attached to many types of other cellular structures. Thus, cytoskeleton can be regarded as a dynamic cytoplasmic matrix, surrounding and embedding other intracellular structures. This matrix may determine the position and movements of other cellular structures as well as the shape of the whole cell. This matrix may also be actively involved in the control of metabolic activity of other organelles and of the whole cell.

Historically, the concept of cytoskeleton is both very old and very young. It is very young because in its modern form the cytoskeleton concept has been developed only during the last two decades. The concept is very old because the first hypotheses about the presence of dynamic fibrillar structures in the cytoplasm were developed more than a century ago, when the first fibrillar

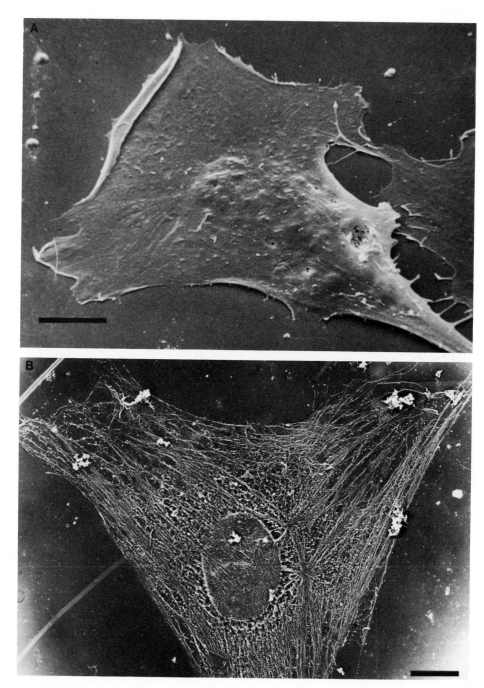

FIG. I.1. Cytoskeleton of cultured fibroblast. **(A)** Intact cultured mouse fibroblast spread on the substrate. Scanning electron micrograph. Scale bar: 10 μm. **(B)** "Model" of cultured fibroblast demembranated by detergent. Cytoskeletal structures filling the cytoplasm and surrounding the nucleus. Electron micrograph of the platinum replica. Scale bar: 10 μm. **(C)** High magnification of the perinuclear area of the model shown in B. Actin filament bundles connected into "stars" and

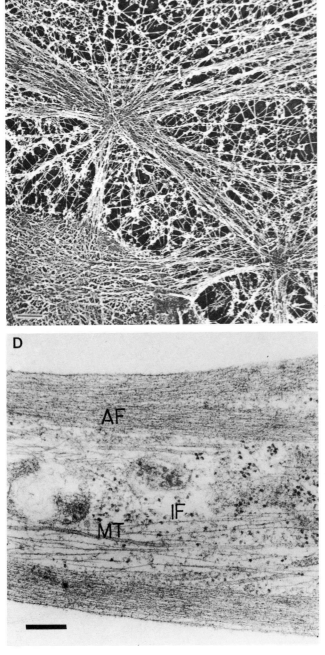

linked to the filaments surrounding the nucleus (left lower corner). Scale bar: 1 μm. (From Svitkina *et al.*, 1984.) **(D)** Three types of cytoskeletal structures in the cytoplasm of fibroblast. Electron micrograph of the peripheral part of the cell. Actin filament (AF) forming the cortical bundles under the cell membrane. Microtubules (MT) and intermediate filaments (IF). Scale bar: 0.2 μm. (From Vasiliev and Gelfand, 1981.)

cytoskeletal structure, mitotic spindle, was discovered. Walter Flemming in 1879 first described the stages of mitosis and introduced this term (the Greek "mitos" means thread). Probably it is not a coincidence that Flemming also proposed the general theory of cell structure, which postulated that protoplasm consists of two components: the fibrillar network and interfibrillar substance. Many other speculations of this type and descriptions of fibrillar or reticular structures in the cytoplasm were published during the last decades of the nineteenth century. However, gradually it became clear that stained and fixed preparations seen under the light microscope can contain many artifactual structures. Therefore, efforts to reveal any fibrillar structures in the cytoplasm of nonmitotic cells without flagella or myofibrils were virtually discontinued for more than half a century.

A number of important observations of living cells were made during this period; these suggested that cytoplasm is a structured body.

For instance, in a number of investigations, N. Koltzoff had found that certain cell parts, e.g., elongated processes of various sperm cells, resist the cell-rounding effect of incubation in medium with low osmotic pressure: he

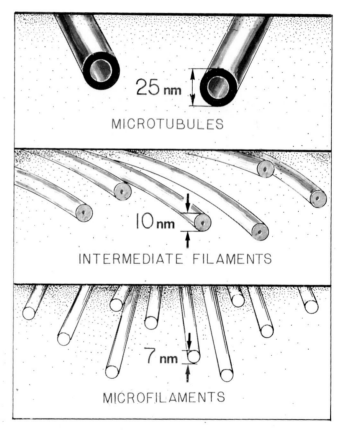

FIG. I.2. Three main types of cytoskeletal fibrils. Diameter of the fibrils is given according to measurements made in the electron micrographs of cell sections.

suggested that these cells contain elastic threads counteracting the effect of pressure. In other experiments it was shown that when the nucleus of a living cell is first displaced by a needle inserted into the cytoplasm and then released, it promptly returns to the original position. Similarly, an iron particle inserted into the cytoplasm and moved by an applied magnetic field recoiled when the field was turned off. The experiments of many authors [A. Heilbronn, L. V. Heilbrunn, G. W. Scarth, F. H. C. Crick, A. F. W. Hughes, and others (see references in Porter, 1984)] indicated that cytoplasm has the properties of elastic gel; the specific nature of the elements responsible for this elasticity is still not clear. Of great importance were the embryological experiments of E. G. Conklin and others which showed that nonfertilized and fertilized egg cells of many animals have a polar organization, which determines the orientation of future divisions and localization of organs of developing embryos. Usually the position of poles can be determined from the localization of pigment granules and other inclusions in the egg cytoplasm. However, displacement of these inclusions by centrifugation has no effect on the plane of divisions and localization of future organs. These data suggested that polar organization essential for future development is associated with some persistent network present in the egg cytoplasm. At present we can guess that this hypothetical network is identical with some cytoskeletal structures, but the concrete nature of these structures is still not known.

These interesting findings, made during the first decades of this century, were not studied further and were virtually forgotten for a long time.

Discovery of cytoskeletal structures was made by other groups of investigators. These were the studies of muscle motility, performed between 1930 and 1960 by A. Szent-Györgyi, V. A. Engelhardt, and others. Major proteins responsible for muscle contraction were discovered, and dependence of the contraction on ATP hydrolysis and ATPase activity of myosin were proven. Next, the main fibrillar structures of the muscle, actin and myosin filaments, were described in early electron microscopic studies, and it was suggested by H. E. Huxley and J. Hanson (1954) and independently by A. F. Huxley and R. Niedergerke (1954) that contraction is due to ATP-dependent mutual sliding of these two types of fibers. This concept of sliding fibers, developed in the late 1950s, is now the cornerstone of all studies of molecular mechanisms of cytoskeletal motility.

The other main type of cytoskeletal fibrils, microtubules, was revealed in electron microscopic studies of cilia and flagella. One type of intermediate filament, cytokeratin filaments associated with desmosomes, was also known at that time under the name of tonofibrils. Thus, in the late 1950s, all the main types of cytoskeletal fibers had already been described, but they were still regarded as specialized organelles characteristic only of certain special cell types. At that time the "typical" nonmitotic cell was usually regarded as a bag filled with some nonstructured fluid, cytosol, in which the nucleus, membranous organelles, and ribosomes were freely floating. Cytoskeletal fibrils were still absent from these generalized schemes of cell structures.

Universality of cytoskeleton structures was proven in the 1960s, when advanced electron microscopic studies using glutaraldehyde fixation and epoxy resin embedment were performed by K. Porter and collaborators, as

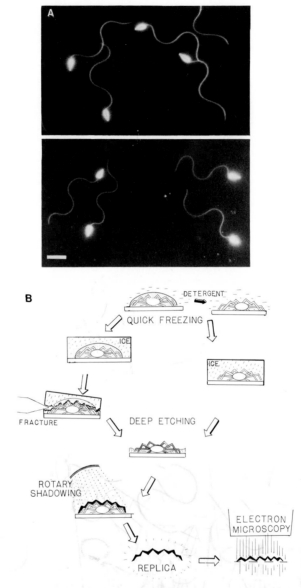

FIG. I.3. **(A)** Dark-field photographs of *Tripneustes* sperm swimming at the bottom surface of an observation dish. Top: Live sperm. Bottom: reactivated demembranated sperm. Scale bar: 10 μm. (From Gibbons, 1982. Courtesy of B. H. Gibbons.) **(B)** Principles of preparation of replicas of cytoskeletal structures for electron microscopy. The whole cells or detergent-treated cytoskeleton preparations from these cells are frozen and freeze-etched, and then the etched surfaces are shadowed with platinum and carbon and the organic structures dissolved with strong acids. **(C)** Principles of immunomorphological examination of cytoskeletal structures. Specific antibody to cytoskeletal protein is attached to the structure containing this protein. Then a second antibody to the first immunoglobulin is attached. The second antibody carries the fluorescent dye molecule; the preparation is then examined by fluorescence microscopy (indirect immunofluorescence method). Alternatively, the second antibody may be bound to colloid gold particles or other labels

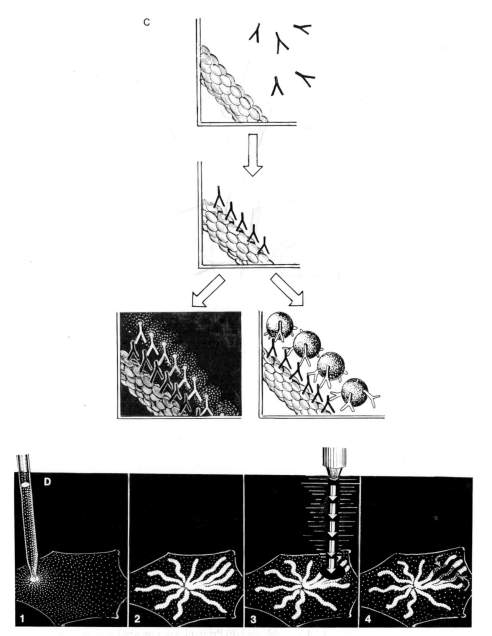

visible at electron microscopy (immunoelectron microscopy). **(D)** Principle of determination of the dynamics of cytoskeletal structures by injection of labeled proteins and photobleaching. From left to right: (1) Cytoskeletal protein labeled by the fluorescent dye is injected into the cell cytoplasm through the micropipette. (2) Some time later labeled protein is incorporated into cytoskeletal structures. (3) Fluorescence in some part of the structure is bleached by the short, strong pulse of the laser beam. (4) Fluorescence in the bleached spot is gradually restored owing to incorporation of nonbleached molecules of labeled protein; degree and rate of restoration can be measured with the aid of a microfluorimeter.

well as by many other investigators. They had shown that, with rare exceptions, all three morphological types of cytoskeletal fibers are present in each eukaryotic cell.

It was also shown that most cytoskeletal structures are retained in the cells from which the membranes and soluble components had been extracted by solutions containing glycerol or nonionic detergents. These so-called "cell models" (Fig. I.3A), when adequately prepared and supplied with ATP, display many motility phenomena, e.g., flagella beating, contraction, mitotic movements of chromosomes. H. Hoffman-Berling was the first investigator to prepare the glycerol-extracted moving models of various cells. The use of cell models facilitated the study of cytoskeleton morphology and function. A combination of extraction procedures and more sophisticated preparation of electron microscopic replicas provided the new method of morphological investigation (Figs. I.1 and I.3B) of the three-dimensional architecture of cytoskeleton. Universality of cytoskeleton was finally established when similar types of cytoskeletal proteins were isolated from various cells, and it was shown that each type of cytoskeletal fiber can be polymerized *in vitro* from corresponding protein subunits: microfilaments from actin, microtubules from tubulin, and intermediate filaments from several types of related proteins, such as cytokeratins and vimentin. Many additional proteins associated with the side surfaces and with the ends of fibrils have been isolated. At present, isolation of cytoskeletal proteins and analysis of their structure is an active area of research. In addition, isolation of purified proteins is the basis of most modern methods used in the study of cytoskeleton. In particular, the processes of formation and function of cytoskeletal structures in the cell are modeled by *in vitro* experiments in which interactions of purified proteins are studied. Polyclonal and monoclonal antibodies against these proteins are used to analyze the molecular composition of various cytoskeletal structures.

In these studies (Fig. I.3C) morphological preparations are first incubated with the antibody against certain cytoskeletal proteins, and then distribution of the first antibody is made visible by the second antibody linked to some label revealed by either fluorescence microscopy (indirect immunofluorescence method) or electron microscopy (indirect immunogold or immunoferritin methods). These techniques, especially indirect immunofluorescence, first used in the study of cytoskeleton by E. Lazarides and K. Weber in 1974, are now the standard methods of investigation of molecular morphology of various cells. Until recently, these "molecular images" remained static, since only nonliving fixed and extracted cells could be examined. However, with the development of special video cameras intensifying the fluorescent images, it became possible to study the molecular dynamics of cytoskeleton in living cells. For this purpose, individual fluorochrome-labeled proteins are injected into living cells; the dynamics of incorporation of the label into various cytoskeletal structures is then followed with the aid of video cameras. Mobility of injected proteins incorporated into various structures can be assessed quantitatively by a special method, so-called photobleaching (Fig. I.3D). Antibodies blocking cytoskeletal proteins and nonfunctional analogs of these proteins injected into the living cell are used to reveal functions of these proteins in various cellular processes. Another older method widely used for the same purpose is incubation of cells with drugs that specifically bind certain cytoskeletal proteins. Colchicine is the best known of these drugs. Its characteristic action on the mitotic spindle was discovered by A. Dustin more than 50 years ago; it is now known that colchicine specifically reacts with tubulin and inhibits polymerization of microtubules.

Development of these methods led to spectacular advances of our knowledge of cytoskeleton during the last two decades. This knowledge has already radically changed the concept of cell organization, and there is reason to hope that we will soon understand the molecular basis of some enigmatic cellular processes, such as directional movements of cells, alterations of cell shape, and mitosis.

In spite of all the achievements, our present knowledge of cytoskeleton structure and function is still very fragmentary. We know very little about many essential molecules of the cytoskeleton and mechanisms regulating cytoskeleton dynamics in the cells. We are only beginning to recognize that besides the three classical groups of cytoskeletal structures, a number of other structures may exist in the cell (see Chapter 4). We do not yet know all the functions of cytoskeleton, especially its role in the regulation of cell metabolism, growth, and differentiation (see Chapter 9).

In this book we review the major facts and problems related to the structure and function of cytoskeleton. In the first part (Chapters 1–4) we describe the main structural components of cytoskeleton: morphology, distribution in the cells, molecular composition, assembly, and disassembly. In the second part (Chapters 5–9) we describe general organization and function of cytoskeleton as well as its reorganization during mitosis and morphogenesis.

Literature Cited

Gibbons, B. H. (1982) Reactivation of sperm flagella: Properties of microtubule-mediated motility, in *Methods in Cell Biology*, Vol. 25, Academic Press, New York, pp. 253–271.

Huxley, H. E., and Hanson, H. J. (1954) Changes in the cross striations of muscle during contraction and stretch and their structural interpretation, *Nature* **173:**973-976.

Huxley, A. F., and Niedergerke, R. (1954) Structural changes in muscle during contraction. Interference microscopy of living muscle fibers, *Nature* **173:**971-973.

Svitkina, T. M., Shevelev, A. A., Gelfand, V. I., and Bershadsky, A. D. (1984) Cytoskeleton of mouse embryo fibroblasts. Electron microscopy of platinum replicas, *Eur. J. Cell Biol.* **34:**64–74.

Vasiliev, J. M., and Gelfand, I. M. (1981) *Neoplastic and Normal Cells in Culture,* Cambridge University Press, Cambridge, United Kingdom.

Additional Readings

General reviews of cytoskeleton structure and function

Alberts, B., Bray, D., Lewis, J., Raff, M., Roberts, K., and Watson, J. D. (1983) *Molecular Biology of the Cell,* Garland, New York, Chapters 10, 11.

Borisy, G. B. G., Cleveland, D. W., and Murphy, D. B., eds. (1984) *Molecular Biology of the Cytoskeleton,* Cold Spring Harbor Laboratory, Cold Spring Harbor, New York.

Cell and Muscle Motility, Vol. 1–5 (1981–1984) Plenum Press, New York.

Fulton, A. B. (1984) *The Cytoskeleton: Cellular Architecture and Choreography*, Chapman and Hall, New York, London.

Lloyd, C. W., ed. (1982) *The Cytoskeleton in Plant Growth and Development*, Academic Press, London.

McIntosh, J. R., ed. (1983) *Modern Cell Biology*, Vol. 2, *Spatial Organization of Eukaryotic Cells*, Liss, New York.

Organization of Cytoplasm (1982) *Cold Spring Harbor Symposium on Quantitative Biology*, Vol. 46, Cold Spring Harbor, New York.

Porter, K., ed. (1984) The cytoplasmic matrix and the integration of cellular functions, *J. Cell Biol.* **99**(no. 1, part 2).

Schliwa, M. (1985) *The Cytoskeleton*, Springer-Verlag, Vienna-New York.

Sheterline, P. (1983) *Mechanisms of Cell Motility, Molecular Aspects of Contractility*, Academic Press, New York.

Stracher, A., ed. (1983) *Muscle and Nonmuscle Motility*, Vols. 1 and 2, Academic Press, New York.

Wilson, L., ed. (1982) *Methods in Cell Biology*, Vol. 24, *The Cytoskeleton*, Part A, *Cytoskeletal Proteins, Isolation and Characterization*, Academic Press, New York.

Wilson, L., ed. (1982) *Methods in Cell Biology*, Vol. 25, *The Cytoskeleton*, Part B, *Biological Systems and in Vitro Models*, Academic Press, New York.

Reviews of the old works related to cytoskeleton

Heilbrunn, L. V. (1956) *The Dynamics of Living Protoplasm*, Academic Press, New York.

Koltzoff, N. K. (1906) Studien über die Gestalt der Zelle, Teil I, *Arch. Mikrosk. Anat. Entwikl. Mech.*, B. 67, S.365–572; Teil II, (1908), *Arch. Zellforsch.*, B.2, S.I-65; Teil III, (1911), *Arch. Zellforsch.*, B.7, S.344–423.

Porter, K. R. (1984) The cytomatrix: A short history of its study, *J. Cell Biol.* **99** (no. 1, part 2):3s–12s.

Wilson, E. B. (1928) *The Cell in Development and Heredity*, 3rd ed., Macmillan, New York.

Articles on cytoskeleton are published mainly in the journals *Cell, Cell Motility and the Cytoskeleton, Experimental Cell Research, European Journal of Cell Biology, Journal of Biological Chemistry, Journal of Cell Biology, Journal of Cell Science, Journal of Molecular Biology, Journal of Ultrastructure Research, Proceedings of the National Academy of Sciences of USA*, and *Tsitologija* (in Russian).

I

Components of Cytoskeleton

Systems of Actin Filaments

I. Introduction: The Main Types of Actin Structures

A. Components of Actin Structures

Actin filaments (Fig. 1.1) are the main components of a large group of cyto-skeletal structures. These filaments, also called microfilaments or F-actin (filamentous actin), are polymerized from globular actin monomers (G-actin).

Polymerized and nonpolymerized actin can interact with many proteins, called actin-binding proteins. These proteins regulate the degree of polymerization of actin and the stability, length, and distribution of actin filaments. Myosin is an actin-binding protein of special importance. Interactions of actin with myosin are accompanied by the consumption of ATP and can lead to the movements of actin microfilaments. These interactions provide the molecular basis for many types of motility. Polymerization of myosin leads to formation of special myosin filaments. Certain fragments of myosin molecule (see Section VIII for details) can attach themselves to actin filaments *in vitro* (Fig. 1.2). In the electron microscope, these attached elongated fragments bound to a single filament look like lateral projections tilted in one direction. These projections are called arrowheads; the direction of the arrowheads makes it possible to distinguish two ends of microfilaments, called pointed and barbed ends. Thus, binding of myosin fragments reveals an inherent polarity of actin filaments.

B. The Main Types of Structures Formed by Actin Filaments

These filaments are grouped together within the cells and form many types of structures. Four large groups of these structures can be distinguished: bundles of parallel filaments with uniform polarity, bundles of filaments with alternate polarities, three-dimensional networks of filaments, and bidimensional submembranous actin–spectrin networks (Fig. 1.3).

Bundles with uniform polarity of filaments usually do not contain myosin. Certain bundles of this type have very densely and regularly packed filaments: in transverse sections these filaments have a hexagonal order. The bundles of this structure form the cores of certain stable, specialized surface

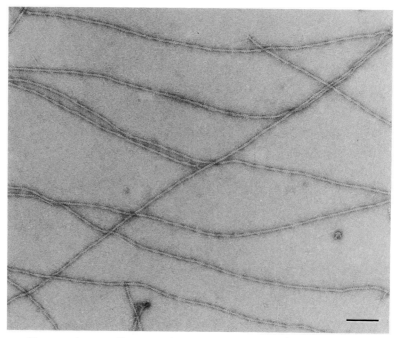

FIG. 1.1. Rabbit muscle actin filaments after negative staining. Electron micrograph. Scale bar: 1.000 Å. (Micrograph courtesy of U. Aebi.)

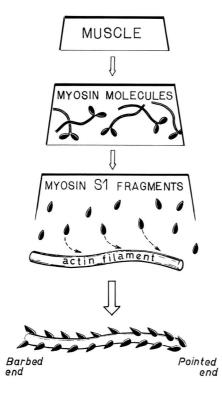

FIG. 1.2. Scheme of determination of polarity of actin filament by attachment with myosin fragments.

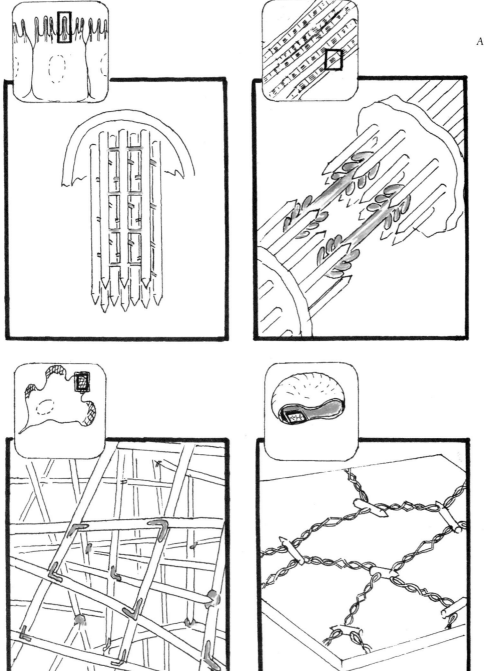

FIG. 1.3. Scheme of various types of structures formed by actin filaments. Upper left: Bundle with uniform polarity in the intestinal brush border; upper right: sarcomere of skeletal muscle; lower left: three-dimensional network in the leading edge of fibroblast; lower right: submembranous actin–spectrin network of erythrocyte.

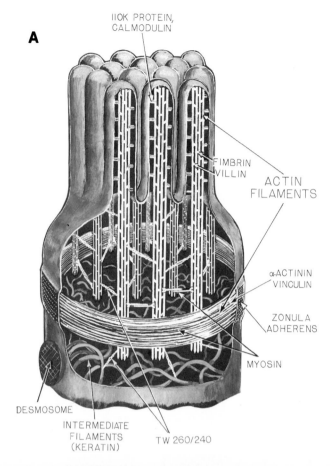

A

IIOK PROTEIN,
CALMODULIN

FIMBRIN
VILLIN

ACTIN
FILAMENTS

αACTININ
VINCULIN

ZONULA
ADHERENS

MYOSIN

DESMOSOME

INTERMEDIATE
FILAMENTS
(KERATIN)

TW 260/240

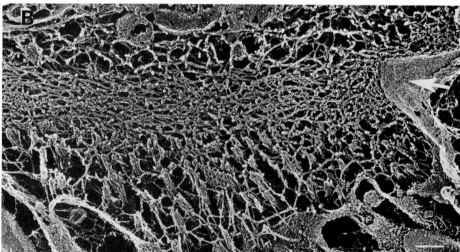

FIG. 1.4. Cytoskeleton of the brush border of intestinal epithelial cells. **(A)** Scheme of the architecture. **(B)** Circumferential bundle of filaments associated with the membrane in the area of zonula adherens (arrow) of the brush border. Glycerinated chicken intestinal cells were incubated in the presence of Mg and ATP; this incubation caused contraction of the bundle; the intercellular junction is split and the zonula adherens pinched in. Platinum replica of quick-frozen

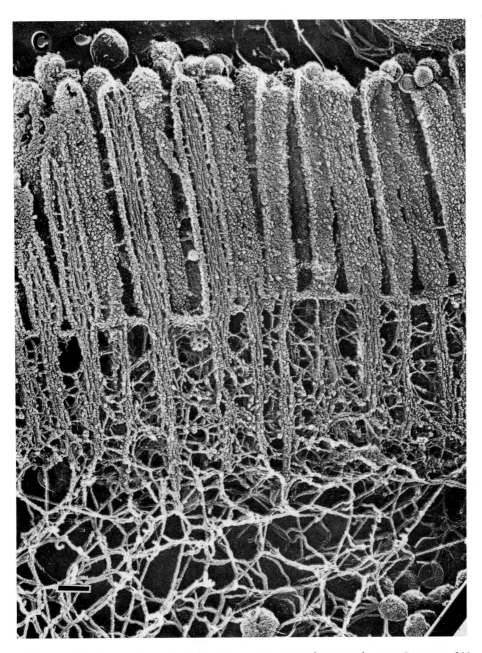

and freeze-etched preparations. Scale bar: 0.1 μm. (From Hirokawa *et al.*, 1983. Courtesy of N. Hirokawa.) **(C)** Tight bundles of actin filaments extend from the microvilli to form straight rootlets. Thin fibrils cross-link the rootlets. Platinum replicas of quick-frozen and freeze-etched cells. Scale bar: 0.1 μm. (From Hirokawa *et al.*, 1982. Courtesy of N. Hirokawa.)

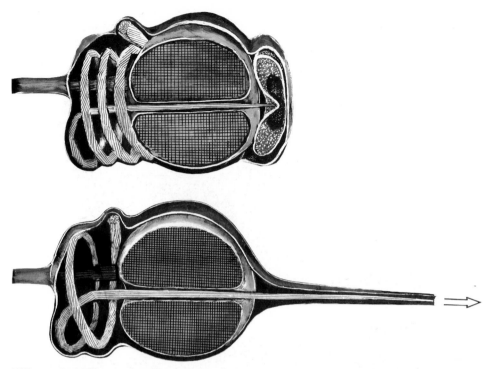

FIG. 1.5. Actin filament bundle within the sperm cell of the horseshoe crab *Limulus*. Top: Non-activated cell with a bundle coiled around the base of the sperm. Bottom: The cell that had formed an acrosomal process; the bundle is extended outward into the process. Scheme based on the work of De Rosier and Tilney (1984).

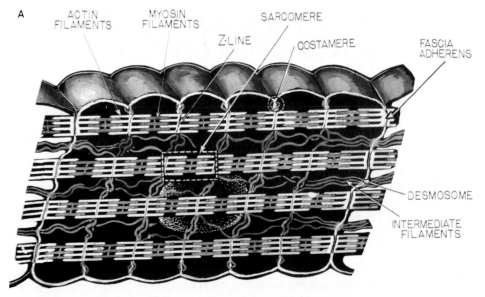

FIG. 1.6. Cytoskeleton of striated muscles. **(A)** Generalized scheme of cytoskeleton of the cardiac striated muscle. **(B)** Electron micrographs of sections of the stretched myofibril of frog skeletal muscle. Left: Thick myosin filaments. Right: Thin filaments attached to the Z line. Inset: Lower

extensions. For instance, these bundles are present within the microvilli of the intestinal brush borders, that is, of the luminal surfaces of intestinal epithelial cells of mammals and birds (Fig. 1.4). These bundles also fill the special large microvilli, so-called stereocilia, present at the apical surfaces of sound-detecting hair cells of the inner ear of higher vertebrates (see Fig. 5.6). Extremely large bundles are found within the acrosomal processes formed by the spermatozoa of the horseshoe crab *Limulus* (Fig. 1.5). Certain other bundles of uniform polarity, e.g., those of lizard stereocilia, are more loosely

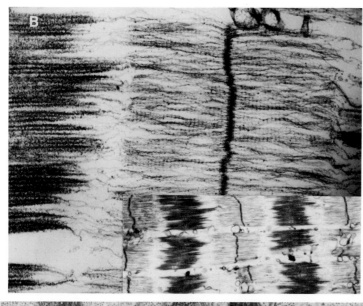

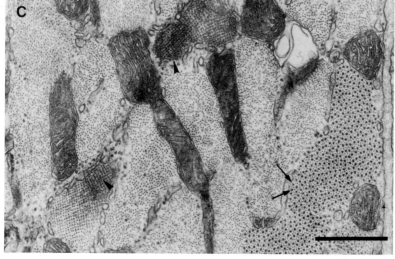

magnification of the same section. Very thin titin filaments (see Chapter 4) connect myosin filaments to actin filaments or Z line. × 52,000; inset × 12,500; reproduced at 65%. (From Maruyama *et al.*, 1984. Micrograph courtesy of K. Maruyama.) **(C)** Cross-section of rat diaphragm muscle. Cross-sections of thick filaments (arrows), of thin filaments, and of Z disks (arrowheads). Scale bar: 0.5 μm. (Micrograph courtesy of L. E. Bakeeva.)

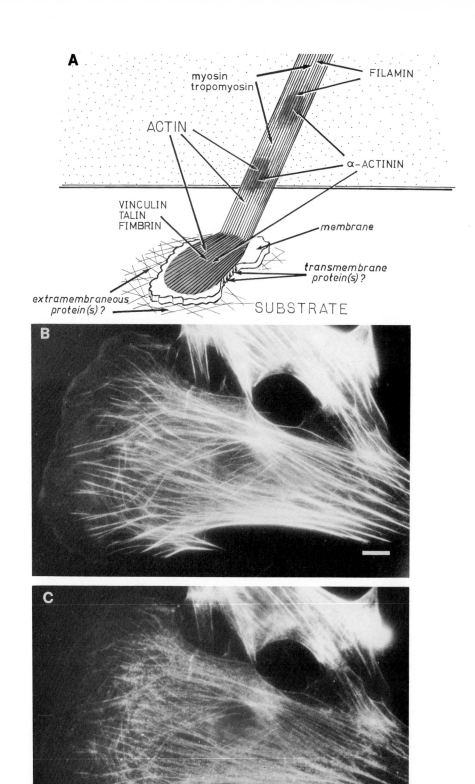

packed and have no hexagonal order of filaments. The barbed ends of filaments in all these bundles are directed toward the cell membrane and away from the cell body.

Bundles of filaments with alternate polarities usually contain myosin. Myofibrils of striated muscle cells have the highest order of organization among the structures of this group. Myofibrils consist of periodically repeating structural units, sarcomeres (Figs. 1.3 and 1.6). Each sarcomere has two arrays of parallel actin filaments with opposite polarities. Barbed ends of the filaments of each array are attached to the structures called Z disks and contain an actin-binding protein, α-actinin. Pointed ends of the filaments are directed toward the middle line of the sarcomere. The central zone of the sarcomere contains the myosin filaments. Actin and myosin filaments can interdigitate in this zone; they slide along one another during muscle contraction.

Stress fiber present in fibroblasts (Fig. 1.7) and other cultured cells is an example of the lower-ordered bundles containing actin and myosin. Filaments with opposite polarities are found in various parts of the bundle. Certain actin-binding proteins are distributed along the bundle in periodic fashion; areas rich in myosin alternate with areas rich in α-actinin. These features suggest that stress fiber has a periodic structure somewhat similar to that of myofibril. However, the degree of similarity is not yet clear. In particular, it is not known whether myosin filaments are present in the stress fiber.

Some myosin-containing bundles of filaments have a ringlike shape. A submembranous contractile ring is formed in all animal cells during division; its contraction seems to be responsible for cytokinesis. Circular bundles of filaments are also present in the interphase epithelial cells (Fig. 1.4).

The actin structure filling the cytoplasm of smooth muscle cells can be regarded as another variant of a bundle with alternate polarities of filaments. Here the barbed ends of filaments are attached to the small, dense bodies containing α-actinin; two groups of filaments radiate in the opposite directions from the two sides of each dense body (Fig. 1.8). Myosin is present in the smooth muscle cell, but its exact localization is not clear. Thick myosin filaments have been observed by some investigators. Possibly these filaments are less stable than those in striated muscles.

Three-dimensional networks consist of filaments crossing one another at various angles and separated from one another by various distances (Fig. 1.9). These networks are present in the submembranous cortex of many types of animal nonmuscle cells and of amebae. The density of the cortical network can vary in different areas of the same cell. The degree of disorder in various parts of the network can also vary. For instance, in certain areas of filament networks, it can become aligned into the bundles of different size (Fig. 1.9).

FIG. 1.7. (A) Scheme of the stress fiber associated with focal contact. **(B, C)** Actin and myosin in cultured mouse fibroblast. The cell doubly stained with rhodamine–phalloidin reacting with actin (B) and with antimyosin antibody (C). Both actin and myosin are present in the stress fibers in the central parts of the cell; actin is continuously distributed in these fibers, while myosin has spotlike periodic localization. Scale bar: 10 μm. (Courtesy of I. S. Tint.)

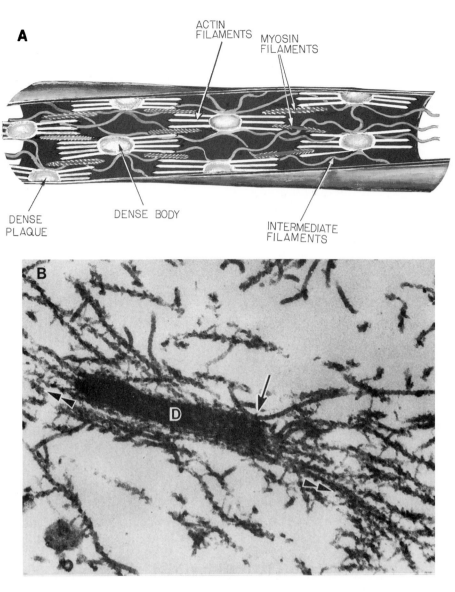

FIG. 1.8. Smooth muscle cytoskeleton. **(A)** Generalized scheme. **(B)** Isolated dense body (D) of chicken gizzard smooth muscle cell. Actin filaments were decorated with HMM fragments of myosin. Intermediate filament (arrow) is not decorated with myosin fragments; it is associated side to side with a dense body. ×80,000. (From Tsukita *et al.*, 1983a. Micrograph courtesy of S. Tsukita.)

Strictly speaking, it is not known whether completely disordered "actin gels" exist in any cells. Filament networks sometimes contain various actin-binding proteins, including myosin. Recently, rod-shaped structures, immunofluorescently stained with myosin antibodies, were found in the cytoplasm of the slime mold *Dictyostelium discoideum*. These findings suggest that more or less stable myosin filaments may be present within some actin networks.

FIG. 1.9. The network of randomly oriented actin filaments and the bundle of parallel filaments in the cortex of the active edge zone of cultured chick fibroblast. Electron micrographs of cytoskeleton negatively stained with sodium silicotungstate. ×102,000; reproduced at 80%. (From Small, 1981. Courtesy of V. Small.)

The **spectrin–actin submembranous network** is the only cytoskeletal structure present within mammalian erythrocyte (Figs. 1.3 and 1.10). It consists of short oligomers of actin which serve as joining points for large elongated molecules of an actin-binding protein, spectrin. These actin–spectrin complexes form a bidimensional fibrillar network under the plasma membrane. This network was also found in avian erythrocytes; it may be responsible for maintenance of the constant shape of erythrocytes.

Recently spectrin-related protein, fodrin, was found in submembranous regions of several types of nucleated cells, such as neurons and fibroblasts. These findings suggest that a spectrin–actin network may be present in some form not only in the erythrocytes, but also in many other cells.

C. Association of Actin Structures with Plasma Membrane

Not only the submembranous actin–spectrin network but also other actin structures described above contact the membrane; often this contact occurs in the specialized area of the membrane. For instance, lateral sides of the bundles in stereocilia and in microvilli of brush borders are linked to the lateral membrane of the microvilli, while their peripheral barbed ends are attached to the submembranous electron-dense plaques at the tips of microvilli (Fig. 1.4). The ends of myofibrils in cardiac muscle are attached to the areas of specialized cell–cell contacts, so called "fascia adherens" (Fig. 1.6). Peripheral

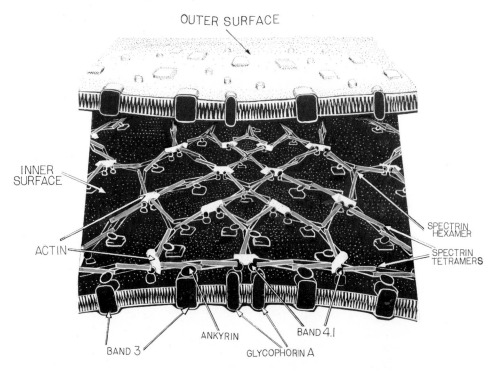

FIG. 1.10. Generalized scheme of a mammal erythrocyte cytoskeleton.

ends of stress fibers in fibroblasts are attached to specialized cell substrate contacts, called focal contacts (Fig. 1.11).

Bidimensional actin–spectrin networks in the erythrocytes are linked to the plasma membrane via special proteins (Fig. 1.10 and Section VI). Three-dimensional networks are usually located under the membrane; the structural basis of their anchoring to the membrane is not yet clear. Thus, association with the membrane is a general feature of actin structures of various types. Concrete forms of this association are different in various cases.

D. Basic Processes Essential for Formation and Function of Actin Structures

We have briefly reviewed morphological features of various types of actin structures. On the basis of this review, we can now list the processes likely to be involved in the formation of these structures. Polymerization of actin filament is the first of these processes. Obviously, this polymerization is a prerequisite for the formation of all types of actin structures. Positions and lengths of actin filaments are strictly determined in the highly ordered actin structures, such as stereocilia or sarcomeres. Thus, formation of these structures requires regulation of initiation and of termination of polymerization of individual filaments. It also requires regulation of interactions of individual filaments with one another during or after their polymerization. Owing to these interactions, associated filaments form various types of bundles or networks. Special processes are needed to establish the association of actin structures with the plasma membrane and with other cellular components. After their formation, many actin structures are able to change their shape reversibly or irreversibly. In particular, contractility is the basic property of the myosin-containing bundles and networks. In the following sections of this chapter we shall discuss each of these basic processes in greater detail.

II. Structure of Globular and Filamentous Actins

A. Amino Acid Sequences of Actins

We begin our discussion of the formation of actin structures by giving a more detailed description of the characteristics of its main component, actin. G-actin is a single polypeptide chain with molecular weight of about 42 kD and isoelectric point of about 5.4. Some variants of actin have 375 amino acid residues; others have 374 residues. At present, the complete sequences of actin genes from many sources and proteins encoded by these genes are known (Fig. 1.12). Comparison of these sequences shows that actin is a remarkably conserved molecule that undergoes only small changes during evolution. For instance, actin from the lamprey muscle differs from mammalian cardiac actin in only 3 of 375 amino acid residues—99% homology. Sea urchin muscle actin has 97% homology with one of the mammalian actins (cytoplasmic β-actin). Even actin from the slime mold *Physarum polycephalum* is very similar to rabbit skeletal muscle actin (Fig. 1.12). Certain

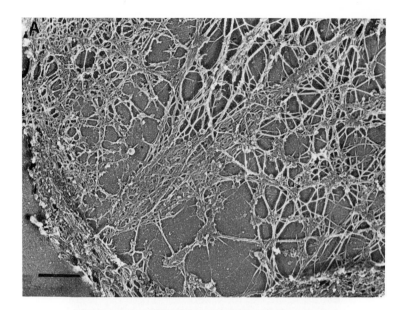

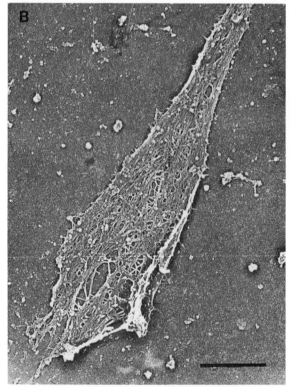

FIG. 1.11. Cytoskeleton structures in the leading lamella of cultured fibroblast. Platinum replicas of critical-point dried preparations of *Triton*-extracted mouse fibroblasts. **(A)** Overall view. Leading edge in lower left corner. Focal contact with attached stress fiber going toward the cell center. Scale bar: 0.5 μm. **(B)** Isolated focal contact. Other parts of the cytoskeleton have been removed from the substrate mechanically by a stream of water. Focal contact and a part of stress fiber remain attached to the substrate. Scale bar: 1 μm. **(C)** Stress fiber (actin filament bundle) at high

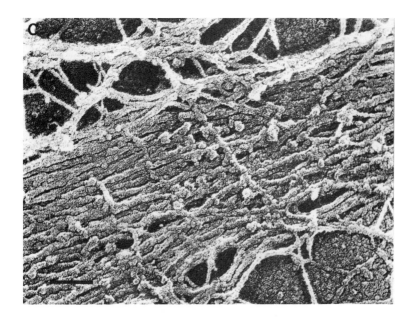

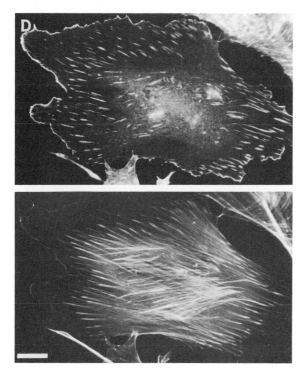

magnification. Scale bar: 0.2 µm. (B and C from Svitkina *et al.*, 1984. Courtesy of T. M. Svitkina.)
(D) Comparison of the distribution of characteristic protein of focal contacts, talin (top) and actin
(bottom) in the same cultured chick fibroblasts. Immunofluorescence microscopy after staining
with antitalin antibody, followed by a rhodamine-labeled second antibody and with fluorescent
derivative of phalloidin. Talin is present in the focal contacts at the ends of actin filament bundles. Talin is also present at the cell margin. Scale bar: 20 µm. (From Burridge and Connell, 1983.
Micrograph courtesy of K. Burridge.)

data suggest, however, that ciliate protozoa may have actins with exceptional structures.

Some lower eukaryotes, such as yeast, have only one actin gene encoding a single protein. In contrast, many higher eukaryotes have several isoforms of actins coded by families of actin genes. Actins of vertebrates fall into three electrophoretic classes: α (the most acidic), β, and γ (Fig. 1.12). Each individual mammal or bird expresses at least six different actins: two nonmuscle actins (β and γ), cardiac muscle α-actin, skeletal muscle α-actin, and two

FIG. 1.12. (A) Primary amino acid sequences of rabbit skeletal muscle actin (upper) and of the actin of *Physarum polycephalum* (lower). Only the residues of *Physarum* actin that are different from those of muscle actin are shown. The blocking group (BL) on the NH₂ terminus of *Physarum* actin is most probably acetyl. [Data from Vandekerckhove and Weber, 1978a (*Physarum* actin) and 1978b (skeletal muscle actin).] **(B)** Scheme of electrophoretic classes of actins in various tissues.

smooth muscle actins (α and γ). The sequences of all these actins are closely related. The highest number of amino acid exchanges is 25 (nonmuscle γ-actin versus skeletal muscle α-actin). No amino acid differences were revealed between the same variants of actins isolated from different species of mammals and birds. Thus, actins of higher vertebrates are tissue-specific and species-independent. Each variant of actin is encoded by a separate gene. The same cell often contains several isoforms of actin, which may be incorporated into different structures. For instance, filaments of skeletal muscle sarcomeres are made of α-actin, while γ-actin is incorporated into submembranous structures of the same cell, costameres (see Section VI).

B. Form of Actin Molecules

To determine the three-dimensional structure of a protein molecule it is necessary to obtain this protein in crystalline form. G-actin molecules are not crystallized. However, there is a protein, bovine DNase I, which forms stable 1:1 complexes with actin. These complexes were crystallized and studied by x-ray crystallographic methods. In other experiments two-dimensional crystalline sheets of G-actin molecules were obtained by addition of gadolinium ions and studied by electron microscopy. Both methods gave similar results. It was found that G-actin is elongated and consists of a larger and smaller domain with a cleft between them (Fig. 1.13).

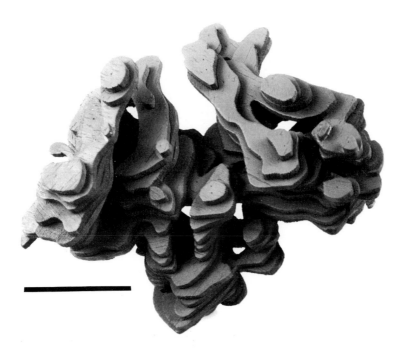

FIG. 1.13. Wood model of the actin subunit at 6 Å resolution. The structure of subunits complexed with pancreatic DNase I has been determined by x-ray crystallographic methods. Scale bar: 20 Å. (Courtesy of D. Suck and W. Kabash.)

A

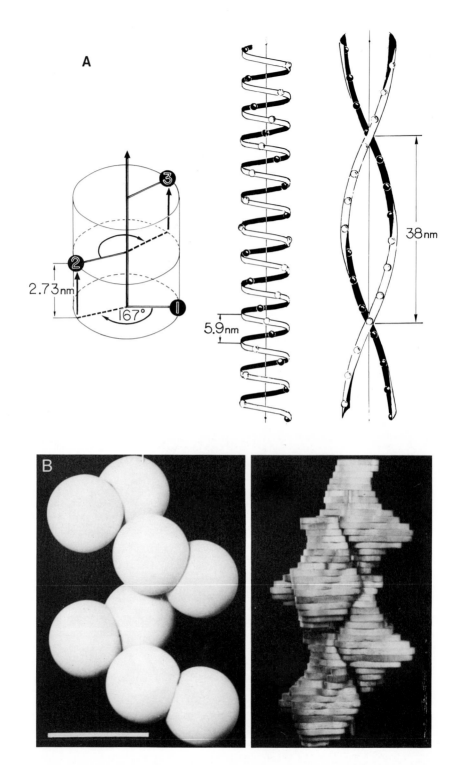

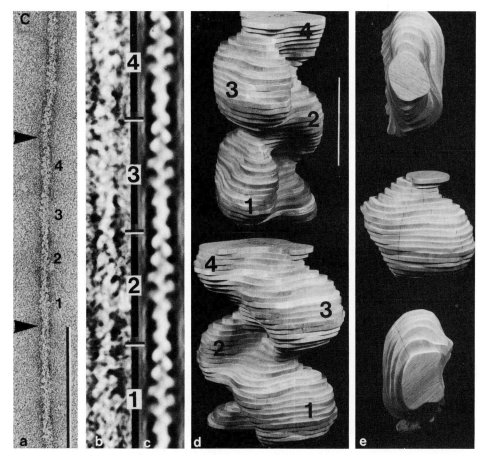

FIG. 1.14. Structure of actin filament. **(A)** Helical symmetry of ideal actin filament. Each actin subunit is represented by one point (e.g., center of gravity). Left: The symmetry operator (the rule according to which the position of each subsequent subunit can be derived from the position of the foregoing subunit). Center: The left-handed 5.9-nm-pitch helix passes through all the subunits. Number of subunits per turn is 2.16. Right: Two right-handed long-pitch helixes that pass through the same subunits. The number of subunits per turn of long-pitch helix is 14. Distance between cross-over points of these two helixes is 38 nm.

(B) Two models of alignment of asymmetrical actin subunits in the actin filament. In the Egelman and De Rosier model (left), the actin subunit consists of two spheres each, of diameter 3.8 nm, separated by 3.0 nm. The long axis of the subunit is tilted by about 20° off perpendicular to the helix axis. This subunit axis does not pass through the helix axis but is skew to it. The distance of the center of the inner sphere from the axis is 1.2 nm. Scale bar: 5 nm. (After Egelman and De Rosier, 1983.) In the Smith *et al.* model (right), the long axis of the subunit is oriented parallel to the filament axis. Here the "best contact" model is shown which reveals extensive contacts between complementary surface features of adjacent subunits. (From Smith *et al.*, 1983. Courtesy of U. Aebi.)

(C) Three-dimensional reconstruction of a molecular actin filament model from negatively stained actin filament. a: Electron micrograph of filament negatively stained with 0.75% uranyl formate. The filament has four helical repeats, marked by two arrowheads. b and c: Two stages of the processing of the filament image shown in a. d: Two views of the wood model representing a four-subunit stretch of filament reconstructed from the data shown in c. e: Three views of the actin subunit as extracted from the reconstruction of filament shown in d. Compare with Fig. 1.13. Bars 1000 Å (a); 50 Å (d). (Courtesy of U. Aebi.)

Many proteases cleave G-actin into two fragments: a small, 9-kD fragment containing the NH$_2$ end of the molecule and a larger, 33-kD COOH end fragment. These fragments may correspond to the small and large domains revealed by structural studies.

Each actin molecule contains one molecule of bound nucleotide (ATP or ADP) and one bound ion (Ca^{2+} or Mg^{2+}). As shown by radioactive tracer experiments, bound nucleotide and cation can undergo rapid exchange with added nucleotides (ADP or ATP) and cations (Mg or Ca), respectively.

The living cell usually contains more Mg than Ca and more ATP than ADP; therefore, most molecules of G-actin *in vivo* are likely to contain bound ATP and Mg.

C. Structure of Actin Filament

Molecules of G-actin form subunits of actin filaments. The structure of these filaments has been studied mainly by analysis of electron micrographs. In negatively stained preparations and in replicas, actin filaments look like strings made of helically arranged subunits (Fig. 1.14). Results of the analysis of these images suggest that the filament is a single-stranded, left-handed helix of subunits with the following parameters. Each subunit is located 2.73 nm above the previous one along the axis. This parameter is called the axial rise per subunit; it does not vary in any filaments. When two neighboring subunits are viewed from the top, they form an average angle of 167°, with a vertex at the axis of filament.

Within the same filament, the angles between subunits can vary; the value of this so-called angular disorder was estimated to be about 10°. The same structure can be described not only as a single helix, but also as a double-stranded, right-handed helix with approximately 14 subunits in each strand per complete turn. Electron microscopic pictures of negatively contrasted filaments often convey the visual impression that the filament is, in fact, a double-stranded helix. Nevertheless, there is no reason to prefer this model to the equivalent single-stranded helix. There are no data indicating that the filament can be dissociated into two strands or that each strand can grow independently of another.

The exact positions of subunits within the filament are not yet clear. According to the model proposed by Egelman and De Rosier (1983), each subunit is so attached to its neighbors that its long axis forms an angle of about 70° with the axis of filament (Fig. 1.14). According to another model proposed by Smith *et al.* (1983), the long axes of subunits are parallel to that of the filament (Fig. 1.14). The common feature of both models is that the contacts between the subunits are assumed to be located along the single-stranded, left-handed helix. The data available are not yet sufficient to choose between the two models.

Both models postulate that the actin filament is a polar structure with two nonequivalent ends (Fig. 1.14). As already mentioned, this polarity can be demonstrated experimentally by decorating the filaments with myosin fragments (Fig. 1.2). Polar morphology of actin filament can also be seen without any decoration in high-quality, negatively stained preparations. The filaments in these preparations have a chevron appearance. The chevrons point toward the barbed end of filament.

One important parameter of the filament, which is not yet exactly known, is its diameter. Experimental estimations of this parameter gave different results depending on the method employed. This width was about 7nm in negatively stained preparations and sections examined by electron microscopy. The width measured in x-ray diffraction studies was larger, about 10 nm. Possibly, the width in electron microscopic preparation is underestimated because the contrasting stain diffuses into the outer part of the filament.

To summarize, the actin filament can be regarded as a helical polar structure made of elongated subunits. Data on angular disorder indicate that the position of each subunit within the filament has a certain degree of freedom. Owing to this structure, the filament has flexibility and ability for torsion.

III. Polymerization of Actin

A. Conditions and Manifestations of Polymerization

In this section we shall discuss the process of actin polymerization; that is, the assembly of filaments from G-actin molecules. G-actin is easily polymerized. To prevent polymerization of monomeric actin, it is necessary to keep it in a solution of low ionic strength. Addition of about 1 mM Mg^{2+} or Ca^{2+} or 100 mM K^+ to this solution induces polymerization.

The usual method of isolation of actin is based on the effect of salts on polymerization. The tissue is first treated with acetone and dried; then actin is extracted by a low ionic strength buffer; next the extract is centrifuged and the salt is added to the supernatant to induce conversion to F-actin. Finally, F-actin is sedimented by centrifugation and dissolved again in low-ionic-strength buffer. Polymerization–depolymerization cycles are repeated several times until 95% pure actin is obtained.

It is not known why the salt promotes polymerization. When the salt is added to the very dilute solution of G-actin, polymerization does not begin, but certain properties of G-actin molecules are changed. These molecules become more resistant to proteases, and light absorption at 212 nm is decreased. These changes, called monomer activation, may be due to some alterations of conformation of G-actin; it is not known whether they are essential for polymerization.

Polymerization is manifested by alterations of many physical parameters of actin solution. In particular, the viscosity of solution, as well as its flow birefringence and light scattering, are increased owing to formation of filaments. For some unknown reason, transformation of G-actin into F-actin is also accompanied by increased absorbance of UV light of 232-nm wavelength. Measurement of each of these alterations, especially viscosity, can be used to assess the progress of polymerization. Centrifugation of filaments from solution is also employed for this purpose.

The two new and more exact methods of measurement of filament assembly use actin monomers labeled by fluorescent dyes, fluorochromes (Fig. 1.15). It was found that fluorescence of actin monomer, labeled with N-pyrenyl iodacetamide, increases considerably after incorporation of this monomer into a filament; this increase is used to assess polymerization. Another method uses the mixture of two types of monomers labeled with different fluorochromes, which are excited by light of different wavelengths. When these two monomers are brought into close proximity by polymerization and fluorescence of the first fluorochrome is excited by light of the appropriate wavelength, part of the excitation energy can be transferred to another fluorochrome. Thus, polymerization reduces the intensity of fluorescence of the first fluorochrome. This reduction is measured to assess polymerization.

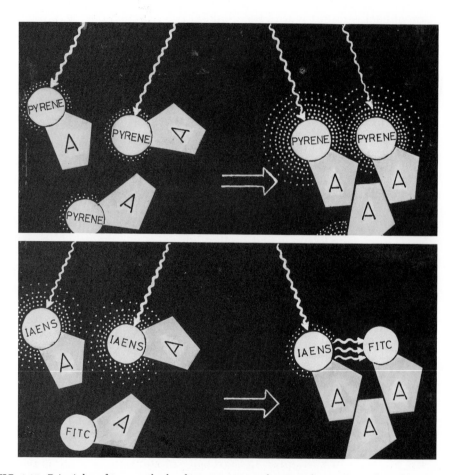

FIG. 1.15. Principles of two methods of measurement of actin polymerization based on the use of monomers (A) labeled by fluorescent dyes. Top: Monomers are labeled by the derivative of pyrene; fluorescence of the label increases after polymerization. (See Cooper *et al.*, 1983.) Bottom: Mixture of monomers labeled by two different dyes: N-(iodoacetamidoethyl)-1-aminonaphthalene-5-sulfonate (IAENS) or fluorescein isothiocyanate (FITC). The fluorescence of the first dye decreases after polymerization owing to the transfer of energy to the molecules of the second dye. (See Taylor *et al.*, 1981, and Spudich *et al.*, 1982.)

B. Two Stages of Actin Polymerization: Nucleation and Elongation

When polymerization is started in the solution of G-actin by the addition of salts, the amount of polymer first increases very slowly and then more rapidly; the rapid increase continues until the amount of polymer reaches a plateau (Fig. 1.16). These data suggest that polymerization has two different

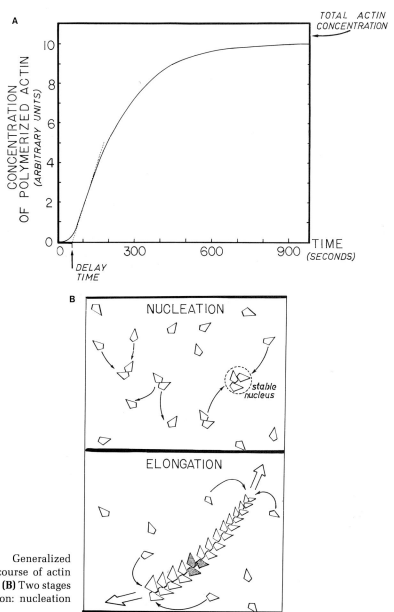

FIG. 1.16. (A) Generalized curve of time course of actin polymerization. (B) Two stages of polymerization: nucleation and elongation.

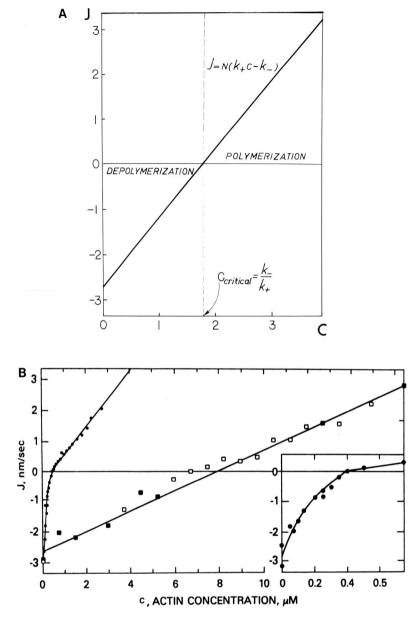

FIG. 1.17. **(A)** Linear dependence of the rate of actin polymerization on monomer concentration as described by equation 1.1. **(B)** Dependence of the rate of filament growth (J) on the concentration of G-actin (c) in real experiments. Polymerization in the presence of 0.2 mM ATP (left curve, ●) or in the presence of 0.2 mM of ADP (right curve, □). The inset shows the $J(c)$ plot at a low concentration of ATP-actin on an expanded scale. The rate of elongation of ADP-actin depends linearly on the concentration of monomer in the entire range of concentration. Elongation proceeds faster in ATP than in ADP. In contrast, the $J(c)$ plot of ATP-actin deviates from a straight line at low concentrations of monomer. **(C)** Correlation between actin polymerization and associated ATP hydrolysis. Polymerization of pyrenyl-labeled G-actin (33 μM) was assessed by measuring the change in fluorescence (solid line). Simultaneously ATP hydrolysis (symbols) was

stages: nucleation and elongation. During the first, slow stage, several actin molecules get together and form a small stable aggregate, the nucleus. Rapid elongation of this small aggregate takes place during the second stage (Fig. 1.16). This interpretation is confirmed by experiments in which small fragments of filaments (so-called seeds) were added to a solution of G-actin; in this case, rapid polymerization began immediately, without the preceding slow stage. The nucleus can be defined as the smallest aggregate of monomers for which the rate of further growth is faster than the rate of disassembly. Several facts indicate that the nuclei, formed at the first stage, are trimers of G-actin. The rate of polymerization at this stage is proportional to the third power of actin concentration. This dependence indicates that, on the average, an encounter of three molecules is needed for the formation of one nucleus.

When actin solution was rapidly fractionated during the first stage of polymerization, the smallest stable aggregates revealed were trimers; dimers were not stable. The time of nucleation was reduced when artifactually produced trimers, i.e., three monomer molecules covalently linked to one another by a special reagent, were added to the solution.

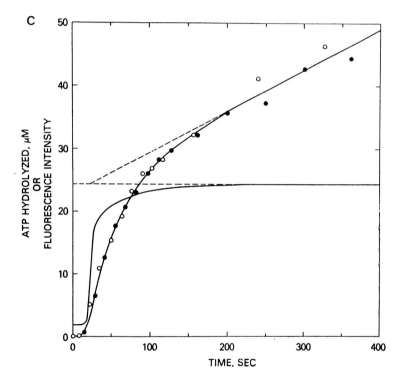

measured. At this high concentration of actin, complete polymerization occurs earlier than the rate of ATP hydrolysis becomes constant. Calculations show that at 10 sec 60–70% of the total F-actin still had bound ATP; this bound ATP was hydrolyzed over the next 200 sec. (B and C are from Carlier *et al.*, 1984. Courtesy of M. F. Carlier.)

C. Elongation: The Role of ATP

To study elongation independently of nucleation, a known number of preformed filaments is added to the monomer solution at the start of polymerization and the initial rates of polymer formation are registered. As spontaneous nucleation of monomers is much slower than elongation, it does not affect measurement results. The rate of elongation is constant in solution with constant concentration of G-actin; it is proportional to the number of added filaments. The dependencies of elongation rates on concentration are substantially different in experiments with solutions of monomers containing bound ADP and bound ATP. In the more simple case of ADP-actin, the rate of elongation (Fig. 1.17) can be described by the linear equation

$$J = EN = (k_+C - k_-)N \qquad [1.1]$$

where J = the rate of change of polymer amount per time unit; N = the number of added filaments, $E = k_+C - k_-$ is the rate of elongation of one filament; C = concentration of monomer; k_+C is the rate of association of monomers to one filament; k_- is the rate of dissociation of monomers from one filament. k_+ and k_- are called association rate constant and dissociation rate constant, respectively.

Critical concentration of monomer, C_{cr}, can be defined as a concentration

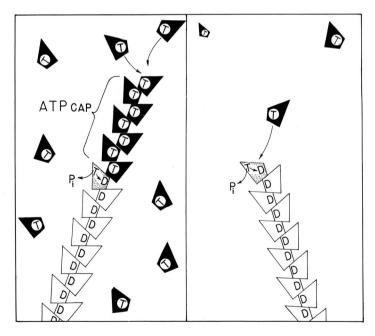

FIG. 1.18. Formation of the ATP cap at the end of the filament at a high concentration of ATP monomer (left); disappearance of the cap at a low concentration of ATP monomer (right). T and D = actin subunits containing ATP or ADP, respectively. T → D = subunits undergoing ATP hydrolysis.

at which there are no alterations of the amount of polymer with time ($J = 0$). This concentration of the ADP-actin is easily obtained from the equation $C_{cr} = k_-/k_+$. If concentration of monomer is lower than critical, net depolymerization takes place ($J < 0$).

Experiments with ATP-actin revealed much more complex dependence of elongation on concentration. At high concentrations of ATP-actin, this dependence is nearly linear. Here the elongation of ATP-actin goes at a much higher rate than that of ADP-actin. For instance, in the experiments with *Acanthamoeba* actin, the association rate constant, k_+, was found to be 10 times higher and the dissociation rate constant, k_-, five times lower for ATP-actin than for ADP-actin. The concentration dependence of the rate of elongation of ATP-actin, in contrast to that of ADP-actin, is not always linear: it shows a sharp bend downward at a low concentration of monomer (Fig. 1.17). This unexpectedly rapid fall of the elongation rate is likely to be caused by alterations of the actin-bound nucleotide within the filament. The polymerization causes two types of alterations in the state of actin-bound ATP. First, ATP bound to actin monomer is hydrolyzed to ADP and inorganic phosphate after monomer incorporation into the filament. Second, the rate of exchange between actin-bound nucleotide and the ATP molecules present in solution becomes very slow, so ADP resulting from hydrolysis remains bound to the actin molecules in the filament. Thus, elongation of a filament consisting of n monomers can be described by the scheme

$$\text{Filament}_n + \text{monomer (ATP)} \underset{\text{dissoc.}}{\overset{\text{assoc.}}{\rightleftharpoons}} \text{filament}_{n+1}\,(\text{ATP}) \rightarrow \text{filament}_{n+1}\,(\text{ADP})$$

$$\searrow$$

$$\text{inorganic phosphate}$$

The ATP hydrolysis is not mechanistically coupled to polymerization. While the rate of ADP hydrolysis inside the filament is independent of the external concentration of ATP–G-actin, the rate of association of new monomers with filament ends is proportional to the concentration of the monomers in solution. When this concentration is high, the rate of addition of new monomers will be faster than the rate of ATP hydrolysis. Therefore ATP monomers will be present at the ends of most filaments forming a so-called "ATP cap"; the more central parts of the filament will contain ADP monomers. In contrast, when the rate of association is low, e.g., in the low concentration of ATP-actin, ATP at the ends of filaments will be hydrolyzed faster than new monomers will be attached (Fig. 1.18). In these conditions the ATP caps will disappear, and ADP will be present at the ends of most filaments. Elongation of ATP-actin proceeds faster than that of ADP-actin. Accordingly, filament ends carrying ATP caps attach new subunits more easily and detach them less easily than uncapped ends. In other words, the population of N-filaments in the solution of ATP-actin consists of T-capped filaments and D-uncapped filaments:

$$N = T + D$$

having different rates of elongation: E_T and E_D ($E_T > E_D$). The rate of change of polymer amount will then be

$$J = TE_T + DE_D$$

As the concentration of monomer is decreased, the uncapped fraction D will progressively grow. At near zero concentration $T = 0$: $N \approx D$, and the average rate of elongation of one filament will approach E_D. Thus, an increase of the fraction of uncapped filaments may explain the rapid decline of the elongation rates at low concentrations of ATP monomers. At critical concentration, the growth of capped filaments is equilibrated by depolymerization of uncapped filaments,

$$J = 0; \ TE_T = -DE_D$$

Thus, at the critical concentration of ADP-actin, each individual filament is in the equilibrium state (Fig. 1.19). In contrast, at the critical concentration of

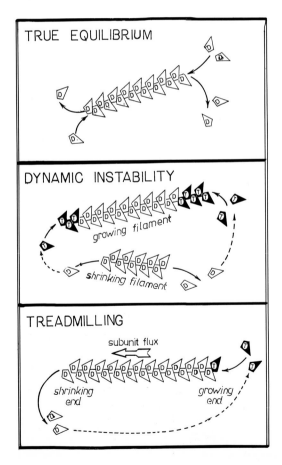

FIG. 1.19. Three possible types of equilibria between polymerized and nonpolymerized actins at critical concentration. Top: assembly and disassembly are equilibrated at each end of each filament. Center: growth of capped filaments at both ends is equilibrated by the shrinkage of others. Bottom: growth of each filament at one end is equilibrated by its shrinkage at the other end.

ATP-actin, only the total amount of polymer is steady, but each individual filament is not in the equilibrium state; some of these filaments grow, while others shrink (Fig. 1.19).

This theoretical model of polymerization, developed by Carlier *et al.* (1984), has not yet been directly confirmed experimentally, but it explains satisfactorily the known data on actin polymerization. According to the model, actin filaments are dynamic structures that may rapidly decrease in length when their ATP caps disappear; they may grow again when new caps are formed at the ends. This property has been termed a built-in dynamic instability. It can be essential for rapid reorganization of actin structures.

D. Elongation: Differences of Two Filament Ends

We have discussed the summarized kinetics of actin elongation at both ends of filaments. In fact, the rates of elongation at the two ends of a filament are different. These different properties of the ends were demonstrated in experiments in which bundles of filaments with uniform polarity, e.g., bundles, isolated from *Limulus* spermatozoa or from brush border microvilli, were incubated in solution containing actin monomers; the rate of elongation of each end of the bundle was measured in the electron microscope.

Measurements made at high concentrations of monomers have shown that the rates of elongation at the barbed ends are substantially faster than those at the pointed ends (Fig. 1.20); this is true both for ADP-actin and for ATP-actin.

Polar growth of a filament is the result of its polar organization; different

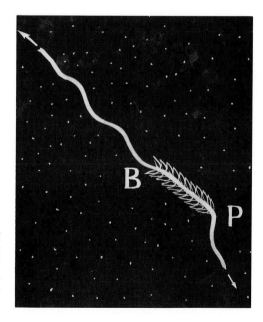

FIG. 1.20. Scheme showing that the rates of elongation at the two ends of the filament are different. The filament decorated by myosin fragments and incubated in actin solution is elongated at a greater rate at the barbed end.

surfaces of two ends of filaments have different affinities for the attachment of new monomers. The exact nature of these differences is not yet clear.

The discovery of the different rates of elongation at the two ends of a filament has led to development of a theoretical model of the special equilibrium state of filaments, called treadmilling. According to this model, not only elongation rates, but also critical concentrations of actin, can be different for the two ends of the filament. When the concentration of actin in solution has a certain intermediate value between these two critical levels, then the barbed end will be elongated, while the pointed end will be depolymerized at the same rate (Fig. 1.19). In these conditions treadmilling will occur; that is, the individual monomers inside the filaments will move from the barbed to the pointed end, while the length of the filament will remain constant. This movement of subunits needs energy that can be supplied only by ATP hydrolysis. Therefore, according to the model, treadmilling is possible only for ATP-actin. It proved to be difficult to find out experimentally whether treadmilling actually occurs. As just discussed, at low concentrations of ATP monomers, some filament ends grow and others shorten. These fluctuations do not exclude treadmilling; it is possible that at certain concentrations of monomer most barbed ends will grow and most pointed ends shrink. However, fluctuations caused by formation and disappearance of ATP caps may overshadow treadmilling, especially if the average length of filaments is not very large. Obviously, polar growth of filaments can be important for the formation and maintenance of many types of actin structures, especially of bundles of uniform polarity.

IV. Agents Regulating Actin Polymerization

Assembly and disassembly of actin filaments can be modified by special actin-binding proteins. Several types of proteins can be distinguished: proteins binding actin monomers, proteins binding to the ends of filaments, and those side-binding along the filaments (Fig. 1.21, Table 1.1).

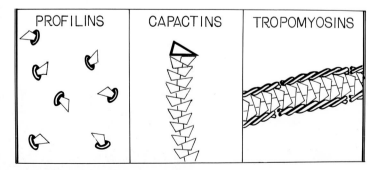

FIG 1.21. Three groups of actin-binding proteins affecting polymerization: profilins, binding to nonpolymerized subunits; capactins, binding to filament ends; and tropomyosins, binding to the lateral sides of filaments.

TABLE 1.1. Actin-Binding Proteins Modifying Polymerization of Filaments[a]

Class	Proteins	Subunit mol. wt. (kD)	Comments
Monomer-sequestering proteins	Profilins		Different proteins with common properties; all profilins form 1:1 complex with G-actin ($K_D \approx 10^{-5}$ M) thus reducing the effective concentration of monomer; profilins do not interact with F-actin
	From mammalian tissues	16	
	From Acanthamoeba	13	
	From Thyone sperm	12.5	
	DNase I	31	Isolated from bovine pancreas; forms very tight 1:1 complex with monomeric actin ($K_D \approx 10^{-10}$ M); formation of the complex leads to total loss of polymerization ability of actin and enzymatic activity of DNase; also binds F-actin and accelerates depolymerization
	Vitamin D-binding protein	52–56	Present in bovine serum; forms 1:1 complex with G-actin and ternary complex with G-actin and DNase I
	Depactin	17	Isolated from starfish oocytes
	Actophorin	17	Isolated from Acanthamoeba
			These G-actin-binding proteins interact also with F-actin, inducing fragmentation of filaments
End-blocking (capping) proteins			
Proteins blocking barbed end	Gelsolin	90	Isolated from macrophages, Ca-dependent capping and severing protein
	Villin	92	Isolated from intestinal microvilli, cross-links filaments in absence of Ca^{2+} but acts as capping and severing protein in the presence of Ca^{2+} (see text)
	Brevin	92	Capping and severing protein from serum, closely related to gelsolin
	Fragmin	42	Three actinlike proteins from Physarum; fragmin has Ca-dependent capping and severing activity; Cap 42a and Cap 42b are capping proteins
	Cap 42a	42	
	Cap 42b	42	
	Severin	40	Isolated from Dictyostelium; Ca-dependent capping and severing protein
	Capping protein from Acanthamoeba castellanii	Two subunits, 29 + 31	Ca^{2+}-independent; no severing activity
Proteins blocking pointed ends	Acumentin	65	Isolated from macrophages; Ca^{2+}-independent
	β-Actinin	34 + 37	Isolated from muscle and kidney cells; Ca^{2+}-independent

(Continued)

TABLE 1.1. (*Continued*)

Class	Proteins	Subunit mol. wt. (kD)	Comments
Side-binding proteins	Tropomyosins	30–40	Inhibits fragmentation of filaments; may also bind troponin (see text); different spectrum of isoforms in different cell types

^aThe selected examples of protein of each group are given here and in Tables 1.2–1.4. The table is based on the data in the review papers of Korn (1982); Craig and Pollard (1982); Weeds (1982); Maruta *et al.* (1984); Pollard and Cooper (1986).

A. Proteins Sequestering Actin Monomers

Profilins are the best-studied representatives of this group. They form 1:1 complexes with G-actin and make it less able to undergo polymerization. More detailed analysis has shown that profilin from *Acanthamoeba* inhibits nucleation to a greater degree than elongation of filaments. Probably, the profilin–actin complex cannot form nuclei, but can bind to the ends of ready-made filaments. This complex is disassembled after binding, detaching the profilin into solution. Profilins do not increase the rate of depolymerization of F-actin. Prevention of actin polymerization is likely to be the main function of profilins in the cell. By accumulating profilin in some place, the cell is able to keep a store of nonpolymerized actin. Surprisingly, one of the pancreatic enzymes, DNase I, can specifically bind actin monomers with very high affinity; the biological significance of this binding is obscure.

B. Proteins Binding the Ends of Filaments (Capping Proteins)

Usually one molecule of this protein binds to one particular end of the filament, preventing further polymerization of actin at that end (Fig. 1.22). Some capping proteins first form a molecular complex with actin monomer; then this complex binds the end of the microfilament and blocks it. Many capping proteins specifically react with the barbed end of the filament, e.g., fragmin, gelsolin, or villin. Few proteins, e.g., acumentin, react with the pointed end. It is interesting that some of the capping proteins (Cap 42b, Cap 42a, and fragmin) have an actinlike molecular structure. They can be regarded as relatives of the actin molecules which, during evolution, have lost the ability to polymerize, but not the ability to attach themselves to the ends of ready-made actin filament. Attachment of this "fake actin" molecule prevents further addition of real actin subunits.

As capping proteins interact only with the ends of filaments, the effective ratio of their concentration to those of actin monomers is very low. Some capping proteins, but not all, can also act on the preformed filaments, fragmenting them into shorter ones; then protein can cap the newly formed end of the fragmented filaments. This severing effect is not accompanied by the depolymerization of actin. Severing proteins can either directly break the filaments or block the ends of filaments broken spontaneously or mechanically.

Some spontaneous fragmentation of filaments always takes place in actin solutions. Capping proteins can also produce certain other alterations of polymerization. Many of them are able to promote nucleation. Possibly, binding of these proteins to one end of actin dimer or trimer stabilizes this nucleus and facilitates its elongation at the opposite, unblocked end. Another possibility is that the severing effect of protein leads to fragmentation of a few filaments formed at the early stage of polymerization and thus increases the number of nuclei. Owing to promotion of polymerization, more filaments can be formed at a given time in actin solution in the presence of capping protein than in its absence; however, owing to the unilateral block of elongation and to the severing effect, the average length of these filaments will be shorter. Calcium dependence is a characteristic feature of certain capping proteins, e.g., gelsolin and fragmin. Other proteins of this type, e.g., acumentin, are Ca-independent.

All the data about the effects of capping proteins *in vitro* show that they

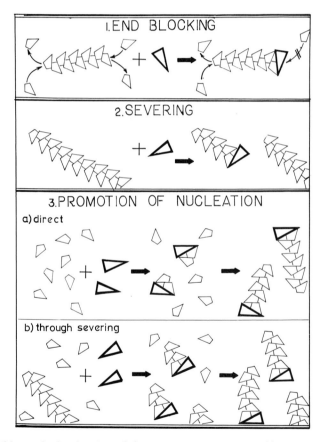

FIG. 1.22. Possible methods of action of the proteins capping actin filaments: (1) Attachment blocks assembly and disassembly at one end; (2) severing action fragments the filaments; (3a) attachment to the nuclei leads to their stabilization; (3b) fragmentation of filaments increases the number of ends undergoing elongation.

can act in a cell as factors regulating the length of filaments. The filaments capped by these proteins at both ends will have a fixed length. Severing proteins can decrease the length of preexisting filaments. In in vitro experiments the molecules of capping protein were attached to the polymer surface and then placed into the actin solution; the filaments nucleated by this protein grew with their pointed ends directed into the solution. Similar polar nucleation of the filaments by the capping proteins attached to some cellular structures can exist in the cell.

C. Proteins Binding along the Filament (Side-Binding Proteins)

Tropomyosins form this group of proteins. Molecules of the best-characterized, skeletal muscle tropomyosin, are dimers consisting of two 33-kD chains. Each of these chains contains numerous α-helical structures. The chains are wrapped helically around each other, forming a so-called coiled coil. This dimeric rod of tropomyosin is about 40 nm long. It binds laterally to actin filament (Fig. 1.21); a row of these rods head to tail forms a helix around the filament. C-ends of tropomyosin chains are directed toward barbed ends of filaments. Binding of tropomyosin inhibits spontaneous fragmentation of filaments as well as shortening of these filaments caused by the severing proteins villin and gelsolin. Thus, tropomyosin appears to stabilize the preformed filaments.

Different molecular variants of tropomyosins were isolated from striated and smooth muscles. Nonmuscle cells contain several special variants of tropomyosin. These molecular isoforms differ somewhat in length and may have different stabilizing effects on filaments.

D. Cytochalasins

Cytochalasins are a group of closely related fungal metabolites (Fig. 1.23) that were found to specifically affect actin polymerization. The effects of cytochalasins are somewhat similar to those of capping proteins. Cytochalasins B and D are the most widely used compounds of this group. Cytochalasins act like capping agents. They reversibly bind the barbed end of the microfilament and inhibit filament growth at this end but not at the pointed end. Possibly, in large concentrations, cytochalasins can also sever preformed microfilaments. Like other capping agents, they can promote nucleation. Certain capping proteins, e.g., the proteins from Acanthamoeba or platelet-derived protein, compete with cytochalasins for the binding to F-actin. This competition suggests that these capping proteins and cytochalasin share the same binding site at the barbed end of the microfilament. Other capping proteins do not compete with cytochalasins.

Cytochalasins are widely used in experiments with cultured and other cells to alter the state of actin microfilaments. This use is justified by the high specificity of their effects on actin. However, in analyzing the results of cell experiments, one should take into account the fact that cytochalasin B, besides its action on actin, also has a side effect: it inhibits sugar transport through the membrane. Cytochalasin D and dihydrocytochalasin B also effi-

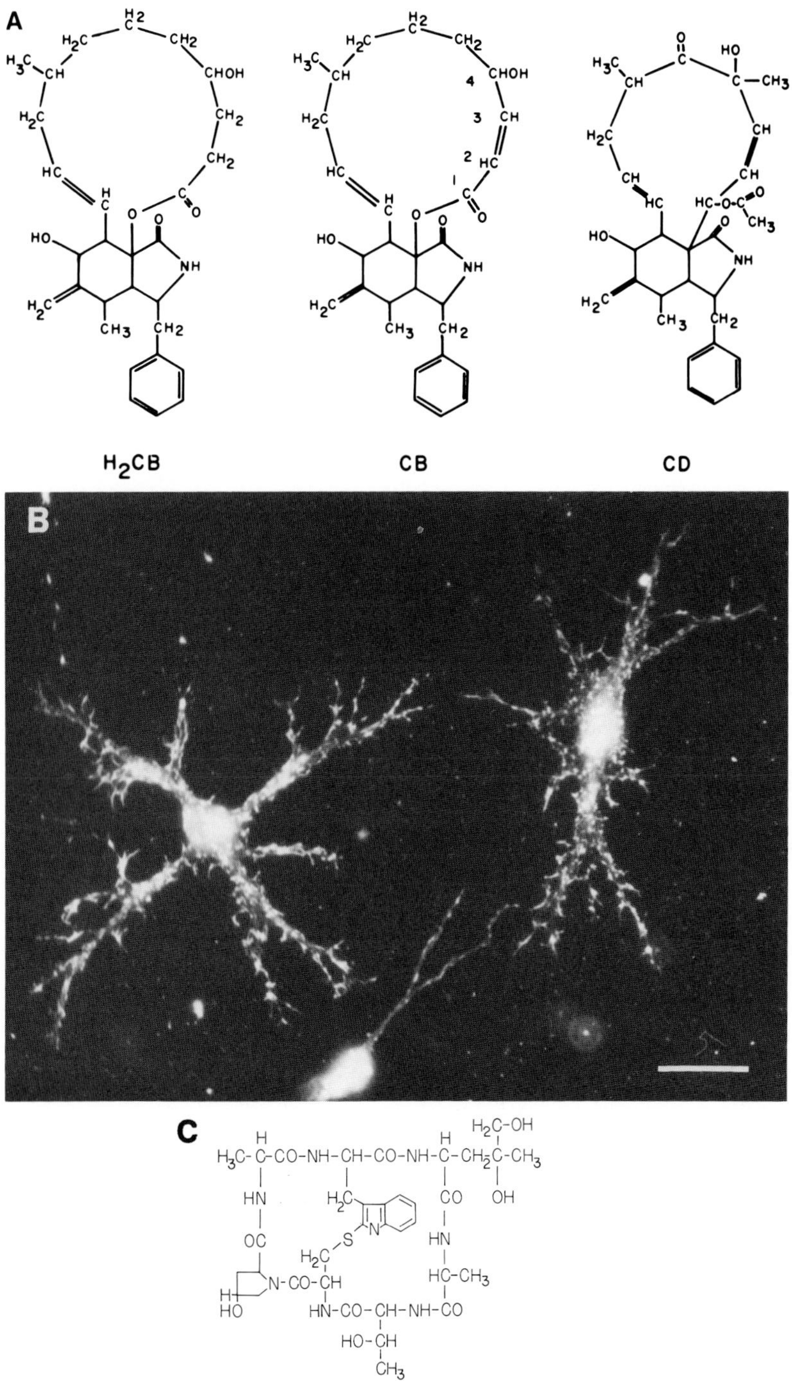

FIG. 1.23. (A) Structure of cytochalasins. (B) Mouse fibroblasts incubated in the medium containing cytochalasin B (10 μg/ml). Immunofluorescence microscopy after antiactin staining. Compare with the cells in control medium (Fig. 1.7B). The cells have aquired arborized shapes; stress fibers have disappeared; actin is accumulated in numerous small needlelike aggregates. Scale bar: 20 μm. (Courtesy of T. M. Svitkina.) (C) Structure of phalloidin.

ciently alter actin polymerization but have much less effect on transport than cytochalasin B.

E. Phalloidin

Phalloidin is a cyclic oligopeptide isolated from toadstools (Fig. 1.23). Like cytochalasins, it specifically affects polymerization of actin. However, its effect is quite different from those of cytochalasins. Phalloidin binds to actin filaments stoichiometrically; that is, one molecule per subunit. Binding of phalloidin decreases the dissociation constants of filament ends and thus increases the rate of elongation. Phalloidin also increases the resistance of filaments toward the action of severing proteins and cytochalasins. The molecular mechanisms of the specific action of phalloidin are not clear.

Fluorochrome-labeled phalloidin is widely used to stain specifically actin filaments.

V. Proteins Cross-Linking Actin Filaments

A. Cross-Linking of Filaments *in Vitro*

We described the formation of individual actin filaments. These filaments are bound together in the cell into various types of structures. The processes of binding can be reproduced to some degree *in vitro*. Filaments in solution can be linked together by special proteins into structures of two types, isotropic gels and bundles. The filaments in the bundles are aligned parallel to one another with different degrees of regularity and with different spacing. Filaments in isotropic gels cross one another at various angles, forming three-dimensional networks. Formation of bundles (bundling) and formation of gel (gelation) can be easily distinguished with the electron microscope. Of course, the structures formed can also be of intermediate character; that is, the gels can contain areas of more or less parallel microfilaments.

Solution containing only actin can be very viscous owing to entanglement of filaments and, possibly, to formation of some weak bonds between these filaments. However, this solution can "creep" when subjected to mechanical stress, thus retaining certain characteristics of liquid. Solid gels, in contrast to liquids, do not creep. Formation of solid gel from actin filaments can take place when special proteins, making stable cross-links between the filaments, are added. At the constant concentration of polymerized actin, the minimal amount of cross-linking protein needed to form the gel is inversely proportional to the average length of the filament. Formation of gel is accompanied by an abrupt increase in the viscosity of the actin solution. In contrast, the bundling of filaments, induced by special proteins, leads to a decrease of viscosity.

B. The Main Groups of Cross-Linking Proteins

Four main groups of these proteins have been isolated and studied: two groups of gel-forming proteins, a group of bundling proteins, and a group of proteins with intermediate properties (Fig. 1.24, Table 1.2).

The first of these groups includes filamin and actin-binding protein (ABP) from the macrophages; these are immunologically related proteins of high molecular weight. Molecules of ABP are dimers, made of two highly flexible, long subunits. These subunits are linked end to end, while their opposite ends bind to F-actin. Owing to their length and flexibility, ABP molecules can link filaments crossing one another at various angles and even filaments perpendicular to one another. Naturally, molecules of this protein are able to induce formation of isotropic gels at low concentrations.

The second group of cross-linkers includes spectrin and related proteins, e.g., fodrin (Fig. 1.25). These are also high-molecular-weight proteins with flexible, elongated molecules. However, their molecular morphology is more complex than that of ABP. Spectrin consists of two different filamentous sub-

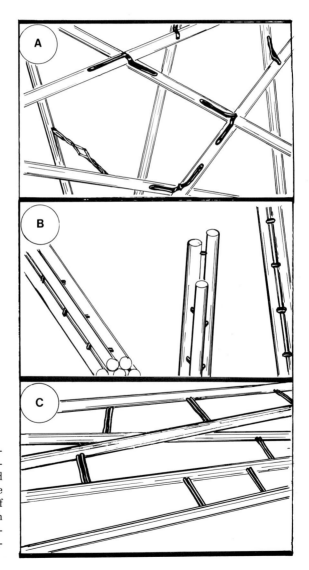

FIG. 1.24. Cross-linking of actin filaments by various groups of proteins. **(A)** Proteins of filamin and spectrin groups with long, flexible molecules. **(B)** Bundling proteins of the fascin and fimbrin group with short molecules. **(C)** Proteins of α-actinin group with rodlike molecules.

units (α and β); they may be bound together forming either the tetramer or more complex multimeric structures. Proteins of this group induce formation of isotropic gels. Cross-linking of actin filaments by spectrin is strongly promoted by another protein isolated from the erythrocyte, protein 4.1. This protein is thought to stabilize the spectrin–actin links.

The third group of cross-linkers includes typical bundling proteins, such

TABLE 1.2. Proteins Cross-Linking Actin Filaments[a]

Class	Protein	Subunit mol. wt. (kD)	Comments
Gel-forming proteins	Myosins	Heavy chains 200; light chains 18–20 and 15–17	Ubiquitous; essential for actin-based motility; forms contractile actomyosin gels with actin
	Actin-binding protein (ABP)	270 (2 subunits)	Isolated from macrophages; Ca-independent
	Filamin	250 (2 subunits)	Isolated from smooth muscle, closely related to ABP
	Spectrin	α-240, β-220	Isolated from the erythrocytes
	TW 260/240	α-240, β-260	Isolated from intestinal brush borders; closely related to spectrin
	Fodrin (calspectin)	α-240, β-235	Isolated from neural tissues; found in most cell types; closely related to spectrin
	Actin-binding protein from *Dictyostelium*	120 (2 subunits)	Ca-independent
	Caldesmon	150 (2 subunits)	Isolated from smooth muscle; cross-linking action is calmodulin-inhibited
Proteins with predominant bundling action	Fascin	58	Isolated from sea urchin eggs
	Fimbrin	68	Isolated from intestinal microvilli; found also in ruffles of cultured cells. Both fascin and fimbrin promote formation of tight paracrystalline bundles
	Villin	92	Isolated from microvilli; cross-linking action only in the absence of Ca^{2+}
	α-Actinins	100 $\times$ 2 (2 subunits)	Present in Z disks of muscle cells and in various nonmuscle cells; *in vitro* promotes formation of loose bundles and networks; nonmuscle α-actinins are inhibited from cross-linking by calcium; muscle proteins are Ca-independent

[a] General reviews: Korn (1982); Craig and Pollard (1982); Weeds (1982); Stossel (1984); Pollard and Cooper (1986). Reviews on spectrinlike proteins: Baines (1984); Speicher (1986).

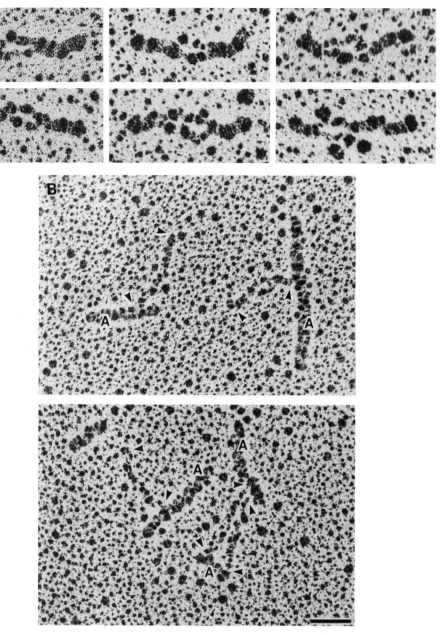

FIG. 1.25. Brain spectrin molecules. Electron micrographs of the rotary-shadowed preparations. **(A)** Tetramers. The separation of two twisted strands in the middle portion was enhanced by incubation in 0.1 M KCl. **(B)** Spectrin tetramers attached to actin filaments (A) with their tail ends (arrowhead). Scale bar: 0.1 μm. (From Tsukita *et al.*, 1983b. Micrographs A and B courtesy of S. Tsukita.) **(C)** Spectrinlike protein in the cultured fibroblast. The same cell stained with antibody to brain spectrin (top) and to actin (bottom). Notice the rather uniform distribution of spectrin in the cell with higher concentration around the nucleus; there is no colocalization of spectrin over actin filament bundles. Scale bar: 20 μm. (C from Burridge *et al.*, 1982. Courtesy of K. Burridge.)

as fascin and fimbrin. These are monomeric globular molecules of much lower molecular weight than gelling proteins of the filamin and spectrin types. Numerous molecules of these proteins form multiple short cross-links between each two filaments (Fig. 1.24). As a result, these filaments are brought together at small, uniform distances; that is, parallel arrays of closely packed filaments are formed.

Proteins of the fourth group, which includes α-actinin and actinogelin, have structure and properties intermediate between the typical gelling and typical binding agent.

These proteins are dimers with molecular weight of subunits about 100 kD. Their molecules have the shape of nonflexible, elongated rods. These proteins added to the filaments induce formation of loose structures, which contain both the gellike zones of criss-crossed filaments and bundlelike groups

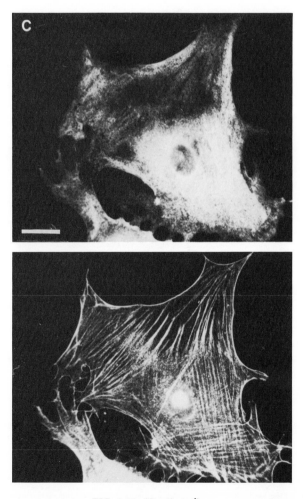

FIG. 1.25. (*Continued*)

of nearly parallel filaments (Fig. 1.24). Electron microscopy reveals the rods of cross-linking proteins between the parallel filaments.

C. Function of Cross-Linking Proteins in the Cell

It is natural to suggest that proteins that cause cross-linking of filaments *in vitro* can also participate in formation of actin structures *in vivo*. This suggestion is strongly supported by the fact that each well-studied group of actin structures was found to contain a characteristic set of cross-linking proteins with appropriate properties. For instance, the brush border bundle contains two proteins with bundling properties, fimbrin and villin. Formation of filament bundles at the periphery of discoid phagocytic cells, coelomyocytes, of the sea urchin is accompanied by accumulation of another bundling protein, fascin. The presence of α-actinin in the structures, binding together the groups of filaments in sarcomeres and other bundles, was mentioned in the Introduction. Gelating proteins are often found in filament networks. For instance, antibody to filamin stains the leading edge and actin bundles in fibroblasts (Fig. 1.26); ABP is a component of the cortical network of macrophages.

Spectrin and protein 4.1 are essential for the formation of bidimensional networks in erythrocytes. Certain genetic defects of spectrin make the cytoskeletons of erythrocytes unstable and prone to lysis so that persons with these defects have hemolytic anemias.

It would, however, be an oversimplification to suggest that each protein with a cross-linking activity has only one function. The type of function performed by a protein can depend on the conditions of its environment, especially on the concentration of calcium. The case of villin is of special interest in this connection. This protein has bundling activity in the absence of calcium, capping activity at low concentrations of calcium (1–5 μM), and severing activity at high concentrations of calcium (10–50 μM). Therefore, when calcium is added to the preparations of brush border cytoskeleton, villin, cross-linking the core bundles, destroys these same bundles. Calcium may also affect the properties of other cross-linking proteins. For instance, micromolar concentrations of calcium prevent the bundling effect of α-actinin extracted from the platelets.

Alterations of the state of cross-linking proteins can cause not only destruction of actin structure, but also changes in its shape. An interesting example is the change of the actin bundle in the *Limulus* spermatozoon during activation of this cell (Fig. 1.5). In nonactivated cells this bundle is coiled posterior to the nucleus. When the sperm cell is activated, by contact with the outer coat of the egg, the bundle uncoils and moves to the anterior into the straight process, which is extended from the pole of the cell and penetrates through the egg coat. The bundle contains, besides actin, the equimolar amount of another protein, called scruin, with mol. wt. 55 kD. The mode of interaction of scruin with actin filaments is still unknown. It had been suggested by De Rosier and Tilney (1984) that scruin is bound to the filaments in such a way that nonactivated bundle is locked in the bent and twisted position. After activation, some ion or small molecule penetrates into the bundle from the surrounding cytoplasm and, altering the state of scruin, uncoils this bundle. This intriguing concept of an actin movement without myosin deserves further detailed study.

VI. Proteins Attaching Actin Filament to the Membrane

Attachment of actin structures to the outer cell membranes is likely to be mediated by a group of special proteins (Table 1.3). The molecular mech-

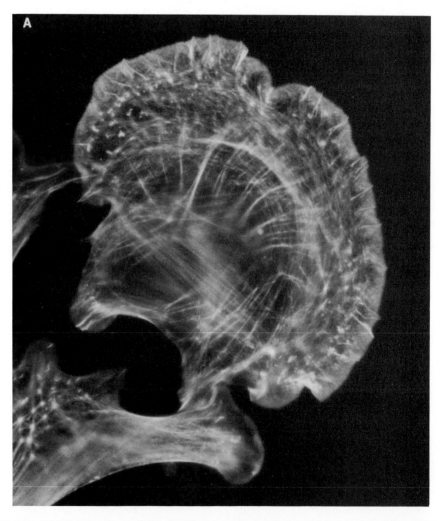

FIG. 1.26. Distribution of actin **(A)** and of the two actin-binding proteins, α-actinin **(B)** and filamin **(C)** in chicken fibroblasts. Actin was labeled by phalloidin; the other proteins were labeled with

anism of this attachment in the erythrocytes is the best known. It involves the protein ankyrin with mol. wt. about 200 kD. Ankyrin binds both to spectrin and to the integral membrane protein, called band 3 protein or anion transporter. Another cytoskeletal component, protein 4.1, can link spectrin and another membrane protein, glycophorin. Thus, the bidimensional actin–spectrin network of the erythrocyte has numerous sites of attachment to the membrane. Possibly, this mechanism of attachment can exist in many cell types, since proteins similar to ankyrin and 4.1 have been found in other cells besides erythrocytes.

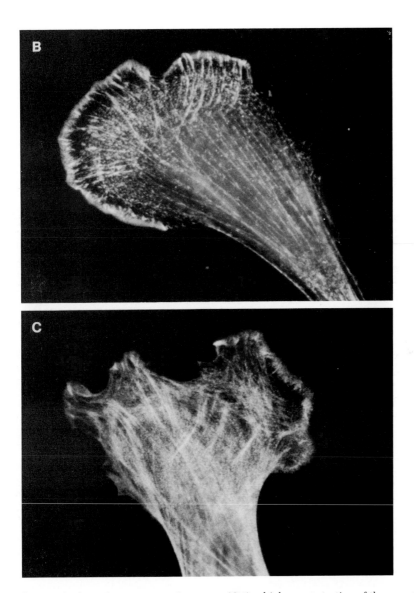

corresponding antibodies. Fluorescence microscopy. Notice high concentration of these proteins within the pseudopods in the active edge zone, and in the stress fibers. (Courtesy of V. Small.)

Special proteins have been isolated from the microvilli of various cells. One of them is a 58-kD protein from the microvilli of mouse tumor cells; it may be involved in the attachment of membrane 75-kD glycoprotein to actin filaments. Another protein of 110 kD forms (together with small Ca-binding protein calmodulin) the side arms connecting the actin bundle with the membrane of brush border microvilli (Fig. 1.4). The 110 K–calmodulin complex binds both to actin filaments and to plasma membrane in an ATP-sensitive manner. Addition of ATP leads to detachment of arms from the actin core and to solubilization of the 110 K–calmodulin complex.

Several characteristic proteins were found in the submembranous termini of the filament bundles of cultured fibroblasts, which are attached to the focal cell substrate contacts. These zones contain vinculin (130 kD) and talin (215 kD) in close proximity to the membrane (Fig. 1.7 and 1.11). Talin binds vinculin in $vitro$ and also binds transmembrane 140-kD glycoprotein, integrin. Vinculin is also accumulated in the submembranous areas of cell–cell contacts associated with filament bundles, e.g., in zonulae adherens of epithelial

TABLE 1.3. Proteins Possibly Involved in the Linkage of Actin Filaments to the Plasma Membrane[a]

Protein	Subunit mol. wt. (kD)	Comments
Spectrin	α-240 β-220	Spectrin, ankyrin, and protein 4.1 are present in the erythrocytes;
Ankyrin (syndein)	200	actin is bound to band 3 membrane protein (anion channel) via spectrin and
Band 4.1 protein	80 (4.1a) 78 (4.1b)	ankyrin (see Fig. 1.10); band 4.1 protein stabilizes the actin–spectrin links and binds to another membrane protein glycophorin; related proteins are present in many other cells
Vinculin	130	Present in focal contacts and other membrane structures, to which actin filaments are linked
Metavinculin	150	Related to vinculin membrane-associated protein; present in smooth muscle
Talin	215	Present in focal contacts; binds vinculin and integrin in $vitro$
110 kD protein of brush border microvilli	110	Forms cross-arms linking lateral sides of the actin bundle of microvillus to the plasma membrane; binds calmodulin
Cytoskeleton-associated glycoprotein (CAG) of ascites tumor cell microvilli	75	Membrane protein isolated in the stable complex with actin

[a]Reviews: Branton et al. (1981); Goodman and Shiffer (1983); (erythrocytes) Geiger (1983); Carraway and Carraway (1984); Mangeat and Burridge (1984); Bennett et al. (1985); Heath (1986); Pollard and Cooper (1986). See also special issue of Cell Motil. **3**(5/6) (1983).

cells and fascia adherens of cardiomyocytes. The striated muscle fibers contain riblike bands of vinculin under their plasma membrane. These bands, costameres, run perpendicularly to the long axis of the fiber (Fig. 1.6). In addition to vinculin, they contain spectrin and γ-actin. It is likely that costameres are the submembranous termini of a system of links which connect laterally the Z disks of the myofibrils with the membrane (see also Chapter 3). The presence of vinculin in several different types of the membrane–cytoskeleton connection strongly suggests that it plays some important, but not yet identified, role in the formation of these connections. Reconstruction of membrane–actin attachments from individual molecular components *in vitro* remains a task for the future.

VII. Myosins

A. Nonpolymerized Myosin

Myosins are components of many actin structures, essential for their motility. Before discussing actin–myosin interactions, we shall describe the properties of myosin molecules and structures formed by polymerization of these molecules.

Myosins are a large family of rather diverse proteins, which nevertheless, share certain common properties; all myosins bind to actin and have ATPase activity, stimulated by actin binding. The skeletal muscle myosin is the best-

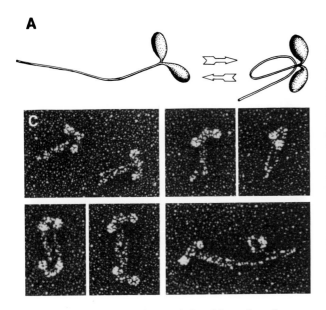
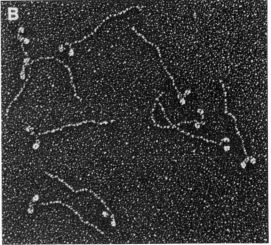

FIG. 1.27. (A) Scheme of extended and looped conformations of the myosin molecule. (B, C) Electron micrographs of platinum-shadowed replicas showing conformational states of myosin molecules. (B) Two-headed molecules with extended tail. (C) Molecules with looped tails. B, C, × 200,000; B reproduced at 75%. (Micrographs from Trybus and Lowey, 1984. Micrographs courtesy of K. M. Trybus.)

studied representative of the family. It can be extracted from the muscle by salt solutions of high ionic strength (0.3–0.6 M KCl). Myosins are very large molecules with mol. wt. about 480 kD. At electron microscopy they have a characteristic shape: a long tail is associated with two pear-shaped heads (Fig. 1.27). Myosin molecules from smooth muscle and nonmuscle cells are able to acquire two different configurations: extended configuration with straight tail and looped configuration with bent tail (Fig. 1.27).

The myosin molecule consists of six polypeptide chains: a pair of heavy chains of mol. wt. about 200 kD and two pairs of light chains of about 20 kD each. The heavy chain consists of the globular head part at the NH_2 end and the rod part at the COOH end. The rod part consists mainly of α-helical domains. The rods of two heavy chains coil around one another, forming a coiled coil structure. This rigid structure forms the tail of the myosin molecule. It contains two short domains, called hinges, where the α-helix structure is interrupted and the molecule is more flexible.

Proteases cut the heavy chains at the hinges, dissecting the myosin molecules into several fragments: S1 (heads), S2 (proximal part of the tail), and LMM (distal part of the tail). Larger fragments can also be obtained: S1 + S2 (called HMM) and S2 + LMM (called rods) (Fig. 1.28). Additional proteolysis can cut the heads (S1 fragment) into three subfragments.

Each head is associated with two nonidentical light chains. One is called the A-chain, because it can be dissociated from the head by alkaline medium. Another chain is called the P-chain, because it can be phosphorylated. Both types of light chain are located in the "neck" region of the myosin molecule, where the pear-shaped head tapers to the rod.

B. Polymorphism of Myosins

Extensive polymorphism of myosins in various tissues of the same organism and at various stages of development of the same tissue has been shown by many methods, including production of specific monoclonal antibodies. For instance, the fast-contracting skeletal muscle contains different isoforms

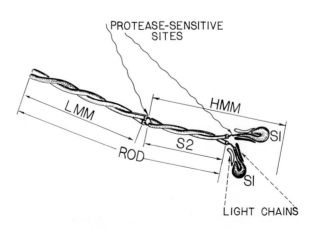

FIG. 1.28. Parts of the myosin molecule.

of the light A-chain: A1 and A2. The myosin molecule can contain either identical chains (A1, A1 or A2, A2) or different chains (A1, A2). Thus, variability of these chains can produce three different variants of myosin molecules within the same cell. The structure of heavy chains can also vary. For instance, cardiac muscles contain two types of heavy chains (HC_α and HC_β). Accordingly, three isotypes of myosins can be present in this tissue: V_1 (α, α homodimer), V_2 (α, β heterodimer), and V_3 (β, β homodimer).

Embryonic forms of light and heavy chains of muscle myosins are replaced by adult forms during development of birds and mammals. The light and heavy chains of adult myosins from fast and slow skeletal muscles, heart muscle, smooth muscle, and the nonmuscle cells are also different.

Despite all their variability, most types of myosin molecules have the same general pattern: they are made of six chains and form double-headed rods. Therefore, one exceptional case is of special interest. Myosins IA and IB isolated from the protozoan *Acanthamoeba castellanii* consist of only one heavy chain and one light chain. These molecules have only one head and no tail. In addition to these single-headed myosins, the same ameba also contains double-headed myosin II molecules of a more usual type. Myosins I and II are coded by different genes. The functions of unusual myosin I in the protozoan cell are still not clear.

C. Polymerization

Skeletal muscle myosin molecules polymerize in buffers of low ionic strength. The tail fragments of myosin also polymerize very efficiently. Thus, the heads are not essential for polymerization. Polymerization of myosin, like that of actin, has nucleation and elongation stages. The tail-to-tail dimers of the myosin molecules are likely to act as nuclei. Polymerization of myosin leads to formation of the two types of filaments, called bipolar and side-polar filaments (Fig. 1.29).

Bipolar filaments are rods of different sizes with two conical poles. Numerous myosin heads are projected from the surfaces near the poles. The surface of the central part of the rod (the bare zone) has no projected heads. Each filament consists of hundreds of myosin molecules with the tails oriented toward the center and the heads toward the periphery.

Side-polar filaments have no central bare zone. Their lateral view has a rhomboid shape, with two parallel bare sides and two sides with projected heads.

The myosins from nonmuscle cells and from smooth muscle usually form smaller filaments than skeletal muscle myosin.

Unusual single-headed myosin I of the *Acanthamoeba* seems to be the only type of myosin that is not polymerized at all, possibly because it has no tail.

D. Phosphorylation of Myosin and Regulation of Polymerization

Phosphorylation of the light P-chains and of heavy chains is an important method for reversible modification of the properties of myosin molecules in the cell.

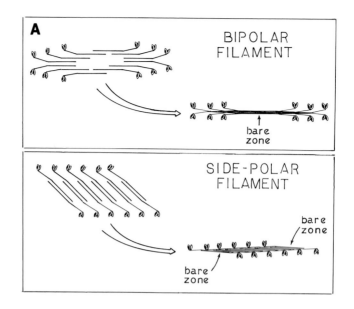

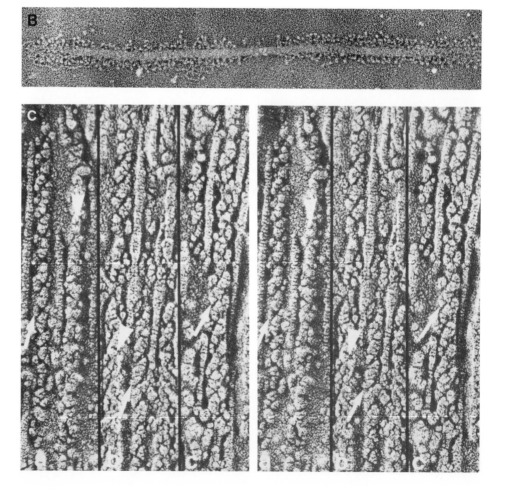

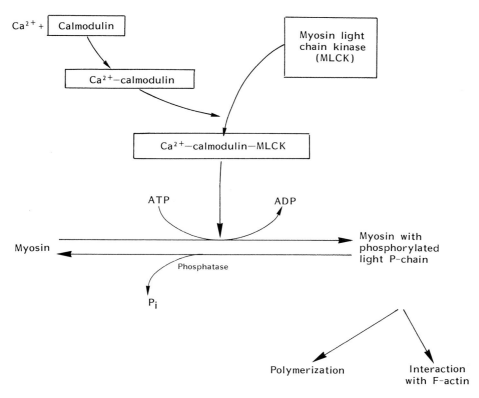

FIG. 1.30. Stages of regulation of phosphorylation of light P-chains of myosin.

Phosphorylation of P-chain in smooth muscle and in nonmuscle cells is performed by a special enzyme, myosin light-chain kinase (MLCK), with a mol. wt. about 100 kD (Fig. 1.30). This enzyme is strictly calcium-dependent. It is activated only when it binds the complex of calcium with another protein, calmodulin. Phosphorylation of the light chains promotes polymerization of smooth muscle and nonmuscle myosin. It also changes the conformation of myosin molecules. Nonphosphorylated molecules usually have a looped shape, while phosphorylated ones often acquire an extended shape. Possibly, the extended conformation is more favorable for polymerization. Another variant of the regulation was revealed in experiments with double-headed myosin II from the *Acanthamoeba*. Here phosphorylation of the heavy

---

←

FIG. 1.29. (A) Scheme of the formation of two types of myosin filaments. **(B)** Myosin thick filament isolated from rabbit psoas muscle, freeze-dried and shadowed with platinum. Notice protruding heads of myosin molecules and the bare zone in the middle of the filament. ×116,000, reproduced at 70%. (From Trinick and Elliott, 1979. Courtesy of J. Trinick.) **(C)** Thick myosin filaments and thin actin filaments in a sarcomere from relaxed fibrils. Platinum replicas of quick-frozen and freeze-etched rabbit psoas muscle. Stereopairs of photographs. Projections at the surface of thick filaments (arrows) are, presumably, myosin heads. ×140,000. (From Ip and Heuser, 1983. Courtesy of W. Ip.)

chains inhibited polymerization. A similar effect was also observed in experiments with myosin of another lower eukaryotic cell, myxomycete *Dictyostellium discoideum*.

E. Formation of Myosin Filaments in the Cell

The thick filament of the striated muscle myofibril is the only conspicuous myosin structure found in cells (Fig. 1.29). The structure of this filament is very similar to that of the bipolar myosin filament formed *in vitro*, so that its assembly in the cell is, obviously, a result of polymerization of myosin molecules. The thick filaments formed in the cell, in contrast to myosin filaments polymerized *in vitro*, have a constant size. It is not yet known how this control of size is achieved. Formation of myosin filaments in certain cells seems to involve not one, but several types of proteins. For instance, polar and central regions of the thick filaments of the nematode muscle contain different isoforms of myosin reacting with different antibodies. The thick filaments of invertebrates contain, besides myosin, another protein, paramyosin. An additional component, C-protein, was also revealed in the thick filaments of vertebrates. The role of these proteins in the formation of filaments is not clear.

As discussed in Section I, it is possible that myosin filaments in smooth muscle and in nonmuscle cells are less stable than these in the striated muscle. Reversible polymerization of myosin in these cells may be controlled by its phosphorylation and dephosphorylation.

VIII. Organization of Actin–Myosin Motile Systems

A. Molecular Interactions of Myosin with Actin and ATP

Interactions of myosin with actin filaments and ATP provide the molecular basis for actin-based movements. Before discussing these triple interactions, we will consider separately reactions of myosin with F-actin and with ATP.

Myosin–actin interactions in the absence of ATP are manifested by formation of stable complexes between myosin heads and actin filaments. These so-called rigor complexes can be formed not only by intact myosin molecules, but also with head-containing fragments (S-1 or HMM). As already mentioned, the complexes have the shape of "arrowheads" with myosin heads attached to the filaments at an angle of about 45° (Fig. 1.2).

Myosin interactions with ATP in the absence of actin are manifested by ATP hydrolysis. The ATPase activity of myosin is localized in the heads. This activity is high in certain nonphysiological solutions, e.g., in solutions containing high concentration of K^+ and lacking bivalent ions (so-called K^+–EDTA activity). In contrast, the ATPase activity of myosin is low in solution containing nearly physiological millimolar concentrations of Mg^{2+} (so-called Mg ATPase activity).

Analysis of the kinetics of ATP hydrolysis in the presence of Mg^{2+} has shown that ATP rapidly forms a complex with myosin and then is rapidly hydrolyzed within this complex, but release of the products of hydrolysis from the complex is slow and limits the rate of the whole process (Fig. 1.31).

When **actin filaments, myosin, and ATP** are brought together, two additional effects are observed in comparison with dual combinations of actin–myosin and myosin–ATP: activation of ATPase activity of myosin by actin (Table 1.4) and detachment of actin from the complex with myosin by ATP.

Actin stimulates only low activity of myosin in the presence of Mg, but not K–EDTA activity. Kinetic evidence shows that inorganic phosphate and ADP are released from the actin-attached myosin much faster than from the unattached myosin. Thus actin increases the Mg^{2+} ATPase activity of myosin by accelerating the rate-limiting stage of the reaction; that is, release of the products of hydrolysis (Fig. 1.31). Addition of ATP also alters the interaction of actin with myosin; it leads to dissociation of the actin–myosin rigor complex. This effect is not dependent on ATP hydrolysis, as it can be produced not only by ATP itself, but also by its nonhydrolyzable analogs.

What is the relationship between the stimulation of ATPase activity by **myosin-associated** actin and **dissociation** of actin from myosin? According to the popular Lymn–Taylor model (1971), myosin interactions with actin and ATP are cyclic (Fig. 1.31). First, ATP binds to the myosin head and detaches this head from actin. Second, ATP is hydrolyzed within the myosin head. Third, the head with the bound products of hydrolysis is reattached to actin. Fourth, reattached actin releases these products from the complex with the head. Then the cycle is repeated again. This scheme is well-supported by the existing experimental data. Some recent evidence indicates that the Lymn–Taylor sequence is not the only existing pathway of ATP hydrolysis. It was shown that myosin heads covalently cross-linked to actin retain high ATPase activity. Kinetic data suggest that some part of the ATP in usual systems is also hydrolyzed directly without the dissociation of actin–myosin complexes. The physiological significance of this additional "direct pathway" is yet to be explained.

At the same time, it is obvious that mutual motility of actin and myosin molecules can be based only on their cyclic associations and dissociations. The force required for motility can be generated in the associated state, while myosin heads, owing to periodic detachments, can change the site of their association with actin filaments.

The exact mechanism of the generation of mechanical force during this cycle is not yet clear. One plausible model of this mechanism is that of swinging myosin heads (Fig. 1.31). According to this model, myosin molecule, containing the products of ATP hydrolysis, is attached to the actin filament at an angle of 90°. Release of these products is accompanied by rotation of the actin–myosin complex to an angle of 45°. This rotation of myosin head in the muscle can pull actin and myosin filaments along one another. However, there is still no direct evidence that the angle of attachment of actin to myosin is changed during contraction of muscle. Another hypothesis (Fig. 1.31) suggests that the moving force is produced not by the rotation of the myosin head, but by the cyclic shortening and relaxation of the proximal part of the

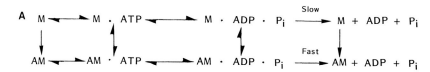

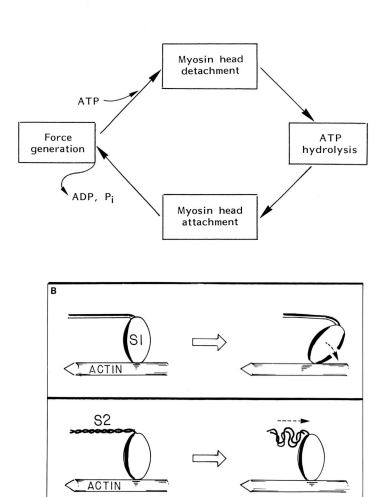

FIG. 1.31. **(A)** Top: Simplified scheme of ATP hydrolysis by myosin heads (M) associated and nonassociated with actin (A). Bottom: Lymn–Taylor model of mechanochemical cycle. **(B)** Two hypothetical models of force generation by actin–myosin interactions: rotation of myosin heads and shortening of S2 region of myosin molecule.

myosin tail (S-2 region). Regardless of the detailed mechanism of the force-producing changes of myosin molecules, it is clear that polarity of the actin filament determines the orientation of myosin attachment. Therefore, myosin molecules during repeated cycles do not randomly wander along the actin filament, but directionally "walk" along it.

TABLE 1.4. The State of Phosphorylation and the Actin-Activated Mg-ATPase Activities of Myosin and HMM[a]

Type of myosin and state of phosphorylation	Phosphate incorporation (mole P_i/mole myosin)	Mg-ATPase (nmole P_i released/min mg)	
		+ Actin	− Actin
Smooth muscle myosin			
Unphosphorylated	0	4	<2
Phosphorylated	1.9	51	<2
Dephosphorylated	0.1	5	
Rephosphorylated	2.0	46	
Smooth muscle HMM			
Unphosphorylated	0	10	<2
Phosphorylated	1.9	357	<2
Dephosphorylated	0.1	20	
Rephosphorylated	2.1	371	
Platelet myosin			
Unphosphorylated	0	5	
Phosphorylated	1.9	89	
Dephosphorylated	0.1	5	

[a] From Sellers *et al.* (1981).

B. Motility of Various Model Systems Containing Actin and Myosin

Movements of actin–myosin structures can be observed not only in the intact cells, but also in demembranated cell models. These models are obtained by extraction of cells with glycerol or detergents. When properly prepared models, containing mostly cytoskeletal structures, are incubated with ATP, they contract. ATP-dependent contraction was observed in experiments with models of striated and smooth muscle and of nonmuscle cells, such as amebae or fibroblasts. The essential role of actin–myosin interactions in these movements has been confirmed by experiments showing that contraction of models can be specifically inhibited by the head-containing myosin fragments (S-1 or HMM) modified by N-ethylmaleimide (NEM). NEM fragments form rigor complexes with actin even in the presence of ATP and thus prevent the attachment of normal myosin heads to actin.

Certain types of motile systems containing actin and myosin can be also prepared *in vitro* from purified proteins. The character of movement in all these systems depends on the geometrical organization of actin and myosin structures. We shall give several examples of actin–myosin systems of various degrees of complexity (Fig. 1.32).

The demembranated skeletal muscle fiber is, certainly, the most highly organized motile actin–myosin system. The lengths and positions of actin and myosin filaments in this system are fixed in such a way that mutual sliding of filaments in the presence of ATP produces highly efficient movements going strictly in one direction.

A more simplified system performing unidirectional walk of myosin molecules along the actin filaments was developed recently be Sheetz and Spudich (1983). They first prepared polymer beads to which myosin molecules

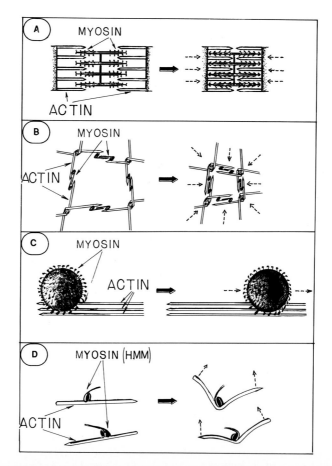

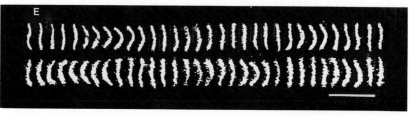

FIG. 1.32. (A–D) Various types of movements based on actin–myosin interactions. (A) Unidirectional contraction of the sarcomere of striated muscle. (B) Three-dimensional contraction of actin gel. (C) Directional sliding of myosin-coated globule along actin filaments. (D) Bending of individual actin filaments caused by interaction with myosin molecules. **(E)** Bending of individual actin filaments in the solution containing heavy meromyosin, Ca^{2+}, and Mg^{2+}-ATP. Sequential micrographs of two filaments (upper and lower rows) taken at 0.1-sec intervals. Filaments were labeled by rhodamine–phalloidin. Bar: 0.5 sec. (E from Yanagida *et al.*, 1984. Micrograph courtesy of T. Yanagida.)

were chemically linked. Then these beads were placed on the surface of actin bundles with uniform polarity of filaments (natural actin bundles from the cytoplasm of the alga *Nitella* or synthetic filament arrays assembled from purified actin *in vitro* can be used). In the presence of ATP and Mg^{2+} the myosin-covered beads moved along the actin filaments. The direction of the movement was always from pointed to barbed end; velocity reached 6 μm/sec. Modification of myosin with NEM blocked the movements.

These experiments show that directional actin–myosin motility can occur in the absence of myosin filaments. Possibly, the system of Sheetz and Spudich (1983) models certain types of organelle movements along the actin filaments (see Chapter 5).

Actin–myosin gel is a motile system with a much lower degree of organization. Its filaments have no definite orientation. These gels can be prepared from purified actin–myosin and other cross-linking proteins or from cell extracts containing all these components. In the presence of ATP and Mg^{2+} the gels contract isotropically; that is, decrease their volume. We do not yet know the exact character of filament movements producing this contraction. It is probable that neighboring filaments slide along one another owing to their interaction with myosin dimers or small polymer fibrils. The essential role of myosin was confirmed by experiments in which contraction of the gel, prepared from the cell extract, was inhibited by the antimyosin antibodies and by NEM-modified myosin fragments. Probably, mechanisms of contraction of these gels *in vitro* are similar to those of three-dimensional actin networks in the cell.

It is interesting that one-headed *Acanthamoeba* myosin I, despite its inability to polymerize, is able to cross-link F-actin and to contract actin gel in the presence of Mg-ATP. This ability seems to be due to the presence of two actin-binding sites on the same heavy chain of this myosin; one of these sites is ATP-sensitive and another ATP-insensitive. Therefore, the single head of this unusual myosin can bind to two actin filaments.

The simplest known model of a motile system consists of actin filaments floating in solution, ATP, and soluble HMM fragments of myosin. Yanagida and collaborators (1984) labeled filaments with fluorochrome–phalloidin and observed their movements in the fluorescence microscope. The thermal bending movements of filaments were seen in a control solution. After addition of ATP and HMM, these bending movements became faster and larger in amplitude. It is likely that attachment and detachment of ATP-activated myosin heads somehow affects the shape of filaments (Fig. 1.32). Obviously, these interactions in solution cannot produce any directional movements of filaments. It is not known whether myosin-induced alteration of the shape of filaments has any biological function in the cell.

Thus, the same molecular force-generating mechanism built into various morphological organizations is able to produce many various types of directional and nondirectional movements, e.g., unidirectional contraction, unidirectional particle movement, isotropic contraction, and deformation of filaments.

IX. Regulation of Contractility in Actin–Myosin Systems

Mutual motility of actin and myosin molecules in the cell is induced or inhibited by reversible alterations of the concentration of calcium ions. Usually the rise from 10^{-7} M to 10^{-6} M acts as the signal permitting the contraction. Although this type of regulation is common for all cells, it is mediated by different molecular mechanisms in various motile systems.

Two large groups of mechanisms are based on calcium-induced modifications of actin–myosin interactions; one of these mechanisms modifies the state of myosin molecules, and another modifies the state of troponin–tropomyosin complex associated with actin filaments. The third group of motility-regulating processes does not affect actin–myosin alterations but modulates the geometry of actin structures; that is, the length and cross-linking of filaments.

A. Regulation of the State of Myosin

Calcium-dependent kinase phosphorylating the light chains of myosin (MLCK, see Section VII.D and Fig. 1.30) is present in smooth muscle and in many nonmuscle cells. Phosphorylation of P-chain by this enzyme increases the actin-activated ATPase activity of myosin and promotes contraction of demembranated cell models and of actin–myosin gels. Phosphorylation does not affect myosin attachment to actin. P-chain, which is also called regulatory light chain, is associated with the myosin head near its junction with the tail; possibly, the nonphosphorylated state of this chain blocks some conformational states of the head essential for the ATPase cycle. Direct binding of calcium to myosin regulatory light chain is another variant of myosin-linked regulation. This variant is characteristic of the scallop muscle and also of some, but not all, vertebrate smooth muscles. Calcium binding is essential for activation of ATPase activity of the myosins of these cells by actin.

Promotion of actin-activated ATPase activity of myosin is often correlated with stimulation of myosin polymerization. Both effects are produced by MLCK-catalyzed phosphorylation of smooth muscle myosin. Phosphorylation of heavy chains of the *Acanthamoeba* myosin II and of *Dictyostelium amebae* myosin inhibits not only its polymerization (see Section VII.D), but also actin-activated ATPase activity. These data suggest that myosin in many motile systems has two different states. Nonpolymerizable myosin poorly interacting with actin can be reversibly transformed by calcium-induced phosphorylation into the molecule, which is easily polymerized and activated by actin.

B. Modulation of the State of Actin-Attached Tropomyosin–Troponin Complex

This regulatory system acts in striated muscle. Its essential components are tropomyosin and another protein, troponin (Fig. 1.33). As previously described (see Section IV), elongated molecules of tropomyosin are side-attached to actin filaments. Troponin consists of three nonidentical subunits. It binds to the tropomyosin–actin complex via two of these subunits. Its third subunit, called troponin C, is a calcium-binding protein. When tropomyosin–troponin complex is attached to the filament, actin interactions with myosin become calcium-dependent. In particular, activation of myosin ATPase by actin is observed only in high calcium ($\geq 10^{-6}$ M) but not in low calcium ($<10^{-7}$ M).

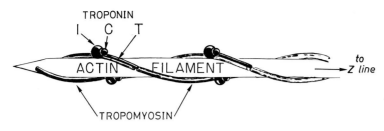

FIG. 1.33. Attachment of tropomyosin–troponin complex to actin.

Analysis of the structure of complexes shows that calcium binding is accompanied by movement of tropomyosin on the filament. The commonly accepted hypothesis suggests that troponin–tropomyosin complex in low calcium sterically blocks actin interaction with myosin by screening filaments from the attachment of myosin heads. Calcium binding to troponin C alters the position of tropomyosin on the filament, so that binding sites for myosin become free.

However, certain data are difficult to explain on the basis of this hypothesis. For instance, Chalovich and Eisenberg (1982) have observed that binding of myosin heads to troponin–tropomyosin-bearing filaments is not changed by calcium, although ATPase activity of these complexes is greatly increased.

Possibly, as suggested by these and other investigators, the tropomyosin position on the actin regulates not the binding of myosin, but the state of the actin-bound myosin; as a result, the rate of certain stages of the ATP-ase cycle, e.g., the rate of release of the products of hydrolysis, can be reversibly changed.

C. Solation and Gelation of Actin Gels

As already discussed, activities of many actin-binding proteins are dependent on calcium. Therefore, calcium can induce reversible alterations in the organization of structures containing these proteins. In particular, the rise of the calcium concentration can partially solate actin gels by activating the severing proteins and by dissociating certain cross-linking proteins from the filaments (Fig. 1.34). This solation can alter the ability of gels for ATP-induced contraction. For instance, when the gel is prepared in the glass capillary and solution containing 1 μM of free Ca^{2+} is added to one end of the capillary, the gel contracts, asymmetrically moving away from this end.

FIG. 1.34. Scheme illustrating hypothetical solation–contraction of actin gel. Interaction with myosin molecules is essential for mutual sliding of actin filaments; this interaction is not shown here.

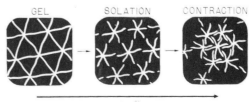

Two alternative hypotheses about the relation of calcium-induced sola-tion to gel contraction in this system have been proposed. The gelation–con-traction hypothesis suggests that maximal contraction takes place in the least solated region of the gel near the nonstimulated end of the capillary. Con-traction of the most gelated region pulls the more solated material away from the calcium-stimulated end. An alternative hypothesis suggests that partial solation of gel promotes contraction because decreased cross-linking of fila-ments makes them more motile. Therefore, both solation and contraction of gel take place at the same stimulated end of the capillary. This solation–con-traction coupling hypothesis, proposed by Taylor and Fechheimer (1982), is supported by their experiments in which stained beads were incorporated into the actin gel and used as markers revealing the movements of various parts of this gel. It was found that gel contraction was initiated at the site of calcium addition, as predicted by the solation–contraction coupling hypoth-esis. It seems probable that solation–contraction relationships can vary depending on the degree of solation. Most rigid gels are unable to contract; partial solation can promote contraction, while full solation can make the contraction impossible.

D. Calcium Regulation of Contraction in Different Cells

The mechanisms regulating contraction in different cells seem to be cor-related with the organization of actin–myosin systems in these cells. Slow contractions of more primitive motile systems, such as three-dimensional networks of actin in nonmuscle cells, are regulated mainly by two groups of mechanisms: the modulations of myosin and the solation of actin gels. These mechanisms, especially the latter, involve considerable reorganization of the structure during each contraction.

The more highly organized contractile system of the smooth muscle is regulated mainly by the first mechanism, myosin modulations. In these cells, reorganization of actin structures during contraction seems to be absent, but reorganizations of myosin structures may occur.

The most highly organized actin–myosin system of striated muscle is reg-ulated by a special tropomyosin–troponin mechanism, which ensures the high rate and frequency of contractions as well as the absence of any detect-able alterations of actin and myosin filaments.

X. Conclusion: Dynamics of Actin Systems

In this chapter we have discussed a number of molecular processes involved in the formation and reorganization of actin structures *in vitro*. In conclusion, we will briefly summarize the role of these processes in dynamic regulation of the actin structures within the cell. Nonpolymerized and poly-merized actin seem to be in a state of dynamic equilibrium in many cells. About half of the actin of cultured nonmuscle cells is nonpolymerized. The concentration of nonpolymerized actin in these cells is much higher than the critical concentration of actin in solution; that is, minimal concentration at which polymerization starts. It is likely that actin-sequestering proteins, such as profilin, are essential for the maintenance of this large pool of nonpolymerized actin in the cell.

Molecules from this pool participate not only in the formation of new actin structures, but also in the renewal of the existing structures. Many, possibly all, actin structures continuously exchange actin molecules with the nonpolymerized pool. This exchange has been demonstrated by various methods, especially by the injection of fluorochrome-labeled nonpolymerized actin molecules into the cells. Labeled actin is incorporated into various preexisting structures, e.g., into the peripheral network and stress fibers of fibroblasts, into the cortical networks of amebae, and even into the myofibrils of myocytes. There is also evidence indicating that the core bundles of the intestinal microvilli undergo continuous renewal. The rate of renewal of different actin structures in the same cell may be unequal; e.g., actin of the peripheral network of fibroblasts was observed to undergo more rapid renewal than that of the stress fibers. Although the data available at present are still incomplete, they suggest that renewal of actin occurs in many types of structures.

Thus, many actin systems have a dynamic composition of filaments, but stable structural organization. How is this combination of dynamics and stability achieved? Continuous renewal is likely to be due to de- and repolymerization of actin at the ends of filaments. The rate of renewal can be increased by the fluctuations causing formation and disappearance of ATP caps at these ends. The rate of renewal can be regulated by the various actin-binding proteins. Stability of the general structure can also be due to the presence of the controlling molecules, which determine the average length and position of filaments. The spectrum of these molecules includes various types of capping proteins, of side-attached and cross-linking proteins, and of the proteins binding filaments to the membrane. These controlling molecules can be associated with special structures, such as Z disks and focal contacts. Filaments can start growing from these structures or, alternatively, stop polymerization by attaching their ends to these structures. Owing to the combined instability of the individual filament and stability of the general pattern, many actin structures can exist and function for long periods of time, continually readjusting their organization to the subtle changes of cell shape, of the position of cellular organelles, and other factors. When intracellular conditions are changed, the existing filaments may be destroyed, and formation of the new structures can be directed by controlling molecules.

Although these general considerations are probably valid, very little is known about the formation and maintenance of specific actin structures. For instance, it is not clear which sequences of molecular events determine the structure and cellular localization of the brush border microvilli, of the stereocilia, of sarcomeres, and of other structures.

Myosin is an essential component of many actin structures, but at present much less is known about its dynamics in the cell than about the dynamics of actin. For example, we do not know the ratio of polymerized to nonpolymerized myosin in the cells or the rate of renewal of the myosin filament. Mutual sliding of actin and myosin is the basis of most motile phenomena in actin systems. In certain systems, e.g., in cortical actin gels, each myosin-dependent movement is coupled with the profound reorganization of the system. In other more specialized systems, e.g., in striated muscle, movements and structural reorganization appear to be completely uncoupled.

One should not forget that certain movements of actin systems, e.g., uncoiling of the bundle in the *Limulus* sperm, can be based on the interactions of filaments not with myosin, but with other actin-binding proteins.

Literature Cited

Baines, A. I. (1984) A spectrum of spectrins, *Nature* **312**:310–311.

Bennett, V., Baines, A. I., and Davis, I. Q. (1985) Ankyrin and synapsin: Spectrin-binding proteins associated with brain membranes, *J. Cell. Biochem.* **29**:157–169.

Branton, D., Cohen, C. M., and Tyler, J. (1981) Interaction of cytoskeletal proteins on the human erythrocyte membrane, *Cell* **24**:24–32.

Burridge, K., and Connell, L. (1983) A new protein of adhesion plaques and ruffling membranes, *J. Cell Biol.* **97**:359–367.

Burridge, K., Kelly, T., and Connell, L. (1982) Proteins involved in the attachment of actin to the plasma membrane, *Phil. Trans. R. Soc. Lond. B* **299**:291–299.

Carlier, M-F., Pantaloni, D., and Korn, E. D. (1984) Evidence of an ATP cap at the end of actin filaments and its regulation of the F-actin steady state, *J. Biol. Chem.* **259**:9983–9986.

Carraway, K. L., and Carraway, C. A. C. (1984) Plasma membrane-microfilament interaction in animal cells, *Bio Essays* **1**:55–58.

Chalovich, J. M., and Eisenberg, E. (1982) Inhibition of actomysin ATPase activity by troponin-tropomysin without blocking the binding of myosin to actin, *J. Biol. Chem.* **257**:2432–3437.

Cooper, J. A., Walker, S. B., and Pollard, T. D. (1983) Pyrene actin: Documentation of the validity of a sensitive assay for actin polymerization, *J. Muscle Res. Cell Motil.* **4**:253–262.

Craig, S. W., and Pollard, T. D. (1982) Actin-binding proteins, *Trends Biochem. Sci.* **7**:88–92.

De Rosier, D. J., and Tilney, L. G. (1984) The form and function of actin. A product of its unique design, in *Cell and Muscle Motility*, Vol. 5, *The Cytoskeleton* (J. W. Shay, ed.), Plenum Press, New York, pp. 139–169.

Egelman, E. H., and De Rosier, D. J. (1983) A model for F-actin derived from image analysis of isolated filaments, *J. Mol. Biol.* **166**:623–629.

Egelman, E. H., and Padron, R. (1984) X-ray diffraction evidence that actin is a 100 Å filament, *Nature* **307**:56–58.

Geiger, B. (1983) Membrane–cytoskeleton interaction, *Biochim. Biophys. Acta* **737**:305–341.

Goodman, S. R., and Shiffer, K. (1983) The spectrin membrane skeleton of normal and abnormal human erythrocytes: A review, *Am. J. Physiol.* **244**:c121–c141.

Heath, J. (1986) Finding the missing links. *Nature* **320**:484–485.

Hirokawa, N., Tilney, L. G., Fujiwara, K., and Heuser, J. E. (1982) Organization of actin, myosin, and intermediate filaments in the brush border of intestinal epithelial cells, *J. Cell Biol.* **94**:425–443.

Hirokawa, N., Keller, T. C. S. III, Chasan, R., and Mooseker, M. (1983) Mechanism of brush border contractility studied by the quick-freeze deep-etch method, *J. Cell Biol.* **96**:1325–1336.

Ip, W., and Heuser, J. (1983) Direct visualization of the myosin crossbridge helices on relaxed rabbit psoas thick filaments, *J. Mol. Biol.* **171**:105–109.

Korn, E. D. (1982) Actin polymerization and its regulation by proteins from nonmuscle cells, *Physiol. Rev.* **62**:672–737.

Lymn, R. W., and Taylor, E. W. (1971) Mechanism of adenosine triphosphate by actomyosin, *Biochemistry* **10**:4617–4624.

Mangeat, P., and Burridge, K. (1984) Actin–membrane interaction in fibroblasts: What proteins are involved in this association? *J. Cell Biol.* **99**:95s–103s.

Maruta, H., Knoerzer, W., Hinssen, H., and Isenberg, G. (1984) Regulation of actin polymerization by non-polymerizable actin-like proteins, *Nature* **312**:424–427.

Maruyama, K., Sawada, H., Kimura, S., Ohashi, K., Higuchi, H., and Umazume, Y. (1984) Connectin filaments in stretched skin fibers of frog skeletal muscle, *J. Cell Biol.* **99**:1391–1397.

Pollard, T. D., and Cooper J. A. (1986) Actin and actin-binding proteins. A critical evaluation of mechanisms and functions, *Annu. Rev. Biochem.* **55**:987–1035.

Sellers, J. R., Pato, M. D. and Adelstein, R. S. (1981) Reversible phosphorylation of smooth muscle myosin, heavy meromyosin and platelet myosin, *J. Biol. Chem.* **256**:13137–13142.

Sheetz, M. P., and Spudich, J. A. (1983) Movement of myosin-coated fluorescent beads on actin cables *in vitro*, *Nature* **303**:31–35.

Small, J. V. (1981) Organization of actin in the leading edge of cultured cells: Influence of osmium tetroxide and dehydration on the ultrastructure of actin meshworks, *J. Cell Biol.* **91**:695–705.

Smith, P. R., Fowler, W. E., Pollard, T. D., and Aebi, U. (1983) Structure of the actin molecule determined from electron micrographs of crystalline actin sheets with a tentative alignment of the molecule in the actin filament, *J. Mol Biol.* **167**:641–660.

Speicher, D. W. (1986) The present status of erythrocyte spectrin structure: The 106-residue repetitive structure is a basic feature of an entire class of proteins, *J. Cell Biochem.* **30**:245–258.

Spudich, J. A., Pardee, J. D., Simpson, P. A., Yamamoto, K., Kuczmarski, E. R., and Stryer, L. (1982) Actin and myosin: Control of filament assembly, *Phyl. Trans. R. Soc. Lond.* **B299**:247–261.

Stossel, T. P. (1984) Contribution of actin to the structure of the cytoplasmic matrix, *J. Cell Biol.* **99**:15s–21s.

Svitkina, T. M., Shevelev, A. A., Bershadsky, A. D., and Gelfand, V. I. (1984) Cytoskeleton of mouse embryo fibroblasts. Electron microscopy of platinum replicas, *Eur. J. Cell Biol.* **34**:64–74.

Taylor, D. L., and Fechheimer, M. (1982) Cytoplasmic structure and contractility: The solation–contraction coupling hypothesis, *Phil. Trans. R. Soc. Lond. B.* **299**:185–197.

Taylor, D. L., Reidler, J., Spudich, J. A., and Stryer, L. (1981) Detection of actin assembly by fluorescence energy transfer, *J. Cell Biol.* **89**:362–367.

Trinick, J., and Elliott, A. (1979) Electron microscope studies of thick filaments from vertebrate skeletal muscle, *J. Mol. Biol.* **131**:133–136.

Trybus, K. M., and Lowey, S. (1984) Conformational states of smooth muscle myosin. Effects of light chain phosphorylation and ionic strength, *J. Biol. Chem.* **259**:8564–8571.

Tsukita, S., Tsukita, S., and Ishikawa, H. (1983a) Association of actin and 10 nm filaments with the dense body in smooth muscle cells of the chicken gizzard, *Cell Tissue Res.* **229**:233–242.

Tsukita, S., Tsukita, S., Ishikawa, H., Kurokawa, M., Morimoto, K., Sobue, K., and Kakiuchi, S. (1983b) Binding sites of calmodulin and actin on the brain spectrin, calspectin, *J. Cell Biol.* **97**:574–578.

Vandekerckhove, I., and Weber, K. (1978a) The amino acid sequence of *Physarum* actin, *Nature* **276**:720–721.

Vanderkerckhove, I., and Weber, K. (1978b) Actin amino acid sequences. Comparison of actins from calf thymus, bovine brain, and SV-40 transformed 3T3 cells with rabbit skeletal muscle actin, *Eur. J. Biochem.* **90**:451–462.

Weeds, A. (1982) Actin-binding proteins-regulators of cell architecture and motility, *Nature* **296**:811–816.

Yanagida, T., Nakase, M., Nishiyama, K., and Oosawa, F. (1984) Direct observation of motion of single F-actin filaments in the presence of myosin, *Nature* **307**:58–60.

Additional Readings

Actin; structure of the filament; assembly *in vitro*

Barden, I. A., Grant, N. J., and dos Remedios, C. G. (1982) Identification of the nucleus of actin polymerization, *Biochem. Intern.* **5**:685–692.

Bonder, E. M., Fishkind, D. J., and Mooseker, M. S. (1983) Direct measurement of critical concentrations and assembly rate constants at the two ends of an actin filament, *Cell* **94**:491–501.

Chang, K. S., Zimmer, W. E., Jr., Bergsma, D. I., Dogdson, I. B., and Schwartz, R. I. (1984) Isolation and characterization of six different chicken actin genes, *Mol. Cell Biol.* **4**:2498–2508.

Egelman, E. H., and De Rosier, D. J. (1983) Structural studies of F-actin, in *Actin; Structure and Function in Muscle and Non-muscle Cells* (C. Dos Remedios and J. Barden, eds.), Academic Press, Sydney, pp. 17–24.

Fyrberg, E. A., Bond, B. I., Hershey, N. D., Mixter, K. S., and Davidson, N. (1981) The actin genes of *Drosophila*: Protein coding regions are highly conserved but intron positions are not, *Cell* **24**:107–116.

Kaine, B. P., and Spear, B. B. (1982) Nucleotide sequence of a macronuclear gene for actin in *Oxytrecha fallax, Nature* **295**:430–432.

Lal, A. A., Brenner, S. L., and Korn, E. D. (1984) Preparation and polymerization of skeletal muscle ADP-actin, *J. Biol. Chem.* **259**:13061–13065.

Mornet, D., and Ue, K. (1984) Proteolysis and structure of skeletal muscle actin, *Proc. Natl. Acad. Sci. USA* **81**:3680–3684.

Ng, S-Y., Gunning, P., Eddy, R., Ponte, P., Leavitt, I., Shows, T., and Kedes, L. (1985) Evolution of the functional human β-actin gene and its multi-pseudogene family: Conservation of non-coding regions and chromosomal dispersion of pseudogenes, *Mol. Cell Biol.* **5**:2720–2732.

Oosawa, F. (1983) Macromolecular assembly of actin, in *Muscle and Nonmuscle Motility*, Vol. 1 (A. Stracher, ed.), Academic Press, New York, pp. 151–216.

Pantaloni, D., Hill, T. L., Carlier, M-F., and Korn, E. D. (1985) A model for actin polymerization and the kinetic effects of ATP hydrolysis, *Proc. Natl. Acad. Sci USA* **82**:7207–7211.

Pardee, J. D., and Spudich, J. A. (1982) Mechanism of $K^{\pm}$ induced actin assembly, *J. Cell Biol.* **93**:648–654.

Pollard, T. (1984) Polymerization of ADP-actin, *J. Cell Biol.* **99**:769–777.

Pollard, T. D., and Craig, S. W. (1982) Mechanism of actin polymerization, *Trends Biochem. Sci.* **7**:55–58.

Pollard, T. D., and Weeds, A. G. (1984) The rate constant for ATP hydrolysis by polymerized actin, *FEBS Lett.* **170**:94–98.

Suck, D., Kasch, W., and Mannherz, H. G. (1981) Three-dimensional structure of the complex of skeletal muscle action and bovine pancreatic DNase I at 6-Å resolution, *Proc. Natl. Acad. Sci. USA* **78**:4319–4323.

Tobacman, L., and Korn, E. D. (1983) The kinetics of actin nucleation and polymerization, *J. Biol. Chem.* **258**:3207–3214.

Vandekerckhove, I., and Weber, K. (1984) Chordate muscle actins differ distinctly from invertebrate muscle actins. The evolution of the different vertebrate muscle actins, *J. Mol. Biol.* **179**:391–413.

Wanger, M., Keiser, T., Neuhaus, J-M., and Wegner, A. (1985) The actin treadmill, *Can. J. Biochem. Cell Biol.* **63**:414–421.

Wegner, A. (1976) Head-to-tail actin polymerization, *J. Mol. Biol.* **108**:139–150.

Wegner, A., and Engel, J. (1975) Kinetics of the cooperative association of actin to actin filaments, *Biophys. Chem.* **3**:215–225.

Woodrum, D. T., Rich, S. A., and Pollard, T. D. (1975) Evidence for the biased bidirectional polymerization of actin using heavy meromyosin produced by an improved method, *J. Cell Biol.* **67**:231–237.

Actin-binding proteins; cytochalasins; phalloidin

Atlas, S. I., and Lin, S. (1978) Dihydrocytochalasin B. Biological effects and binding to 3T3 cell, *J. Cell Biol.* **76**:360–370.

Bonder, E., and Mooseker, M. S. (1983) Direct electron microscopic visualization of barbed end capping and filament cutting by intestinal microvillar 95-kdalton protein (villin). A new actin assembly assay using the *Limulus* actosomal process, *J. Cell Biol.* **96**:1097–1107.

Bretscher, A. (1984) Smooth muscle caldesmon. Rapid purification and F-actin cross-linking properties, *J. Biol. Chem.* **259**:12873–12880.

Bretscher, A., and Weber, K. (1980) Fimbrin, a new microfilament-associated protein present in microvilli and other cell surface structures, *J. Cell Biol.* **86**:335–340.

Burridge, K., and Feramisco, J. R. (1981) Non-muscle α-actinins are calcium-sensitive actin-binding proteins, *Nature* **294**:565–567.

Coluccio, L. M., and Tilney, L. G. (1984) Phalloidin enhances actin assembly by preventing monomer dissociation, *J. Cell Biol.* **99**:529–535.

Cooper, J. A., and Pollard T. D. (1985) Effect of capping proteins on the kinetics of actin polymerization, *Biochemistry* **24**:793–799.

Di Nubile, M. J., and Southwick, F. S. (1985) Effects of macrophage profilin on actin in the presence and absence of acumentin and gelsolin, *J. Biol. Chem.* **260**:7402–7409.

Gifford, R. G., Weeds, A. G., and Spudich, J. A. (1984) Ca^{2+}-dependent binding of severin of actin: A one-to-one complex is formed, *J. Cell Biol.* **98**:1796–1803.

Glenney, J. R., Jr., and Glenney, P. (1983) Fodrin is the general spectrin-like protein found in most cells whereas spectrin and TW protein have a restricted distribution, *Cell* **34**:503–512.

Grumet, M., and Lin, S. (1980) A platelet inhibitor protein with cytochalasin-like activity against actin polymerization *in vitro*, *Cell* **21**:439–444.

Kwiatkowski, D. J., Janmey, P. A., Mole, J. E., and Yin, H. L. (1985) Isolation and properties of two actin-binding domains in gelsolin, *J. Biol. Chem.* **260**:15232–15238.

MacLean-Fletcher, S., and Pollard, T. D. (1980) Mechanism of action of cytochalasin B on actin, *Cell* **20**:329–341.

Markey, F., Larsson, H., Weber, K., and Lindberg, U. (1982) Nucleation of actin polymerization from profilactin. Opposite effects of different nuclei, *Biochem. Biophys. Acta* **704**:43–51.

Matsudaira, P., Jakes, R., and Walker, J. E. (1985) A gelsolin-like Ca^{2+}-dependent actin-binding in villin, *Nature* **315**:248–250.

Matsumura, F., and Yamashiro-Matsumura, S. (1985) Purification and characterization of multiple isoforms of tropomyosin from rat cultured cells, *J. Biol. Chem.* **260**:13851–13859.

Payne, M. R., and Rudnick, S. E. (1984) Tropomyosin as a modulator of microfilaments, *Trends Biochem. Sci.* **9**:361–363.

Sobue, K., Muramoto, Y., Fijita, M., and Kokinchi, S. (1981) Purification of a calmodulin binding protein from chicken gizzard that interacts with F-actin, *Proc. Natl. Acad. Sci. USA* **78**:5652–5655.

Southwick, F. S., and Hartwig, J. H. (1982) Acumentin, a protein in macrophages which caps the "pointed" end of actin filaments, *Nature* **297**:303–307.

Speicher, D. W., and Marchesi, V. T. (1984) Erythrocyte spectrin is comprised of many homologous triple helical segments, *Nature* **311**:177–180.

Stossel, T. P., Hartweg, J. H., Yin, H. L., Zaner, K. S., and Stendahl, O. I. (1982) Actin gelation and the structure of cortical cytoplasm, *Cold Spring Harbor Symp. Quant. Biol.* **46**:569–577.

Sutoh, K., and Hatano, S. (1986) Actin–fragmin interactions as revealed by chemical cross-linking, *Biochemistry* **25**:435–440.

Tilney, L. G., Bonder, E. M., Coluccio, L. M., and Mooseker, M. S. (1983) Actin from thyone sperm assembles on only one end of an actin filament: A behavior regulated by profilin, *J. Cell Biol.* **97**:112–124.

Vandekerckhove, J., Debobin, A., Nassal, M., and Wieland, T. (1985) The phalloidin binding site of F-actin, *EMBO J.* **4**:2815–2818.

Verkhovsky, A. B., Surgucheva, I. G., and Gelfand, V. I. (1984) Phalloidin and tropomyosin do not prevent actin filament shortening by the 90 kD protein-actin complex from brain, *Biochem. Biophys. Res. Commun.* **123**:596–603.

Verner, K., and Bretscher, A. (1985) Microvillus 110K-calmodulin: Effects of nucleotides on isolated cytoskeletons and the interaction of the purified complex with F-actin, *J. Cell Biol.* **100**:1455–1465.

Weber, K., and Glenney, Jr., J. R. (1982) Calcium-modulated multifunctional proteins regulating F-actin organization, *Cold Spring Harbor Symp. Quant. Biol.* **46**:541–551.

Wegner, A. (1982) Kinetic analysis of actin assembly suggests that tropomyosin inhibits spontaneous fragmentation of actin filaments, *J. Mol. Biol.* **161**:217–227.

Weihing, R. R. (1985) The filamins: Properties and functions, *Can. J. Biochem. Cell Biol.* **63**:397–413.

Wieland, T. (1977) Modification of actins by phallotoxins, *Naturwissenschaften* **64**:303–309.

Yonezawa, N., Nishida, E., and Sakai, H. (1985) pH control of actin polymerization by cofilin, *J. Biol. Chem.* **260**:14410–14412.

Adelstein, R. S., Pato, M. D., Sellers, J. R., de Lanerolle, P., and Conti, M. A. (1982) Regulation of actin–myosin interaction by reversible phosphorylation of myosin and myosin kinase. *Cold Spring Harbor Symp. Quant. Biol.* **46**:921–928.

Bagshaw, C. R. (1982) *Muscle Contraction*, Chapman and Hall, London, New York.

Cande, W. Z., Tooth, P. J., and Kendrick-Jones, J. (1983) Regulation of contraction and thick filament assembly–disassembly in glycerinated vertebrate smooth muscle cells, *J. Cell Biol.* **97**:1062–1071.

Chen, T., Applegate, D., and Reisler, E. (1985) Cross-linking of actin to myosin subfragment 1: Course of reaction and stoichiometry of products, *Biochemistry* **24**:137–144.

Craig, R., Smith, R., and Kendrick-Jones, J. (1983) Light chain phosphorylation controls the conformation of vertebrate non-muscle myosin and smooth muscle myosin molecule, *Nature* **302**:436–439.

Crow, M. T., and Stockdale, F. E. (1984) Myosin isoforms and the cellular basis of skeletal muscle development, *Exp. Biol. Med.* **9**:165–174.

Eisenberg, E., and Hill, T. L. (1985) Muscle contraction and free energy transduction in biological systems, *Science* **227**:999–1006.

Engelhardt, V. A., and Ljubimova, M. N. (1939) Myosin and adenosinetriphosphatase, *Nature* **144**:668–669.

Epstein, H. F., Miller, D. M., III, Ortiz, I., and Berliner, G. C. (1985) Myosin and paramyosin are organized about a newly identified core structure, *J. Cell Biol.* **100**:904–915.

Flicker, P. F., Phillips, Jr., G. N., Cohen, C. (1982) Troponin and its interactions with tropomyosin, an electron microscope study, *J. Mol. Biol.* **162**:495–501.

Flicker, P. F., Peltz, G., Sheetz, M. P., Parham, P., and Spudich, J. A. (1985) Site-specific inhibition of myosin-mediated motility *in vitro* by monoclonal antibodies, *J. Cell Biol.* **100**:1024–1030.

Fujusaki, H., Albanesi, J. P., and Korn, E. D. (1985) Experimental evidence for the contractile activities of *Acanthamoeba* myosins IA and IB, *J. Biol. Chem.* **260**:11183–11189.

Goody, R. S., and Holmes, K. C. (1983) Crossbridges and the mechanism of muscle contraction, *Biochem. Biophys. Acta* **726**:13–39.

Harrington, W. F. (1979) On the origin of the contractile force in skeletal muscle, *Proc. Natl. Acad. Sci. USA* **76**:5066–5070.

Huxley, H. E. (1983) Molecular basis of contraction in cross-striated muscles and relevance to motile mechanisms in other cells, in *Muscle and Nonmuscle Motility*, Vol. 1 (A. Stracher, ed.), Academic Press, New York, pp. 1–104.

Huxley, H. E., and Hanson, H. J. (1954) Changes in the cross striations of muscle during contraction and stretch and their structural interpretation, *Nature* **173**:973–976.

Huxley, A. F., and Niedergerke, R. (1954) Structural changes in muscle during contraction. Interference microscopy of living muscle fibers, *Nature* **173**:971–973.

Kiehart, D. P., and Pollard, T. D. (1984) Inhibition of *Acanthamoeba* actomyosin-II ATPase activity and mechanichemical function by specific monoclonal antibodies, *J. Cell Biol.* **99**:1024–1033.

Kuczmarski, E. R., and Pardee, J. D. (1983) Actin and myosin from *Dictyostelium amoebae*, in *Cell and Muscle Motility*, Vol. 4 (R. M. Dowben and J. W. Shay, eds.), Plenum Press, New York, London, pp. 269–316.

Leavis, P. C., and Gergely, J. (1984) Thin filament proteins and thin filament-linked regulation of vertebrate muscle contraction, *CRC Crit. Rev. Biochem.* **16**:235–305.

Mahdavi, V., Chambers, A. P., and Nadal-Ginard, B. (1984) Cardiac α- and β-myosin heavy chain genes are organized in tandem, *Proc. Natl. Acad. Sci. USA* **81**:2626–2630.

Masuda, H., Owaribe, K., Hayashi, H., and Hatano, S. (1984) Ca^{2+}-dependent contraction of human lung fibroblasts treated with triton X-100: A role of Ca^{2+}-calmodulin-dependent phosphorelation of myosin 20,000-dalton light chain, *Cell Motil.* **4**:315–331.

Miller, D. M., Ortiz, I., Berliner, G. C., and Epstein, H. F. (1983) Differential localization of two myosins within nematode thick filaments, *Cell* **34**:477–490.

Nag, A. C., Cheng, M., and Zak, R. (1985) Distribution of isomyosin in cultured cardiac myocytes as determined by monoclonal antibodies and adenosine triphosphate activity, *Exp. Cell Res.* **158**:53–62.

Ngai, P. K., and Walsh, M. P. (1984) Inhibition of smooth muscle actin-activated myosin Mg^{2+}-ATPase activity by caldesmon, *J. Biol. Chem.* **259:**13656–13659.

Onishi, H., and Wakabayashi, T. (1984) Electron microscopic studies on myosin molecules from chicken gizzard muscle. III. Myosin dimers, *J. Biochem.* **95:**903–905.

Pepe, F. A. (1983) Macromolecular assembly of myosin, in *Muscle and Nonmuscle Motility*, Vol. 1 (A. Stracher, ed.), Academic Press, New York, pp. 105–149.

Reines, D., and Clarke, M. (1985) Immunochemical analysis of the supramolecular structure of myosin in contractile cytoskeletons of *Dictyostellium amoebae*, *J. Biol. Chem.* **260:**14248–14254.

Reisler, E., Cheung, P., Borochov, N., and Lake, J. A. (1986) Monomers, dimers and minifilaments of vertebrate skeletal myosin in the presence of sodium pyrophosphate, *Biochemistry* **25:**326–332.

Rosenfeld, S. S., Taylor, E. W. (1984) The ATPase mechanism of skeletal and smooth muscle acto-subfragment 1, *J. Biol. Chem.* **259:**11908–11919.

Sellers, R. (1985) Mechanism of the phosphorylation-dependent regulation of smooth muscle heavy meromyosin, *J. Biol. Chem.* **260:**15815–15819.

Sellers, J. R., Spudich, J. A., and Sheetz, M. P. (1985) Light chain phosphorylation regulates the movement of smooth muscle myosin on actin filaments, *J. Cell Biol.* **101:**1897–1902.

Spudich, J. A., Kron, S. J., and Sheetz, M. P. (1985) Movement of myosin coated beads on oriented filaments reconstituted from purified actin, *Nature* **315:**584–586.

Squire, J. M. (1981) *The Structural Basis of Muscular Contraction*, Plenum Press, New York.

Squire, J. M. (1983) Molecular mechanisms in muscular contraction, *Trends Neurosci.* **6:**409–413.

Squire, J. M. (1986) Muscle myosin filaments: Internal structure and crossbridge organization, *Comments Mol. Cell. Biophys.* **3:**155–177.

Waller, S., and Lowey, S. (1985) Myosin subunit interactions. Localization of the alkali light chains, *J. Biol. Chem.* **260:**14368–14373.

Weydert, A., Daubas, P., Lazaridis, I., Barton, P., Garner, I., Leader, D. P., Bonhomme, F., Catalan, J., Simon, D., Guenet, J. L., Gros, F., and Buckingham, M. E. (1985) Genes for skeletal muscle myosin heavy chains are clustered and are not located on the same mouse chromosome as a cardiac myosin heavy chain gene, *Proc. Natl. Acad. Sci. USA* **82:**7183–7187.

Actin structures; dynamics of actin in the cell

Bennett, V. (1985) The membrane skeleton of human erythrocytes and its implications for more complex cells, *Annu. Rev. Biochem.* **54:**273–304.

Bershadsky, A. D., Gelfand, V. I., Svitkina, T. M., and Tint, I. S. (1980) Destruction of microfilament bundles in mouse embryo fibroblasts treated with inhibitors of energy metabolism, *Exp. Cell Res.* **127:**421–429.

Bond, M., and Somlyo, A. V. (1982) Dense bodies and actin polarity in vertebrate smooth muscle, *J. Cell Biol.* **95:**403–413.

Byers, H. R., White, G. E., and Fujiwara, K. (1984) Organization and function of stress fibers in cells *in vitro* and *in situ*: A review, in *Cell and Muscle Motility*, Vol. 5, *The Cytoskeleton* (J. W. Shay, ed.), Plenum Press, New York, London, pp. 83–137.

Byers, T. J., and Branton, D. (1985) Visualization of the protein associations in the erythrocyte membrane skeleton, *Proc. Natl. Acad. Sci. USA* **82:**6153–6157.

Craig, S. W., and Pardo, J. V. (1983) Gamma actin, spectrin, and intermediate filaments proteins colocalize with vinculin at costameres, myofibril-to-sarcolemma attachment sites, *Cell Motil.* **3:**449–462.

Geiger, B., Avnur, Z., Rinnerthaler, L., Kinssen, H., and Small, V. J. (1984) Microfilament-organizing centers in areas of cell contact: Cytoskeletal interactions during cell attachment and locomotion, *J. Cell Biol.* **99:**83s–91s.

Glacy, S. D. (1983) Pattern and time course of rhodamine–actin incorporation in cardiac myocytes, *J. Cell Biol.* **96:**1164–1167.

Hartwig, J. H., Niederman, R., and Lind, S. E. (1985) Cortical actin structures and their relationship to mammalian cell movements, in *Subcellular Biochemistry*, Vol. 11 (D. B. Roodyn, ed.), Plenum Press, New York, Chapter 1, pp. 1–49.

Heuser, J. E., and Kirschner, M. W. (1980) Filament organization revealed in platinum replicas of freeze-dried cytoskeletons, *J. Cell Biol.* **86**:212–234.

Ishikawa, H. (1983) Fine structure of skeletal muscle, in *Cell and Muscle Motility*, Vol. 4 (R. M. Dowben and J. W. Shay, eds.), Plenum Press, New York, pp. 1–84.

Ishikawa, H., Bischoff, R., and Holtzer, H. (1969) Formation of arrowhead complexes with heavy meromyosin in a variety of cell types, *J. Cell Biol.* **43**:312–328.

Kreis, T. E., Geiger, B., and Schlessinger, J. (1982) Mobility of microinjected rhodamine–actin within living chicken gizzard cells determined by fluorescence photoblenching recovery, *Cell* **29**:835–845.

Langanger, G., De Mey, J., Moeremans, M., Daneels, G., DeBrabander, M., and Small, J. V. (1984) Ultrastructural localization of α-actinin and filamin in cultured cells with the immunogold staining (IGS) method, *J. Cell Biol.* **99**:1324–1334.

Lazarides, E. (1976) Two general classes of cytoplasmic actin filaments in tissue culture cells: The role of tropomyosin, *J. Supramol. Struct.* **5**:531(383)–563(415).

Mooseker, M. S., Bonder, E. M., Conzelman, K. A., Fishkind, D. J., Howe, C. L., and Keller, III, C. S. (1984) Brush border cytoskeleton and integration of cellular functions, *J. Cell Biol.* **99**:104s–112s.

Sanger, J. W., Mittal, B., and Sanger, J. M. (1984) Analysis of myofibrillar structure and assembly using fluorescently labelled contractile proteins, *J. Cell Biol.* **98**:825–833.

Schlessinger, J., and Geiger, B. (1983) The dynamic interrelationships of actin and vinculin in cultured cells, *Cell Motil.* **3**:399–403.

Schliwa, M. (1982) Action of cytochalasin D on cytoskeletal networks, *J. Cell Biol.* **92**:79–91.

Stidwill, R. P., Wysolmerski, T., and Burgess, D. R. (1984) The brush border cytoskeleton is not static: *In vivo* turnover of proteins, *J. Cell Biol.* **98**:641–645.

Tilney, L. G., and Tilney, M. S. (1984) Observation on how actin filaments become organized in cells, *J. Cell Biol.* **99**:76s–82s.

Tilney, L. G., Bonder, E. M., and De Rosier, D. J. (1981) Actin filaments elongate from their membrane-associated ends, *J. Cell Biol.* **90**:485–494.

Wang, Y-L., Lanni, F., McNeil, P. L., Ware, B. R., and Taylor, D. L. (1982) Mobility of cytoplasmic and membrane-associated actin in living cells, *Proc. Natl. Acad. Sci. USA* **79**:4660–4664.

Wehland, J., and Weber, K. (1980) Distribution of fluorescently labelled actin and tropomyosin after microinjection in living tissue cultured cells as observed with TV image intensification, *Exp. Cell Res.* **127**:397–408.

2

Systems of Microtubules

I. Introduction: Microtubular Structures in the Cell

A. Components of Microtubular Structures

Microtubules, like actin filaments, are universal components of all the eukaryotic cells (Fig. 2.1). Microtubules have the largest diameter of all cytoskeletal fibrils; this diameter is usually about 25 nm. The wall of microtubules is about 5 nm wide. This wall is made of the single protein tubulin. Other proteins are usually attached to the outer surface of the wall of the microtubule: they are called microtubule-associated proteins (MAPs). Dynein is an MAP of special importance because its interaction with tubulin is essential for the motility of cilia and flagella.

Microtubules, like actin filaments, are polar structures with unequal ends, called plus and minus ends. These ends have a number of different properties; in particular, the growth of microtubules in tubulin solution is usually faster at the plus ends than at the minus ends. Later we shall describe methods of determination of the polarity of microtubules.

One or both ends of cellular microtubules are often associated with special structures called microtubule-organizing centers (MTOCs). There are many different morphological types of MTOCs, including centrosomes, basal bodies, and kinetochores. Microtubules linked to one another and/or to MTOCs form many types of arrays in the cells.

B. Morphology of Microtubular Structures

Regular parallel arrays and radial arrays form two large groups of microtubular structures (Fig. 2.2A). Regular **parallel arrays** are the most highly organized structures; the distance between microtubules, the pattern of the array in cross-section, and often the number of microtubules in the array are strictly fixed in these structures. Many variants of these arrays are found in Protozoa. For instance, the feeding organelle of ciliates, the pharyngeal basket, consists of microtubular bundles that have a hexagonal pattern in cross-section. The microtubular array of the long cytoplasmic projections of Heliozoa consists of microtubules forming two concentric spirals in cross-section

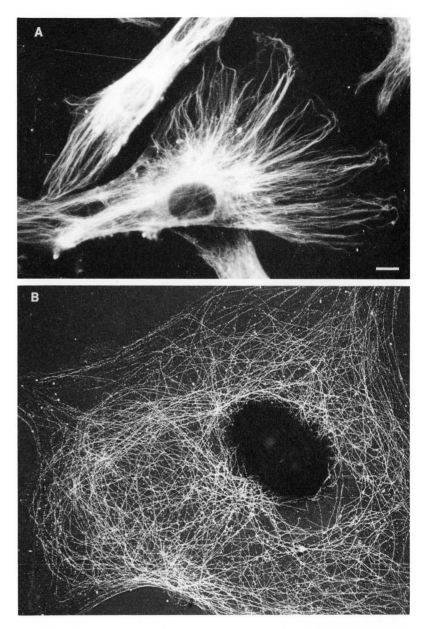

FIG. 2.1. **(A)** Microtubules radiating from the perinuclear zone into the peripheral parts of cultured mouse fibroblast. Immunofluorescence microscopy using antitubulin antibody. (Courtesy of I. S. Tint.) Scale bar: 5 μm. **(B)** Distribution of cytoplasmic microtubules in cultured African green monkey kidney epithelial cell (CV-I line). The specimen was incubated with gold-labeled antibodies, then examined in the photoelectron microscope; specimens were illuminated by UV light, and the electrons emitted by the gold particles were visualized. Immunophotoelectron microscopy. ×3000; reproduced at 55%. (From Birrell *et al.*, 1985. Courtesy of G. B. Birrell.)

(Fig. 2.2B, C, D). Mitotic half-spindles of certain lower eukaryotes, diatoms, also have nearly parallel microtubules. Axoneme of the cilia and flagella is the most common type of regular array, which is found in most groups of eukaryotes (Fig. 2.2E). This axoneme is a cylinder composed of nine doublets of fused microtubules (outer doublets) and a pair of central microtubules. Each outer doublet consists of the complete A-microtubule and incomplete B-microtubule. Both central microtubules are complete. The plus ends of all the microtubules are at the distal tips of the axoneme. Several types of transverse projections are attached to the walls of axonemal microtubules. Two curved sidearms are attached to the A-microtubules of each outer doublet; these sidearms consist of dynein, and their interaction with the walls of B-microtubules of adjacent doublets is the basis of motility of cilia and flagella. Three other types of projections are (1) the links connecting adjacent outer doublets with one another, (2) the radial spokes going from the A-microtubules of outer doublets toward the central pair, and (3) the central sheath projections radiating from the walls of the central pair.

The ciliary axoneme is a direct continuation of its MTOC, the basal body. This MTOC itself is a highly regular parallel array of microtubules. It is a cylinder made by nine triplets of fused microtubules. Each triplet has one complete A-microtubule and two incomplete B- and C-microtubules. A- and B-microtubules of the outer doublets of the axoneme grow from A- and B-microtubules of the basal body triplets, respectively. Central microtubules of the axoneme have no counterparts within the basal body. This complex organization of the MTOC is characteristic of cilia and flagella, but not of other types of parallel arrays. For instance, microtubules of Heliozoa grow from the outer parts of nuclear membrane or from the clumps of electron-dense material without any visible organization. In all the parallel arrays, microtubules always have the same polarity; their plus ends are directed away from MTOCs.

Radial arrays have a central MTOC, from which microtubules radiate with their plus ends out. Mitotic half-spindles in cells of higher eukaryotes and interphase arrays of cytoplasmic microtubules in melanocytes and other animal cells are the two best-known examples of these structures (Figs. 2.1 and 2.2F, G, H, I). The MTOCs of both these arrays, centrosomes, usually consist of a pair of centrioles surrounded by dense pericentriolar material, which sometimes forms more discrete foci, called satellites (Fig. 2.2J, K). The structure of the centriole is identical to that of the basal body. The central ends of radiating microtubules seem to be associated not directly with the centriole, but with the surrounding pericentriolar material. The MTOC of certain radial arrays, e.g., of the mitotic spindles of plant cells, consists only of the dense material without any centriole. Plus ends of microtubules in radial arrays are usually free; the only exception is the association of certain spindle microtubules with the kinetochore. Radial patterns of the interphase and mitotic arrays may be modified and altered in various ways. In particular, certain groups of microtubules of these arrays can become nearly parallel to one another, forming bundlelike structures (Fig. 2.2L). These nearly parallel groups are most often seen at the cell periphery, especially within the cytoplasmic projections of various types, e.g., within the processes of neural cells. These parallel groups of microtubules have a much looser pattern than the

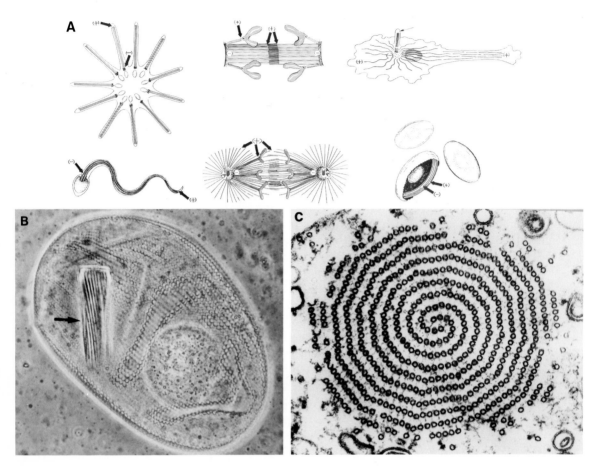

FIG. 2.2A–I. **(A)** Various types of structures made of microtubules. Parallel arrays in the projections of a heliozoan, in a sperm flagellum, and in the mitotic spindle of a diatom. Radial arrays in animal mitotic spindle and in animal interphase cell. Part of an interphase radial array in the cell projection is aligned into the bundlelike structure. Primary cilium of the interphase cell is marked by an arrow. Submembranous microtubular bundle of the avian erythrocyte. Polarities of microtubules are marked by + and −.

(B–D) Microtubular arrays of Protozoa. Cytopharyngeal basket of the ciliate *Nassula*. (B) Cytoskeleton of the *Nassula* extracted with digitonin. The cytopharyngeal basket is indicated by an arrow. Phase contrast. Notice the rods forming the basket. (From Tucker, 1971. Courtesy of J. B. Tucker.) (C) Cross-section of the microtubule bundle in the axopodium of the heliozoan *Echinosphaerium nucleofilum*. ×80,000, reproduced at 85%. (From Tucker,

1984.) (D) Microtubules forming the rod of the basket in cross-section. (From Tucker, 1979. Courtesy of J. B. Tucker.)

(E) Structure of flagellum. Only a sector of the cross-section is seen. Protofilaments of the microtubules are labeled by numbers.

(F–I) Radial microtubular system in pigment cells (erythrophores) of the squirrel fish *Holocentrus ascensionis*. (F) Indirect immunofluorescence. ×700. (G–I) A whole-mount cytoskeleton seen in high-voltage electron microscopy. Microtubules were revealed by immunogolding (incubation with antitubulin antibody was followed by incubation with colloidal gold particles attached to protein A specifically reacting with immunoglobulin molecules). Black pigment granules remain attached to the microtubules. G ×2800, H, I ×10,000. (F–I are from Ochs and Stearns, 1981. Courtesy of M. E. Stearns.)

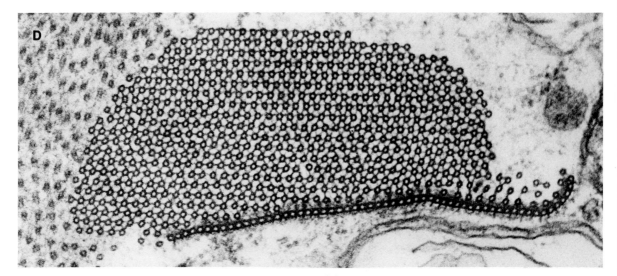

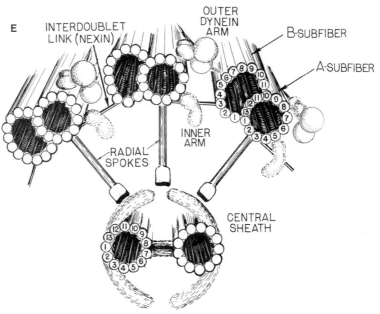

INTERDOUBLET
LINK (NEXIN)

OUTER
DYNEIN
ARM

B-SUBFIBER

A-SUBFIBER

RADIAL
SPOKES

INNER
ARM

CENTRAL
SHEATH

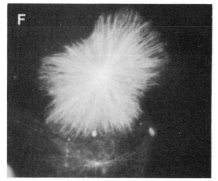

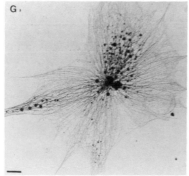

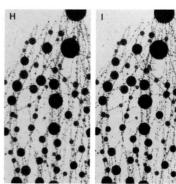

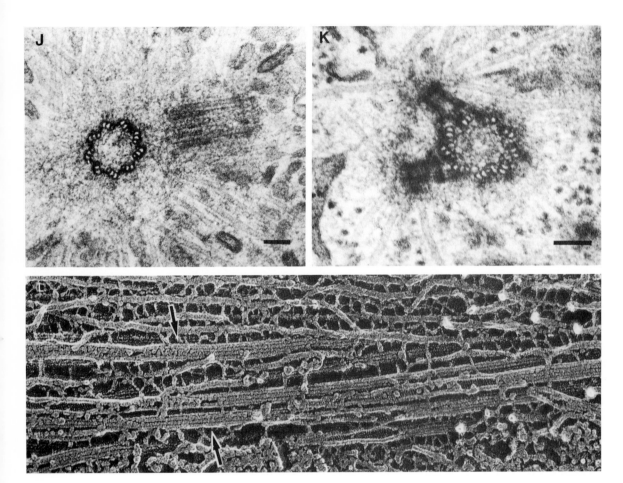

FIG. 2.2. (*Continued*) (**J, K**) Structure of centrosome. (J) Centrosome of mitotic cultured cell (pig kidney embryo PE line). Centriole pair and pericentriolar halo with radiating microtubules are seen. Electron micrograph of the sectioned cell. (K) Centrosome of interphase cultured cell (PE line). Cross-sectioned centriole is surrounded by dense pericentriolar material; this material forms two extensions (satellites); microtubules radiate from these satellites. (Courtesy of I. A. Vorobjev.)

(**L**) Microtubules (arrows) in the cytoplasm of frog axon. Electron micrograph of the platinum-shadowed replica; axons were first saponin-treated, then quick-frozen and freeze-etched. Notice the protofilament structure of microtubules. Intermediate filaments are parallel to the microtubules. Thin fibrils cross-link adjacent microtubules and intermediate filaments. ×180,000. (From Hirokawa, 1982. Courtesy of N. Hirokawa.)

(**M**) Parallel submembranous microtubules beneath the membrane folds in the unicellular alga *Distigma proteus*. Notice that each microtubule has 13 protofilaments in cross-section. Electron micrographs of the section of cell cortex at three magnifications: ×16,500, ×218,000, and ×84,000, all reproduced at 80%. (From Murray, 1984. Courtesy of J. M. Murray.)

(**N, O**) Primary cilium of cultured cells. (N) Primary cilium and centriole (arrow) revealed by immunofluorescence microscopy using antitubulin antibody. Mouse fibroblast preincubated with colcemid to destroy all the other microtubular structures. Scale bar: 10 μm. (O) Primary cilium at the surface of cultured mouse kidney epithelial cell. Scanning electron microscopy. Scale bar: 1 μm. (Courtesy of I. S. Tint.)

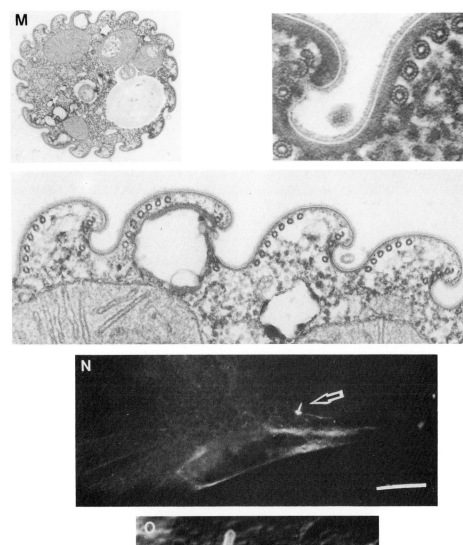

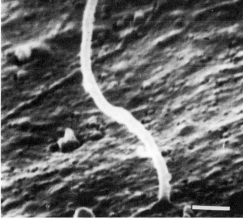

FIG. 2.2. *(Continued)*

highly regular parallel arrays described earlier; the distances between the individual microtubules of the group may vary considerably. Presence of microtubules not attached to the central MTOC is another possible modification of radial array. For instance, most microtubules present within the long processes of neurons, in contrast to the microtubules of the cell body, seem to be unconnected to perinuclear centrosome. However, the polarity of these microtubules is similar to that of the other interphase microtubules.

Submembranous microtubules can be regarded as a special group of structures (Fig. 2.2M). Numerous parallel microtubules ar present under the membranes of many types of cells of higher plants. The circular submembranous bundles of microtubules are characteristic of two types of blood cells having discoid shapes, bird erythrocytes and mammal platelets. All these submembranous structures are rather loosely organized and do not contain any visible MTOCs.

Various arrays present within the same cell can interact with one another, forming "superarrays." The most obvious example is formation of the mitotic spindle from the two half-spindles (Fig. 2.2A). Formation of the radial array of microtubules around the centrosome of the interphase cell can be combined with growth of a single modified cilium from the end of the centriole present within this centrosome (Fig. 2.2A). This so-called primary cilium has no central microtubules ("9 + 0 pattern"). Primary cilium is present in many types of cultured mammal cells (Fig. 2.2N, O).

Particularly diverse and complex combinations of microtubular arrays are found in ciliates.

C. Basic Processes Essential for the Formation and Function of Microtubular Systems

Morphologies of actin structures and of microtubular structures have many similarities. The systems of both types are formed by polar fibrils with associated proteins. Naturally, many questions arising during the analysis of formation and function of microtubular systems are similar to those already discussed in Chapter 1 with regard to actin systems. It is necessary to understand how the microtubules are formed, what factors regulate their length and positions in various structures, how these structures move, and other factors. However, microtubular systems also have certain specific features, which have no counterpart in actin systems. In particular, most microtubular systems are closely associated with special organizing centers. These centers have diverse structures; some of them are very complex. It is necessary to understand how MTOCs organize microtubular structures and how the MTOCs themselves are formed.

II. Tubulin

A. Amino Acid Sequences

Tubulin is the main building block of all the microtubular systems. Brain tissue has an especially high concentration of tubulin and is used most often for its isolation.

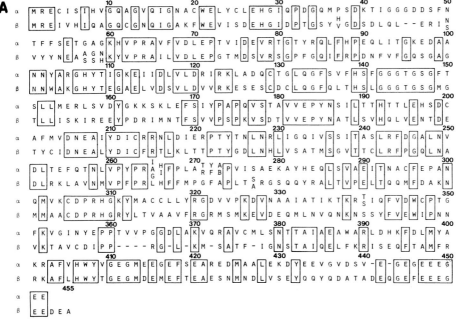

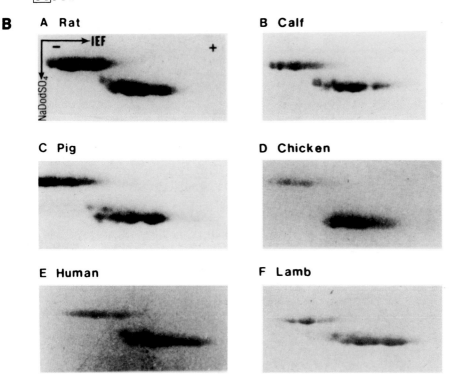

FIG. 2.3. (A) Comparison of α- and β-tubulin sequences from porcine brain. Boxed areas indicate identical residues. A (Ala); C (Cys); D (Asp); E (Glu); F (Phe); G (Gly); H (His); I (Ile); K (Lys); L (Leu); M (Met); N (Asn); P (Pro); Q (Gln); R (Arg); S (Ser); T (Thr); V (Val); W (Trp); Y (Tyr). (From Krauhs *et al.*, 1981. Courtesy of H. Ponstingl.) **(B)** Numerous isoforms of brain tubulins. Purified tubulins of different vertebrates were separated by two-dimensional sodium dodecyl sulfate electrophoresis and isoelectric focusing (IEF). Several isoforms of the α-tubulins (upper line) and of the β-tubulins (lower line) are resolved in each preparation. (From Field *et al.*, 1984. Courtesy of J. C. Lee.)

The tubulin molecule is composed of two nonidentical polypeptide chains, α and β.

α-Tubulin (pI 5.3) is slightly more basic than β-tubulin (pI 5.1). Brain α-tubulin contains 451 amino acid residues; β-tubulin is somewhat shorter, containing 445 residues. These two subunits have about 40% of homologous residues (Fig. 2.3A).

Tubulins were highly conserved during evolution. Considerable homologies are found between the amino acid sequences of tubulins from very distant organisms and between the nucleotide sequences of corresponding genes. For example, the chicken tubulin gene hybridizes strongly with the tubulin genes of the alga *Chlamydomonas*. Owing to the conservation of tubulin structure, microtubules of very different organisms can often be stained by the same antibody. However, some exceptional results have also been reported. Thus, only minimal homologies were found between the tubulin gene of the protist *Naegleria* and chicken β-tubulin gene. This finding indicates that the tubulin structure may have undergone considerable changes in certain branches of the evolutionary tree. This possibility requires further study.

The same organism and even the same tissue usually contain multiple forms of α- and β-tubulins distinguishable by isoelectric focusing, peptide mapping, and other methods. For instance, recent analysis revealed six isoforms of α-tubulin and 12 isoforms of β-tubulin in brain tissue (Fig. 2.3B) Some of these isoforms are likely to be coded by different genes. The chick genome has four genes for each α- and β-tubulin. Obviously, the number of isoforms in this case is larger than the number of genes. This suggests that posttranslational modifications can generate at least a part of the tubulin heterogeneity. In fact, several types of these modifications have been revealed in various organisms: tyrosination, acetylation, and phosphorylation. Detyrosination and tyrosination are unique modifications of tubulin; no other protein is known to undergo these modifications. α-Tubulin polypeptide is synthesized with a tyrosine at its C-terminus. This tyrosine can be removed by a specific carboxypeptidase. This nontyrosinated α-tubulin is a substrate for addition of tyrosine by another specific enzyme, tubulin tyrosine ligase. Recently it was found that various microtubules of the same cultured cell contain either tyrosinated or nontyrosinated tubulin. These results suggest that tyrosination–detyrosination can be used for the formation of functionally different gorups of microtubules, the nature of which is not known. α-Tubulin present in the flagella of *Chlamydomonas* appears to be an acetylated form of the α-tubulin present in the body of this unicellular organism. This is the only case in which specific posttranslational modification of tubulin seems to be correlated with the special function of microtubules.

B. Properties of Tubulin Dimers

α- and β-Tubulins have molecular weights of about 50 kD. Molecules of nonpolymerized tubulin have a molecular weight twice as large, about 100 kD. Treatment of the tubulin solution with cross-linking agents covalently binds α- and β-subunits. Thus, each tubulin molecule is a heterodimer consisting of one α- and one β-chain. The dissociation constant of tubulin is about 10^{-6}, so that small amounts of free α- and β-chains are usually present in solution.

Tubulin dimer can bind two molecules of GTP. One of the bound molecules is rapidly exchanged for another similar GTP or GDP molecule, present in the solution, but the exchange of the second molecule is very slow. Exchangeable GTP is bound to the β-subunit; nonexchangeable nucleotide is likely to be bound to the α-subunit.

The specific property of the tubulin molecule is the ability to bind colchicine and related organic substances, derived from plants (see Section VI). Tubulin can also bind calcium and magnesium. In the presence of zinc ions, tubulin molecules are precipitated as sheets, which are used for electron microscopic structural studies. These studies have shown that tubulin consists of elongated, wedge-shaped subunits. Neighboring subunits in the sheets, which probably correspond to α- and β-chains, have slightly different shapes.

Thus, nonpolymerized tubulin molecules are heterodimeric; their structure is more complex than that of monomeric actin molecules. Polymerization of actin can lead to formation of only one structure, the filament. As discussed in the next paragraph, several variants of microtubules, as well as some other structures, can be built from the more complex tubulin molecules.

III. Structure of the Microtubule

A. Protofilament

The wall of the microtubule is made of protofilaments, that is, of linearly arranged pairs of alternating α- and β-subunits (Fig. 2.4A–C). The protofilament is about 5 nm in width. By varying the conditions of polymerization *in vitro*, one can obtain other morphological structures from tubulin, besides microtubules, e.g., hoops up to 1000 nm in diameter or rings about 30 nm in diameter (Fig. 2.4D). All these forms of tubulin polymers are composed of protofilaments, put together in different ways. The protofilament is not a geometrical abstraction, but an existing structure. Fragments of single protofilaments can sometimes be seen in the preparations of polymerizing tubulin.

B. Variants of Microtubules

Protofilaments forming the wall of microtubules are parallel to its axis. Individual subunits in the wall of microtubules can be seen in preparations pretreated with tannic acid (Fig. 2.4E). The number of these subunits in the cross-section of the microtubule corresponds to the number of protofilaments.

Examination of tannic acid preparations from various sources has shown that microtubules can have different numbers of protofilaments. Microtubules in the brain and other tissues of most animals have 13 protofilaments. The complete A-microtubules of cilia and flagella also have 13 protofilaments,

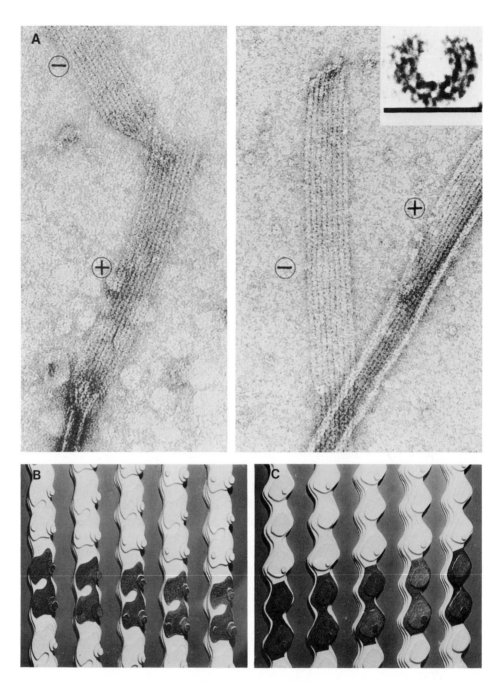

FIG. 2.4. (A) Incomplete microtubule walls (sheets) consisting of protofilaments. Electron micrographs of negatively stained preparation. The inner (+) or outer (−) side is facing the viewer.

(B, C) Model of the protofilaments in the sheet viewed from inside (B) and from outside (C). The model is a three-dimensional reconstruction of the structure of sheets computed from the electron microscopic images. One row of α–β heterodimers is shaded. They are aligned along the same line (B-lattice). (A–C are from Schulteiss and Mandelkow, 1983. Micrograph courtesy of E. Mandelkow.)

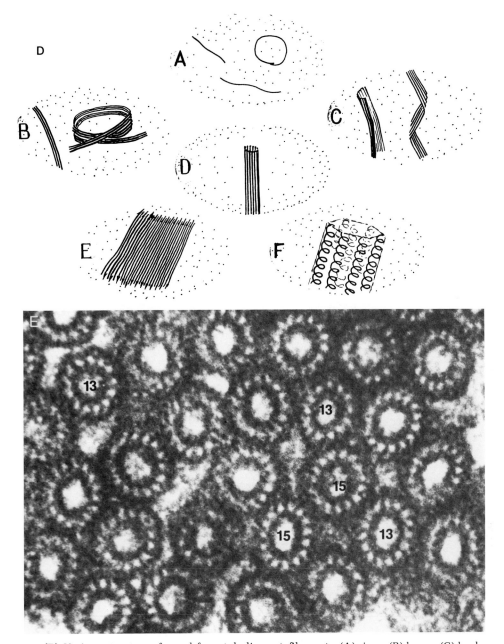

(D) Various structures formed from tubulin protofilaments; (A) rings; (B) hoops; (C) hooks and helical ribbons; (D) microtubule; (E) Zn-induced sheet; (F) vinblastin paracrystals.

(E) Cross-section of microtubules in the dividing micronucleus of the ciliate *Paramecium tetraurelia*; the cells were fixed in the presence of tannic acid. Microtubules containing 13 and 15 protofilaments are present. (From Tucker *et al.*, 1985. Courtesy of J. B. Tucker.)

(F) Helical structure of microtubule consisting of 13 protofilaments. "Three-start" helix is shown by the arrows. The real helix may not be right-handed, as shown, but left-handed.

(G) Two types of lattices (A and B) forming the microtubular wall. One type of subunit is shadowed, the other is unshadowed. Discontinuity of the B-lattice is shown by the arrow.

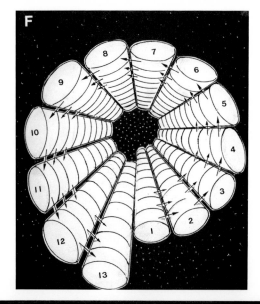

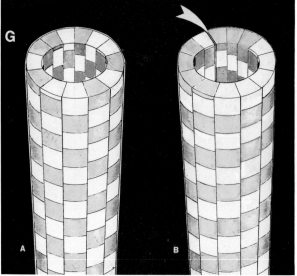

FIG. 2.4. (Continued)

whereas incomplete B-microtubules have 10 or 11 protofilaments. There are, however, certain types of complete microtubules that have not 13, but 11, 14, 15, or 16 protofilaments. For instance, microtubules of most cells of the nematode *Caenorhabditis elegans* have 11 protofilaments; certain neurons of this animal have microtubules with 15 protofilaments. The number of protofilaments in the microtubules in the cells is constant. In contrast, microtubules assembled *in vitro* from brain tubulin or from flagellar tubulin have varying numbers of protofilaments, with 14 predominating. These data suggest that the number of protofilaments within the microtubule is not predetermined by the structure of tubulin, but is controlled by some special cellular mechanism. We shall return later to the possible mechanism of this control.

Structural studies have shown that protofilaments composing the microtubule are slightly staggered; subunits of neighboring protofilaments are shifted about 1 nm along the axis of the microtubule, so that these subunits can be connected by a helical line (Fig. 2.4F). When this line makes a complete revolution around the axis, it moves a distance of three subunit lengths. The protofilament consists of alternating nonidentical subunits. Therefore, there can be two types of regular lattices forming its wall: A-lattice, with α- and β-subunits alternating along the helical line, and B-lattice, with helical lines composed of all α-subunits or of all β-subunits (Fig. 2.4G).

Geometrical considerations permit a continuous A-lattice to be formed only if the number of protofilaments is odd. In fact, structural studies revealed this type of lattice in the 13-protofilament A-microtubules of flagella. B-lattice cannot be folded into a continuous self-closing structure, regardless of the protofilament number. However, geometry permits this type of lattice in incomplete microtubules. In fact, B-microtubules of flagella were found to have this type of lattice.

The 14-protofilament structure of microtubules polymerized in vitro is geometrically incompatible with either A-type or B-type continuous lattices. Examination of the structure of these microtubules has shown that they have a mixed discontinuous lattice with areas of both A and B types. The lattice of the microtubules of other types has not yet been studied.

Variations of the lattice structure can be important for the function of the microtubule in the cell. Dynein molecules are differently bound to A- and B-microtubules of flagella (see Section VIII.B). Possibly, other MAPs also distinguish the type of lattice of the microtubule wall.

C. Polarity of Microtubules

As shown by structural studies, each protofilament is a polar structure with all the asymmetrical subunits oriented in one direction. All the protofilaments forming the microtubule wall have the same polarity, so that the whole microtubule is also polar and has unequal plus and minus ends.

Two methods are used to determine the polarity of the microtubule: decoration by dynein and formation of hooks.

Dynein molecules are attached to the microtubule by their ends and tilted with regard to its axis. In cross-section the asymmetrical dynein arms are oriented clockwise or counterclockwise. This orientation is used to detect polarity (Fig. 2.5A–D). Another method involves decoration of microtubules with curved ribbon structures consisting of tubulin and called hooks. In certain buffer solutions, tubulin is polymerized into hooks. Under these conditions, when lysed cells are incubated with exogenous tubulin, hooks are formed on the cellular microtubules. All hooks attached to the microtubule are curved in the same direction, indicating its polarity. When the microtubule is viewed from the plus to minus end, all the hooks are curved clockwise (Fig. 2.5E–H).

Thus, microtubules are complex polar structures, which can have different numbers of protofilaments and different geometrical structure of the wall.

IV. Polymerization and Depolymerization of Tubulin

A. Conditions of Polymerization

Tubulin assembly is very sensitive to many environmental factors. It is inhibited by low temperature and by the presence of calcium ions; it requires

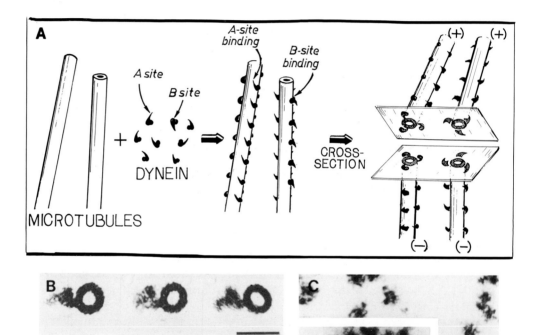

FIG. 2.5. **(A)** To determine the polarity of microtubules they can be incubated with dynein molecules. Depending on the structure of the microtubule wall, these asymmetrical molecules are attached to the microtubules either by A ends or by B ends. Orientation of the attached molecules indicates the polarity of microtubules.

(B–D) Transverse sections of microtubules preincubated with dynein. (B) A-type attachment. Microtubules polymerized *in vitro*. Scale bar: 50 nm. (C) B-type attachment. Microtubules of the meiotic spindle of the surf clam, *Spisula solidissima*. (D) Dynein arms cross-bridging microtubules polymerized *in vitro*. Scale bar: 50 nm. (C and D are from Haimo, 1982. B is from Haimo *et al.*, 1979. Micrographs courtesy of L. T. Haimo.)

the presence of GTP and of magnesium ions. Brain tubulin preparations, containing MAPs, assemble into the microtubules at 37°C in solutions holding millimolar concentrations of Mg^{2+} and of GTP, as well as the calcium-chelating agent, EGTA. However, when MAPs are removed from the preparation, assembly of purified tubulin does not take place under these "physiological" conditions. Purified tubulin without MAPs can be forced to form microtubules only by certain additional alterations of the medium, such as an

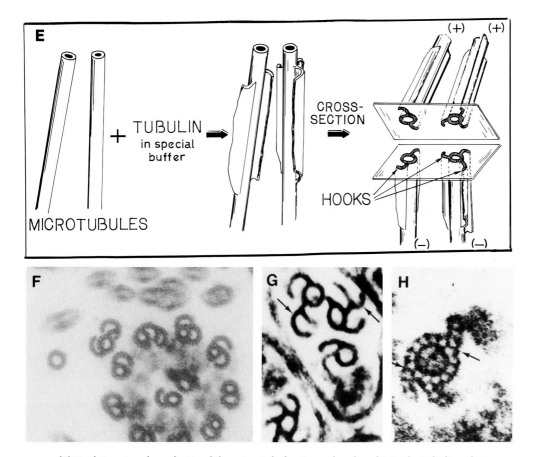

(E) To determine the polarity of the microtubule, it may be placed into the tubulin solution in special assembly buffer containing a high concentration (0.5 M) of piperazine N,N'-bis 124 (2-ethane sulfonate) and 2.5% of dimethylsulfoxide. In these conditions tubulin is polymerized into curved sheets, called hooks, attached to preexisting microtubules. Orientation of the curvature of the hooks indicates the polarity of the microtubules.

(F–H) Cross-section of *Tetrahymena* cilium (F) and of the frog olfactory nerve (G, H) with attached hooks. The cilium is viewed looking toward the basal body; the nerve, toward the peripheral end of the axon. In F, all hooks are curved clockwise; in G and H, counterclockwise. Thus, in both cases microtubules have similar polarities with their plus ends distal to the cell centers. F and G ×200,000, H ×500,000. In H, individual protofilaments can be seen. (F from Euteneuer and McIntosh, 1981. Courtesy of U. Euteneuer. G and H from Burton and Paige, 1981. Courtesy of P. R. Burton.)

increase of magnesium concentration up to 10 mM, or addition of 3–4 M of glycerol.

Conditions of polymerization depend also on the type of tubulin used. There are some exceptional tubulins that are easier to polymerize than brain tubulin. For instance, purified tubulin from sea urchin eggs was observed to form microtubules in minimal "physiological" conditions described earlier. By varying environmental conditions, one can repeatedly polymerize and depolymerize microtubules. This method is used for purification of tubulin from tissue homogenates (see description of similar method of actin purification in Chapter 1). A special drug, taxol (see Section VI), is often used to stabilize the polymerized microtubules in the course of tubulin and MAP purification. Polymerization of tubulin, like that of actin, can be assessed by measuring certain physical parameters of tubulin solution, e.g., light scattering or viscosity.

B. Nucleation of Microtubules

The time course of polymerization of tubulin, like that of actin, is described by a sigmoidal curve: the stage of growth of polymer concentration is preceded by a lag period. Probably, formation of polymerization nuclei takes place during the lag period and their elongation during the stage of growth. In fact, when preformed fragments of microtubules are added to the tubulin solution, rapid increase of the polymer concentration begins immediately without the lag period. Nucleation seems to be more sensitive to environmental conditions, and to removal of MAPs, than elongation. Elongation of preformed microtubules in the presence of purified brain tubulin can take place in "physiological" conditions without glycerol or high magnesium.

The nature of the nuclei formed during the lag period is not clear. Electron microscopic studies had shown that small fragments of several laterally associated protofilaments are formed at the beginning of polymerization. These data suggest that the early stage of microtubule assembly is a two-dimensional process. It includes both the formation of fragments of protofilaments and lateral association of several such fragments (Fig. 2.6). Further progress of these two processes leads to assembly of the cylindrically closed fragments of the microtubule, which grow later only by elongation of protofilaments.

Obviously, this multistage process, going in two dimensions, is more complex than formation of the one-dimensional nucleus of actin filaments. The more complex nucleation process demands more complex conditions. A correct relationship between elongation of protofilaments and their lateral association is essential for nucleation of microtubules. If this relationship is upset by altered conditions of polymerization, abnormal polymer structures can be formed. For instance, when elongation of protofilament fragments is more effective than their lateral association, hoops will be formed. These considerations give a reasonable explanation of nucleation, but at present they remain hypothetical.

C. Elongation

To measure the rate of elongation one can place the preformed microtubules into a solution of tubulin and then determine the alterations of their length and of the amount of polymer. The kinetics of elongation in the presence of purified tubulin has many similarities with that of actin filaments (see Chapter 1). As already mentioned, each tubulin dimer contains two bound molecules of GTP or of GDP. One of these molecules, GTP at the exchangeable site, plays the same role in regulation of elongation of microtubules as bound ATP in elongation of actin filaments. In both systems attachment of subunits to the end of the fibril is followed by hydrolysis of the bound nucleotide. Bound exchangeable GTP in the wall of microtubule is hydrolyzed into GDP.

In the presence of high concentrations of GTP-tubulin, elongation is very efficient; in these conditions, elongation depends linearly on the concentration of tubulin in solution. In contrast, in the presence of similar concentrations of GDP-tubulin, elongation proceeds very slowly, if at all. These differences in the efficiency of polymerization in the presence of the bound nucleoside triphosphate and of diphosphate are greater for tubulin than for actin.

At low concentration of GTP-tubulin the rate of elongation is sharply decreased (Fig. 2.7A); dependence of this rate on the subunit concentration

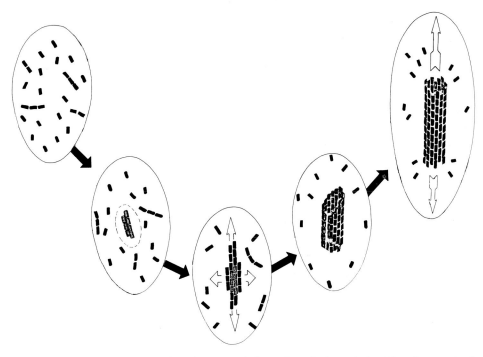

FIG. 2.6. Hypothetical scheme of nucleation and elongation of microtubules. First, fragments of protofilaments are formed; next, the bidimensional fragment of the microtubule wall; and finally, a cylindrically closed short microtubule growing from its ends.

becomes nonlinear. In experiments with tubulin, this downward bend of the rate of elongation is even more conspicuous than in experiments with actin. It is likely that these rapid decreases in the polymerization rates of tubulin and actin have similar bases. Decrease of the concentration of tubulin diminishes the rate of attachment of GTP subunits to the microtubule, but does not affect the rate of hydrolysis of GTP after attachment (Fig. 2.7B). As a result, when GTP-tubulin concentration is low, many microtubules will have GDP-subunits at their ends; in other words, these microtubules lose "GTP caps" (Fig. 2.7C). Without GTP caps these ends cannot continue elongation and, in contrast, begin to depolymerize. Loss of GTP caps at low concentrations of tubulin may explain the sharp fall of the elongation rates leading to depolymerization.

This "capping model" of elongation (see also Chapter 1) predicts that the population of microtubules at low concentrations of GTP-tubulin consists of two fractions: growing capped microtubules and shrinking uncapped microtubules. Disassembly of uncapped microtubules can lead to their complete disappearance. Experiments have confirmed that at the critical concentration of GTP tubulin, the number of microtubules in solution decreases with time, while the average length of remaining microtubules increases (Fig. 2.7D). The amount of polymerized tubulin in these conditions is not changed because the growth of some microtubules is compensated by the shrinkage of others. Even when the concentration of tubulin is lower than critical, some microtubules continue to grow, although most microtubules will be depolymerized.

Polarity of elongation is another similar feature of microtubules and actin filaments. This polarity is best revealed when the flagellar axonemes are placed in the tubulin solution. Two ends of the axoneme can be distinguished morphologically; the plus end looks more frayed. At a high concentration of GTP-tubulin, the rate of elongation at the plus ends is much higher than that at the minus ends (Fig. 2.7E). It has been suggested that critical concentrations

---

FIG. 2.7. **(A)** Dependence of the rate of elongation (J) of microtubules assembled from pure tubulin on the concentration of tubulin dimer (c) in the presence of GTP. C_c = critical concentration of tubulin. The plot $J(c)$ does not consist of a single straight line intersecting the concentration axis at $c = C_c$, but exhibits a change in slope at the critical concentration. (From Carlier et al., 1984. Courtesy of M.-F. Carlier.) **(B)** Effect of tubulin concentration on the rate of polymerization and concomitant GTP hydrolysis. Ordinates: Fraction of the maximal polymerization (solid line) and of the maximal rate of GTP hydrolysis (symbols). Tubulin concentrations were 34.5 μM (a, ●) and 17.2 μM (b, ○). The two processes are closer to each other at low concentrations of tubulin. They seem to be disconnected at high tubulin concentrations; in this case only 10% of GTP is hydrolyzed when 90% microtubules are formed. (From Carlier and Pantaloni, 1981. Courtesy of M.-F. Carlier.) **(C)** Microtubule with GTP cap. T = GTP; D = GDP. **(D)** Scheme showing the time course of the alterations of the numbers and lengths of microtubules in solutions containing critical concentrations of tubulin. Capped microtubules 1–4 grow, while uncapped microtubule 5 shrinks and eventually disappears. As a result, the number of microtubules decreases with time, while the average length of each microtubule increases. By definition, the total concentration of polymerized tubulin remains constant at critical concentration. **(E)** Scheme of experiment showing polar elongation of isolated flagellar axonemes placed in the tubulin solution.

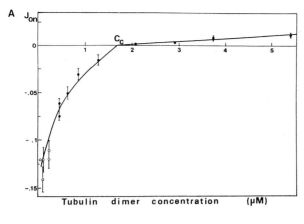

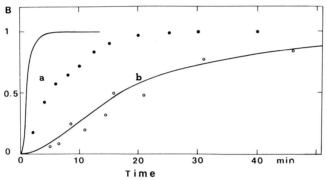

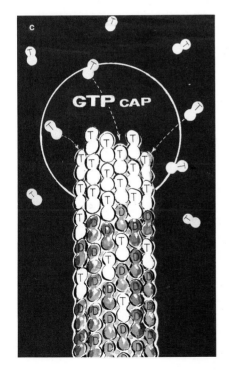

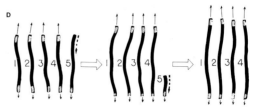

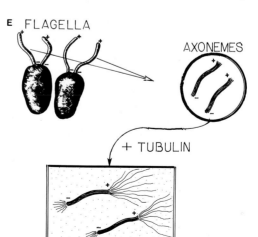

of tubulin can be different at the plus and minus ends. As a result, at certain intermediate concentrations of tubulin the microtubules will undergo tread-milling; that is, they will grow at one end and depolymerize at another. In fact, concurrent addition and loss of tubulin subunits from the opposite ends of microtubules was recently observed in experiments in which chicken erythrocyte microtubules were incubated in a solution containing brain tubulin; immunoelectron microscopy was used to distinguish two types of tubulin. Fluctuations of the state of microtubule ends at the low concentrations of tubulin may overshadow treadmilling. Therefore, experimental demonstration of microtubule treadmilling is difficult in many situations for the same reasons as that of actin treadmilling (see Chapter 1).

To summarize, dynamic instability is a characteristic feature of microtubules made of purified tubulin. This instability is even more dramatic than that of actin filaments. The "solid flesh" of microtubule can melt readily when it loses its GTP cap.

V. Microtubule-Associated Proteins

A. The Main Types of MAPs

MAPs are attached to microtubules *in vivo* and coisolated with tubulin. The preparations of tubulin obtained by repeated polymerization–depoly-merization contain as much as 20% MAPs. Tubulins can be dissociated from MAPs by ion-exchange chromatography. The best-studied MAPs of nonfla-gellar microtubules are those isolated from the brain. They form three main groups: MAP-1 with mol. wt. 300–350 kD, MAP-2 with mol. wt. 270–285 kD, and tau group with mol. wt. about 60 kD (Fig. 2.8A,B). *In vitro* all these MAPs readily reattach themselves to the walls of microtubules polymerized from pure tubulin.

MAP-2 is the best-investigated brain MAP; it has two related variants, MAP-2A and MAP-2B (Fig. 2.8A,B). MAP-2 is remarkably heat stable; it is not denatured by heating up to 100°C. Molecules of MAP-2 have very elongated shapes. When these molecules are reattached to the microtubule wall, they look like filamentous projections 80–100 nm long (Fig. 2.8C–E). Filamentous MAP-2 molecules can be split by chymotrypsin into longer and shorter fragments. Only the shorter fragment can be attached to the microtubule; thus, it contains the microtubule-binding domain of MAP-2.

As shown by Vallee and collaborators (see Vallee and Bloom, 1984; Vallee *et al.*, 1984), the MAP-1 group consists of at least three proteins: MAP-1A, MAP-1B, and MAP-1C (Fig. 2.8A,B). Reattached MAP-1 proteins form projections on the wall of microtubules (Fig. 2.8F). The tau group can be resolved into four or five components in one-dimensional electrophoresis and into more than 20 components in two-dimensional electrophoresis; these proteins are closely related to one another. When they are reattached to microtubules, projections are not formed, but the surface of the microtubule looks less smooth than in control preparations.

The tissues, other than brain, are poorer sources of microtubules, and

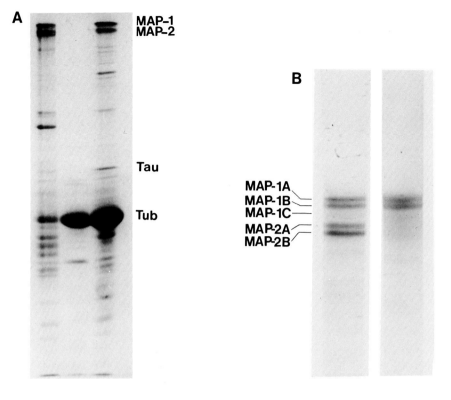

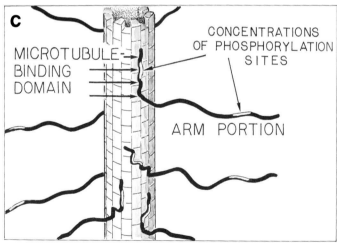

FIG. 2.8. **(A)** Sodium dodecyl sulfate–polyacrylamide gel electrophoresis (SDS–PAGE) of a MAP-containing tubulin preparation of bovine brain (right); MAP (left) and tubulin (center) fractions obtained from this preparation. **(B)** Electrophoresis resolving several fractions in MAP-1 and MAP-2 groups. Only the high molecular portion of the gel is shown. (Courtesy of V. I. Rodionov and S. A. Kuznetsov.) **(C)** Scheme of interaction of MAP-2 molecule with the microtubules. **(D, E)** Microtubules with (D) and without (E) MAP-2 molecules. Microtubules were assembled from purified tubulin in the presence of MAP-2 or with no MAP-2 added. Electron micrographs of preparations shadowed with platinum. ×9000. (From Voter and Erickson, 1982. Courtesy of W. A. Voter.)

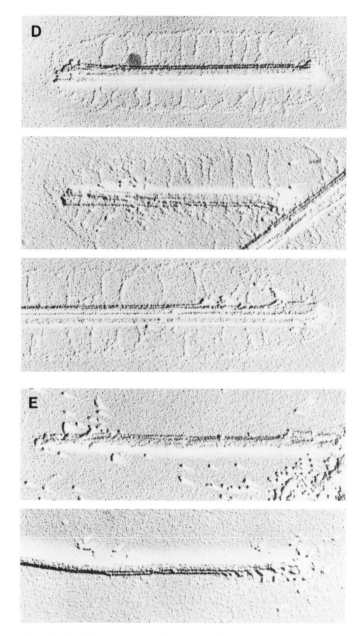

FIG. 2.8. (*Continued*) **(F)** Microtubules with attached MAP-1 molecules. (Courtesy of V. I. Rodionov and S. A. Kuznetsov.) **(G)** Effects of MAPs on the polymerization of tubulin in buffer with EGTA and GTP. Ordinates: Amount of polymer (optical density at 330 nm); abscissas: time. Pure tubulin is not polymerized; polymerization takes place in the presence of MAP-1 (left) or MAP-2 (right). The tubulin concentration in both experiments was 0.8 mg/ml. The concentration of MAPs (mg/ml) is indicated above each curve. (Courtesy of S. A. Kuznetsov.)

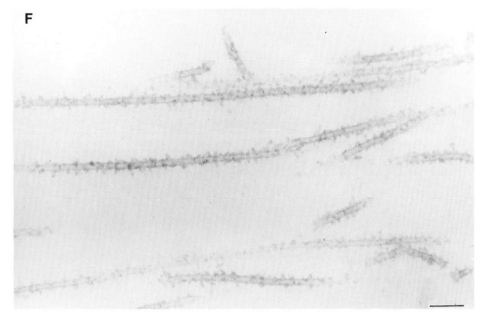

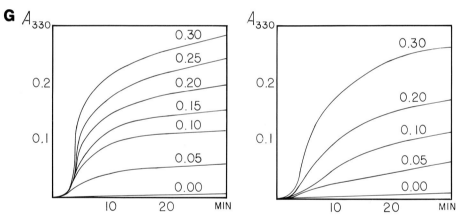

FIG. 2.8.   (*Continued*)

their MAPs are more difficult to isolate. The major groups of MAPs in cultured cells of various origin seem to be different from those in the brain (see Table 2.1). Structural relations between all these proteins are still not clear.

B. Effects of MAPs on Polymerization and Depolymerization of Tubulin *in Vitro*

All types of MAPs added to tubulin solution promote the formation of microtubules. After addition of a small amount of any MAP, polymerization occurs readily under "physiological" conditions, so there is no need to add glycerol or increase the concentration of Mg^{2+}

MAPs strongly promote nucleation of microtubules. They also increase the net rate of elongation and decrease the critical concentration of tubulin (Fig. 2.8G). Kinetic studies have shown that MAPs do not increase the rate of association of subunits, but diminish the rate of their dissociation from the ends of microtubules. It has been also suggested that MAPs can diminish the differences between the rates of dissociation from capped and uncapped ends and damp out fluctuations of microtubule length at low concentrations of tubulin. In other words, it is possible that MAPs diminish dynamic instability of microtubules.

MAPs isolated from the tissues, other than brain, e.g., 210-kD MAP from cultured epithelial human cells, have also been observed to promote microtubule polymerization. Thus, stabilization of microtubule seems to be a common function of many different proteins that attach themselves to the wall of this organelle.

C. Regulation of the State of MAPs

MAPs change the stability of microtubules. In turn, the state of MAPs themselves can be reversibly changed by phosphorylation and dephosphorylation. The best-studied MAP-2 molecule can acquire up to 20 phosphate residues. Phosphorylation of MAP-2 can be catalyzed by several kinases. One of them is activated by cAMP, another by calcium–calmodulin. cAMP-depen-

TABLE 2.1. Microtubule-Associated Proteins

Group	Subunit mol. wt. (kD)	Comments
MAP-1	Three isoforms ~350	Isolated from brain, projected from microtubular wall; promotes assembly
Proteins associated with MAP-1 (light chains of MAP-1)	30 28	Associated with MAP-1 at stoichiometry 1 mole of each per mole of high-molecular-weight polypeptide
MAP-2	Two closely related isoforms ~270	Isolated from brain, projected from microtubular wall; promotes assembly
Proteins associated with MAP-2	70 54 + 39	Function unknown Subunits of cAMP-dependent protein kinase
Tau	Family containing several isoforms 55–62	Isolated from brain; promotes assembly
MAPs isolated from cultured cells of human HeLa line	Triplet centered at ~210; 125	Promotes assembly
MAP-3	180	Present in axons and glial cells
MAP-4	Triplet of related proteins 215–240	Isolated from mouse neuroblastoma cells; occurred in different mouse tissues

[a] Reviews: Vallee and Bloom (1984); Vallee et al. (1984); Vallee (1984). For HeLa MAPs: Bulinski and Borisy (1979). For tau proteins: Cleveland et al. (1977). For MAP-3: Huber et al. (1986). For MAP-4: Parysek et al. (1984).

dent kinase is bound to the projected domain of MAP-2. Phosphorylation of MAP-2 and of tau proteins decreases their ability to promote tubulin polymerization and to stabilize microtubules. Decrease of the polymerization-promoting ability correlates with the degree of phosphorylation. Possibly, phosphorylated MAPs are less readily bound to the microtubules and, of course, cannot stabilize them.

Microtubules in the cell are very sensitive to calcium, especially in the presence of calcium-binding protein, calmodulin. Intracellular injection of calcium–calmodulin complex rapidly leads to depolymerization of microtubules. It is likely that calcium–calmodulin complex acts on the microtubules via MAPs. As already mentioned, this complex can activate one of the MAP-phosphorylating kinases. There are also data indicating that calcium–calmodulin can bind directly to tau proteins and to other MAPs. Possibly, this binding can modify the state of MAPs in such a way that they lose the ability to stabilize microtubules. Thus, MAP-induced stabilization of microtubules can be abolished by certain reversible modulations of the state of MAP molecules.

D. Functions of MAPs in the Cell

Effects of MAPs *in vitro* suggest that *in vivo* these proteins can perform a number of common functions. MAPs can control the stability of microtubules and act as the targets of intracellular regulatory signals, such as altered concentration of cAMP or of calcium.

Besides stabilization of microtubules, certain MAPs perform some other functions. The most obvious example is dynein, which is essential for force generation in cilia and flagella. Specific additional functions of other MAPs are not yet identified. It seems likely that some of these proteins can be necessary for attachment of various organelles to the microtubules. Certain MAPs isolated from sea urchin spindles promote the formation of microtubule bundles *in vitro*; possibly, they participate in microtubule–microtubule interactions in the cell.

Specificity of functions of various MAPs is suggested by the differences of the spectra of these proteins in various cell types. Within the same cell, some MAPs are specifically localized on certain microtubules. For instance, staining of brain cells with antibodies against MAP-2 has shown that this protein is concentrated on the microtubules of dendritic processes but absent from the microtubules of axonal processes. Possibly, the state of microtubules in various neuronal processes is regulated by different signals and these signals are mediated by various MAPs.

In this connection it is worth mentioning that various microtubules within the same cell can be remarkably heterogeneous in their reaction to external factors. For instance, most cytoplasmic microtubules in cultured cells are rapidly depolymerized by low temperature ($+4°C$), but some microtubules can survive this temperature for a long time. It is natural to suggest that these cold-resistant microtubules have a particular spectrum of MAPs. This suggestion remains to be tested.

Thus, different MAPs may have both common and specific functions. Further study of these functions is of great interest because it may reveal an

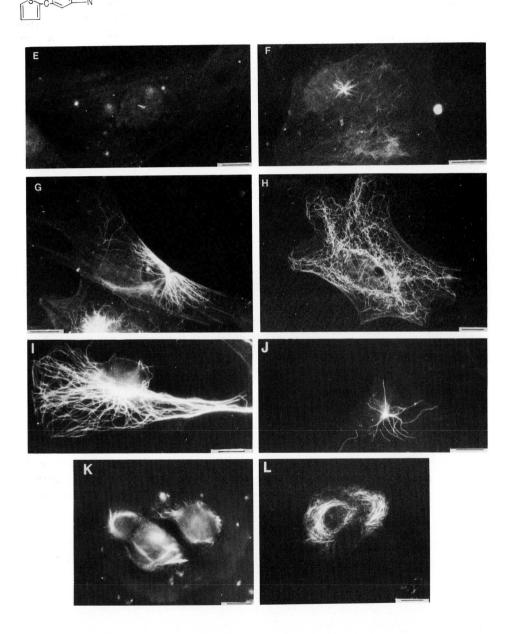

important group of mechanisms controlling the assembly–disassembly of microtubules in the cells.

VI. Effects of Specific Drugs on Polymerization of Tubulin

Several groups of drugs can bind to tubulin and affect its polymerization *in vitro* and *in vivo*. The effects of these drugs deserve special comment because they are widely used in the study of microtubules. Colchicine, isolated from the plants *Colchicum autumnale* or *Colchicum speciosum*, is the best-studied tubulin-specific drug (Fig. 2.9A). Colchicine inhibits the polymerization of tubulin *in vitro* and causes rapid disassembly of most types of cellular microtubules *in vivo* (Fig. 2.9E).

One tubulin dimer is able to bind one colchicine molecule with high affinity. Colchicine is capable of inhibiting microtubule polymerization at concentrations that are considerably below the concentration of free tubulin molecules. For example, considerable inhibition was observed at ratios as low as 1 colchicine molecule per 20–50 tubulin dimers. This phenomenon has been termed "substoichiometric poisoning." The action of colchicine on polymerization can be divided into two stages. At the first stage colchicine forms molecular complexes with nonpolymerized tubulin; at the second stage these complexes bind to the end of the microtubule and inhibit its polymerization. The exact nature of the effect of tubulin–colchicine complex on the microtubule end is not clear. It had been suggested that these complexes selectively bind to the plus ends of microtubules, but not to their minus ends. However, recent experimental data suggest that colchicine–tubulin can bind to both ends of the microtubule and alter them in such a way that the rate of association of tubulin with these ends is decreased, while the rate of dissociation is not altered. Possibly, this decrease of association rate promotes disappearance of the GTP caps from the ends of microtubules. As a result, further detachment of colchicine–tubulin complexes from the end will be followed by rapid depolymerization. These suggestions need further testing.

Several other drugs, such as colcemid, podophyllotoxin, and nocodazole

---

FIG. 2.9. **(A)** The structure of colchicine (R = CO:CH$_3$) and of colcemid (R = CH$_3$). **(B)** The structure of nocodazol. **(C)** The structure of vinblastine. **(D)** The structure of taxol. **(E–H)** Restoration of a radiating microtubular system destroyed by colcemid. Mouse fibroblasts were incubated for 24 hr in the medium containing colcemid (E) and then transferred into a colcemid-free medium and fixed 30 min (F), 60 min (G), and 2 hr (H) later. Immunofluorescence microscopy using anti-tubulin antibody. Notice the gradual growth of microtubule from the perinuclear center. Free microtubules not connected with the center are also seen in the cytoplasm at the intermediate stages of restoration (F and G). (Courtesy of I. S. Tint.) **(I, J)** Inhibition of colcemid action by azide. Mouse fibroblasts exposed to 10 μM of colcemid for 30 min with (I) or without (J) 20 mM sodium azide. No destruction of microtubules is observed in I. (From Bershadsky and Gelfand, 1981.) **(K)** Paracrystals of tubulin in the cultured fibroblast incubated with vinblastine. Immunofluorescence microscopy using antitubulin antibody. (Courtesy of I. S. Tint.) **(L)** Taxol-treated mouse fibroblast containing numerous microtubules not associated with centrosome in the peripheral parts of the cytoplasm. (Courtesy of V. I. Rodionov.)

(Fig. 2.9B), bind to the same site of the tubulin molecule as colchicine. Their effects on tubulin polymerization are also qualitatively similar to those of colchicine.

Vinblastine (Fig. 2.9C) and vincristine form another group of drugs inhibiting microtubule polymerization. These substances bind to other sites of the tubulin molecule than colchicine. Each tubulin dimer has two vinblastine binding sites. Vinblastine, like colchicine, substoichiometrically inhibits polymerization of tubulin. Exposure of cells to large concentrations of vinblastine or vincristine leads not only to the disassembly of microtubules, but also to the formation of crystallike aggregates of tubulin in the cytoplasm (Fig. 2.4D and 2.9K).

Taxol (Fig. 2.9D,L) is derived from the bark of the western yew, *Taxus brevifolia*. Its effects on the microtubules are in many respects opposite to those of other microtubule-binding drugs. Taxol enhances both the yield and rate of microtubule assembly. It decreases the critical concentration of tubulin required for assembly and promotes polymerization in unfavorable conditions, e.g., in the absence of MAPs or of GTP and at low temperature. Incubation of preformed microtubules with taxol makes them resistant to depolymerization by calcium. Taxol is maximally efficient when it is bound to microtubules stoichiometrically; that is, one molecule per one dimer of tubulin. Kinetically, taxol stabilizes the microtubule, decreasing the rate of dissociation of subunits.

An intriguing problem related to the study of microtubule-specific drugs is the possible physiological role of high-affinity binding sites for colchicine and vinblastine on the tubulin molecule. One may suggest that these drugs are toxic analogs of some normal regulatory substances binding to the same sites and controlling assembly–disassembly of microtubules. As yet, no such physiological "colchicinelike" or "vinblastinelike" molecules have been found in animal tissues.

Mechanisms controlling the sensitivity of cellular microtubules to specific drugs present another interesting problem. Full-grown ciliary microtubules, in contrast to those of most other structures, are resistant to colchicine. Cytoplasmic microtubules in usual conditions are highly sensitive to colchicine. However, when the level of ATP in the cell is reversibly decreased by metabolic inhibitors, cytoplasmic microtubules become resistant to colchicine (Fig. 2.9I,J). Possibly, the sensitivity depends on the spectrum of MAPs present on the microtubule. ATP deprivation can alter the state of these MAPs, e.g., inhibit their phosphorylation.

VII. Function and Formation of Microtubule-Organizing Centers

A. The Main Types of MTOCs

Microtubules formed in tubulin solution are randomly oriented. Parallel or radial arrays of microtubules are formed in these solutions only in the presence of an additional structure, microtubule-organizing centers (MTOCs).

As mentioned in Section I, MTOCs present in the cells have many diverse morphologies (Fig. 2.10A–G). Two large morphological groups of MTOCs can be distinguished. The structures of the first group, basal bodies and centrioles, not only organize the microtubules, but are themselves highly regular parallel arrays of microtubules. In contrast, the MTOCs of the second group do not contain microtubules. They often consist of accumulations of electron-dense amorphous material having laminated, ringlike, spherical, or irregular shapes. These accumulations are sometimes associated with the nuclear envelope. Kinetochore consists of several laminae of electron-dense material associated with a special region of the chromosome, the centromere. Only a few types of MTOCs, namely, centrosomes, basal bodies, and kinetochores, have been isolated from the cells. Biochemical studies of these MTOCs have just begun. Specific antibodies staining centrosomes have been described; other antibodies are specific for kinetochores. Antigens reacting with antikinetochore serum have been identified as four proteins of 14, 20, 24, and 34 kD. Thus, various MTOCs apparently contain specific proteins.

B. Nucleation and Stabilization of Microtubules by MTOCs *in Vitro*

When tubulin solution is added to isolated centrosomes, polymerization of microtubules is rapidly nucleated. Minus ends of the centrosome-nucleated microtubules are attached to centrosomes, while their free plus ends radiate into the medium. Nucleation of microtubules on centrosomes occurs with high efficiency even in conditions when spontaneous nucleation is strongly inhibited, e.g., in solutions of purified tubulin without MAPs or glycerol.

Microtubules nucleated on centrosomes in low concentrations of pure tubulin, like free microtubules, vary considerably in length; some of them continue to grow, while others shrink and disappear. There are, however, important differences between the free and centrosome-nucleated microtubules in these conditions. Depolymerization of a free microtubule usually leads to its irreversible disappearance, but a depolymerized centrosome-associated microtubule may be succeeded by a new one because its nucleation site at the MTOC is preserved. Centrosome continuously renucleates microtubules, so that the radial array remains stable, although each individual microtubule of this array is unstable (Fig. 2.10I). Owing to facilitated nucleation, centrosomal microtubules often win the competition with free ones for limited tubulin resources. For instance, when both free and centrosome-associated microtubules are first formed in solution with a high concentration of tubulin and this solution is then diluted, free microtubules gradually disappear, while centrosomal microtubules are retained.

The ability to nucleate microtubules *in vitro* is characteristic not only of centrosomes, but also of other types of MTOCs, such as basal bodies and kinetochores of metaphase chromosomes. The minimal concentration of tubulin needed for polymerization on kinetochores is much higher than that on centrosomes. While all the centrosome-nucleated microtubules grow with their plus ends out, at least some kinetochore-nucleated microtubules have opposite polarity; that is, they grow with their minus ends out. This polarity is similar to that of kinetochore-associated microtubules in the cell.

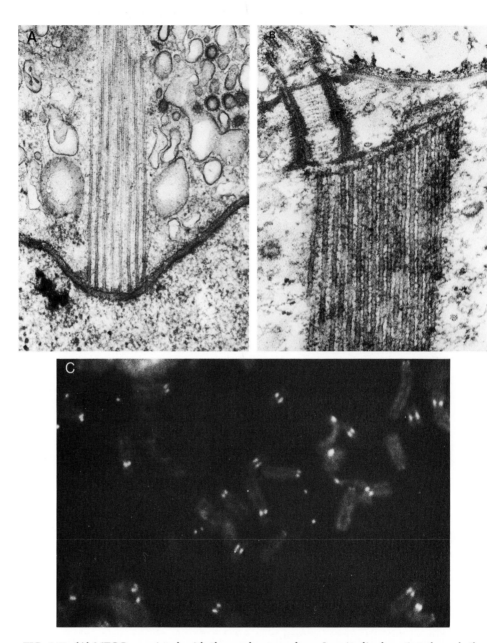

FIG. 2.10. **(A)** MTOC associated with the nuclear envelope. Longitudinal section through the base of the axoneme of the heliozoan *Actinophrys sol.* ×51,000, reproduced at 75%. (From Jones and Tucker, 1981. Courtesy of J. B. Tucker.) **(B)** Electron micrograph of laminated flat MTOC of the microtubules of the rod of a cytopharyngeal basket. ×130,000, reproduced at 50%. (From Tucker, 1971. Courtesy of J. B. Tucker.) **(C)** Kinetochores of mitotic chromosomes of cultured cells (CHO line) revealed by specific antibody. Immunofluorescence microscopy. ×2065, reproduced at 85%. (From Valdivia and Brinkley, 1984. Micrograph courtesy of B. R. Brinkley.)

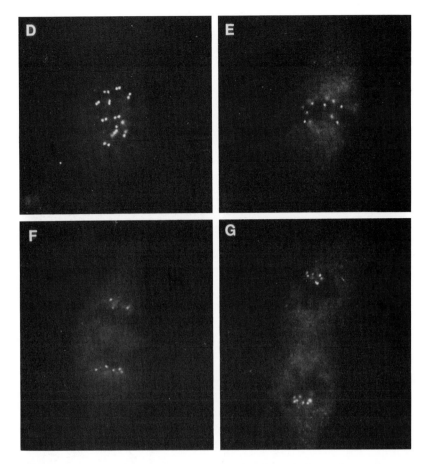

(**D–G**) Cultured cells of PTK$_2$ line doubly stained with antibodies to kinetochores. (D) Prophase. (E) Metaphase. Kinetochores are seen as pairs of fluorescent spots. (F) Anaphase. (G) Telophase. (D and E ×1800, F ×2600, reproduced at 65%. Micrographs courtesy of B. R. Brinkley.)
(**H**) Cultured cells stained to antibodies specific to centrosomes. Left, interphase cells; right, mitotic cells. (Courtesy of B. R. Brinkley.)
(**I**) Scheme comparing polymerization of MTOC-associated and free microtubules in tubulin solution. Fluctuations of polymerization process can lead to the shrinkage and disappearance of some microtubules of both types. Free microtubules are not restored later. In contrast, after disappearance of MTOC-associated microtubule, their nucleation sites are preserved on the MTOC and new microtubules, marked by asterisks, can start growing from these sites. Eventually only the MTOC-associated microtubules remain in solution. (**J**) Capture of growing microtubule by the kinetochore. Isolated centrosomes and chromosomes are placed into the tubulin solution. Microtubules grow outward from the centrosome; one of them is attached to the kinetochore of the chromosome. Then the solution is diluted by buffer. Decrease of the tubulin concentration leads to depolymerization of all the centrosome-attached microtubules with the free outer ends; only the microtubules, with the outer ends captured by the kinetochores, remain stable. (Scheme based on Mitchison and Kirschner, 1984.)

Another interesting property of kinetochores is their ability not only to nucleate microtubules, but also to capture microtubules nucleated by other MTOCs. This capture was demonstrated in experiments in which microtubule polymerization was first initiated on centrosomes and then chromosomes were added. The mixture of chromosomes and microtubule asters was

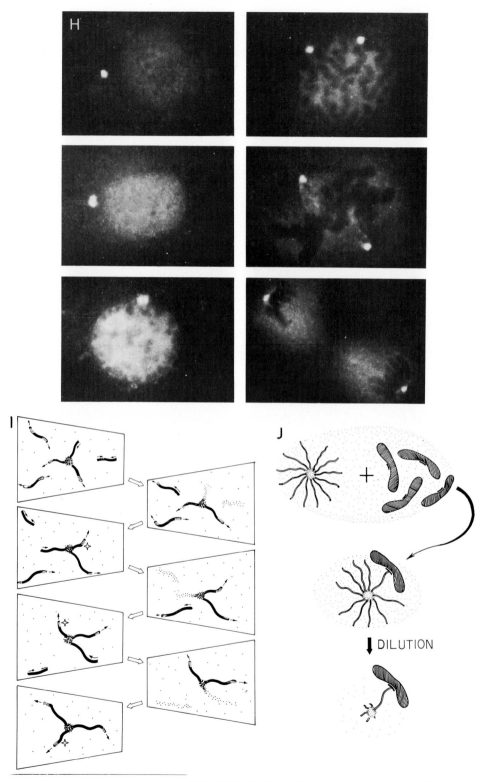

FIG. 2.10. (*Continued*)

incubated for some time in the tubulin solution to allow interaction. Then the complexes of centrosomes and kinetochores were formed; namely, the plus ends of some microtubules growing from centrosomes became attached to the kinetochores of chromosomes. When tubulin solution around these complexes was diluted by the buffer, rapid depolymerization of the microtubules emanating from the centrosome and not captured by the kinetochore took place. Captured microtubules were, however, much more stable (Fig. 2.10J). The microtubules linking the centrosomes with the chromosomes remained intact when all the other microtubules around the centrosome disappeared. This elegant experiment shows that MTOCs capping both ends of the microtubule stabilize it and prevent its depolymerization in the low concentration of tubulin. The obvious suggestion is that kinetochore microtubules of the mitotic spindle *in vivo* are stabilized in a similar way.

Thus, MTOCs can nucleate the polar growth of microtubules and stabilize these microtubules by capping their ends.

C. Determination of the Structure of Microtubular Arrays by MTOCs

Experiments *in vitro* show that MTOCs not only nucleate microtubules, but also control the structure of these microtubules and of their arrays. In particular, different arrays are nucleated by basal bodies and by centrosomes not only *in vivo,* but also *in vitro.* Centrosomes *in vitro* nucleate the asterlike array with microtubules radiating in all directions. Microtubules nucleated by the basal body grow from its two opposite ends and are approximately parallel to one another.

The maximal number of microtubules within the array is also determined by the MTOC. When the tubulin concentration in solution is gradually increased, the average number of microtubules growing from each centrosome increases, until a certain maximum is reached. This saturation effect suggests that the number of nucleating sites on each centrosome is limited.

Self-assembly in brain tubulin solutions gives predominantly microtubules with 14 protofilaments. However, under exactly the same conditions, microtubules growing off centrosomes usually have 13 protofilaments. Obviously, the nucleation sites of centrosomes determine the number of protofilaments in the microtubule.

Thus, certain aspects of the controlling role of MTOCs in the formation of microtubule arrays can be reproduced *in vitro.* However, *in vitro* we are still unable to form most regular and complex microtubular structures. For instance, *in vitro* the basal body nucleates not the highly regular axoneme with its outer and inner doublets, but a rather loose parallel array with frayed ends. Obviously, in this case some additional factors within or outside the MTOC are needed for more exact specification of the numbers, structures, and positions of microtubules within the array.

The mechanisms of determination of array structures by MTOCs *in vitro* and *in vivo* are still completely obscure. It is possible that basal bodies act as preformed fragments of microtubules, from which corresponding microtubules of the axoneme elongate. However, the structure of the axoneme corresponds only partially to that of the basal body, and it is not clear how these differences are specified.

The situation is even more complex when microtubules seem to be nucleated by apparently structureless dense material. For instance, radiating arrays nucleated by centrosome grow not from the centriole, but from diffuse pericentriolar material. Many other types of MTOC contain nothing but this material. At present we have no idea how this seemingly uniform material of unknown composition can determine the number, polarity, and lattice structure of the nucleated microtubules.

As mentioned in Section I, certain arrays of microtubules *in vivo* are associated not with discrete MTOCs, but with certain areas of the plasma membrane. These membrane–microtubule complexes of two cell types, of chick erythrocytes and of the alga *Distigma proteus*, were successfully reassembled *in vitro* from the membrane fragments of these cells and nonpolymerized tubulin. Thus, certain areas of the membrane apparently have nucleating properties. In contrast to discrete MTOC, these membrane sites may be associated not only with the ends of growing microtubules, but also with their side walls, so that growing microtubules are anchored to the membrane along their entire length. The chemical nature of these special membrane sites is unknown.

D. Formation of MTOCs

Morphological observations of the cells at various stages of MTOC formation are the only source of information about these processes. These observations suggest that there are several methods of formation of MTOCs.

The kinetochore is the only known type of MTOC whose localization and formation are directly determined by specific genetic material, centromeric DNA, located at a certain site of each chromosome, centromere. This DNA, like other DNAs, is replicated during each cell cycle. Multilaminar kinetochores are formed at the two sides of each centromere at the beginning of mitosis. As already mentioned, the kinetochore contains several apparently specific proteins. The specific centromere DNA of yeast was recently isolated and cloned in bacteria. Thus, it may soon be possible to directly study specific DNA–protein interactions involved in the formation of the kinetochore. Centrioles and basal bodies can be formed either near the preexisting structure of the same type (so-called centriolar pathway) or independently of these structures (acentriolar pathway) (Fig. 2.11). Formation of new centriole within the centrosome of an animal cell is an example of a centriolar pathway. In this case the daughter structure is formed within the dense material surrounding the mother centriole. The daughter centriole does not directly contact the mother structure and is usually oriented at a right angle to it. An acentriolar pathway had been observed in the epithelial cells of monkey oviducts, where numerous basal bodies are rapidly formed after hormonal stimulation. An acentriolar pathway begins with development of the focal aggregates of dense material in the cytoplasm. Then new centrioles appear near the surface of these aggregates.

Thus, in both pathways centrioles are formed in association with dense material. In both cases this material undergoes characteristic local changes during the formation of centrioles. It is arranged into a circular plate or a ring.

Often a so-called cartwheel is formed; that is, the central dense cylinder with dense spokes going to the outer ring. Microtubules of a new centriole appear sequentially in the ring of dense material, and then each microtubule is elongated. The mechanisms for these processes are unclear. Formation of centrioles can be regarded as a special variant of the enigmatic process discussed in the previous paragraph, the formation of a complex microtubular array from apparently structureless material.

Morphological observations suggest that focal aggregates of this material can be formed *de novo* in the cytoplasm. Preexisting centrioles are not directly involved in the formation of new centrioles. Why, then, is the centriole so often present within the centrosome? Perhaps the centriole can induce the formation of foci of dense material around itself. This material, in turn, can nucleate both the radiating array of microtubules and the daughter centriole. In other words, the centriole may be able to perform two different types of nucleating functions. It may nucleate the ciliary axoneme directly in

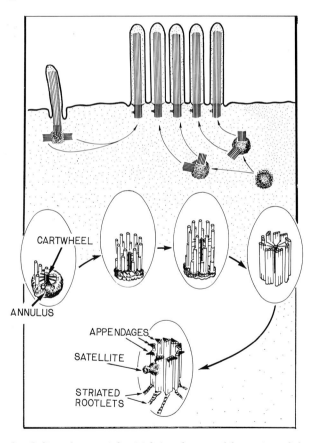

FIG. 2.11. Centriolar (left) and acentriolar (right) pathways of formation of the basal bodies of cilia. Bottom: Stages of the basal body formation from the aggregates of dense material connected or not connected with centriole. (Scheme based on Anderson and Brenner, 1971, and Dirksen, 1971.)

one stage. It may also nucleate the radiating arrays of microtubules and daughter centriole indirectly, via induction of an aggregation of dense material. This dual nucleating function of the centriole can provide a means of strict control of relative positions of several microtubular arrays formed in the same cell. There are many examples of this high-level control. For instance, the primary cilium of animal cells is always located in the center of the radiating cytoplasmic array (see Introduction). In all probability, the cilium is nucleated directly by the centriole, and the radiating array by the pericentriolar material. Ciliata, e.g., *Paramecium*, have numerous rows of cilia aligned parallel to one another. During cell growth each row of cilia is replicated in such a way that the position of each cilium and its orientation are exactly reproduced. Possibly this strictly controlled replication of cilia is achieved by a series of nucleating events. A preexisting basal body can nucleate an accumulation of dense material; this material can nucleate a new basal body; this body can nucleate the new cilium.

All these considerations are at present purely hypothetical as we do not know the nature of dense material and have no means of studying the MTOC formation *in vitro*. Recently, antibodies specifically reacting with pericentriolar material of centrosomes were found; this finding suggests that isolation and identification of the mysterious "dense material" may soon be possible.

VIII. Motility of Microtubular Systems

A. Dyneins

Motility is a characteristic feature of many types of microtubular systems, such as cilia and flagella, the mitotic spindle, and pharyngeal baskets of Infusoria. Flagella and cilia are the only systems in which the basic molecular mechanism of microtubular motility is more or less understood. Mutual sliding of microtubules in these organelles is caused by cyclic interactions of dynein arms with the microtubular walls. Before discussing these interactions, we shall briefly describe the properties of dyneins.

Dyneins are multimolecular complexes forming outer and inner sidearms attached to the walls of A-microtubules of the cilia and flagella (Fig. 2.12A–C). Dyneins are the major microtubule-associated ATPases. Dynein arms can be selectively extracted from the axonemes. These isolated dynein arms have molecular weights about 1000 or 2000 kD (the molecular weight of myosin in only 500 kD). Study of the complex molecular composition of dyneins was facilitated by genetic analysis of the motility of a unicellular alga, *Chlamydomonas*. Normal *Chlamydomonas* actively swims waving its two flagella. Numerous mutants of the organism with decreased motility were selected and analyzed. Some of these mutants have only the inner dynein arms; others have only the outer arms. There are also certain completely immobile mutants, which lack both types of arms. Loss of the outer or the inner arms was found to be correlated with disappearance of certain groups of polypeptides extracted from the flagellar axonemes. These correlations helped to elucidate the composition of dynein arms. The better-studied outer

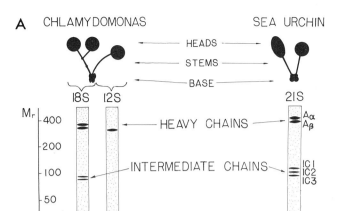

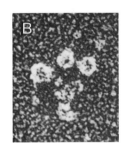 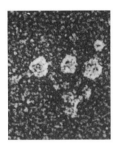  

FIG. 2.12. (A) Structure and polypeptide composition of outer-arm dyneins. Scheme showing the molecular shapes and the results of SDS–PAGE of the two best-studied dyneins. **(B)** Three-headed dynein molecule of *Tetrahymena* cilium. Platinum replica. ×450,000. (From Goodenough and Heuser, 1984. Courtesy of U. W. Goodenough.) **(C)** Dynein molecules attached to the walls of A-microtubules of *Tetrahymena* cilium. B, B-microtubules. (From Goodenough and Heuser, 1984. Courtesy of U. W. Goodenough.)

arms of *Chlamydomonas* contain at least 14 polypeptide chains: three noni-dentical heavy chains, two intermediate chains, and nine light chains. The molecular weight of one heavy chain, estimated by the sedimentation equilibrium method, proved to be as high as 450 kD. The compositions of the outer arms of other animals, of the ciliate *Tetrahymena* and of sea urchins, were found to be very similar to those of *Chlamydomonas*, except that sea urchin dynein has only two heavy chains. The morphology of dynein arms is as complex as their polypeptide composition. Each dynein arm consists of several globular heads and of slender threads. The outer arm of *Tetrahymena* and of *Chlamydomonas* has three heads, while that of the sea urchin has only two heads. Thus, each heavy chain corresponds to one head. The two heads of the *Chlamydomonas* outer arm have a diameter of about 12 nm each; the third head is larger (14 nm). The inner arms contain unique sets of heavy, inter-mediate, and light chains; the inner-arm fraction of *Chlamydomonas* contains up to six distinct heavy-chain types. The inner arms of *Chlamydomonas* are dimorphic: some are two-headed, and others are three-headed.

According to the most widely used model, the dynein arm is a bouquet of two or three globular heads connected by threads to a rootlike base. Iso-lated dynein arms have Mg-stimulated ATPase activity, which seems to be located in the heavy chains.

Thus, the structure of dynein has many similarities with its counterpart in the actin system, myosin. Both myosin and dynein have globular heads in which ATPase activity is localized. However, the structure of dynein is more complex than that of myosin. While myosin has two similar heads, dynein has two or three nonidentical heads, each made by a special heavy chain. The number of diverse intermediate and light chains in dynein is also much greater than that of light chains of myosin molecules. Are dyneins present in any other structures besides cilia and flagella? This question is still contro-versial. The dyneinlike protein isolated from sea urchin eggs has ATPase activity; it contains a heavy chain and several intermediate chains; it binds to microtubules in an ATP-dependent manner. It is not clear whether this protein is a special precursor of ciliary dynein or a special protein that per-forms some function outside the cilia, e.g., in the mitotic spindle. Certain antibodies against ciliary dyneins were found to stain the mitotic spindles, but the identity of the stained molecules remains to be determined.

B. Dynein Interactions with Microtubules

When isolated dynein arms are added to the microtubules from which they have been previously extracted, these arms are reattached to the micro-tubules. Each reattached arm, like that in the intact axoneme, can contact A- and B-microtubules of two neighboring doublets (Figs. 2.5 and 2.12B). Various parts of the arm specifically interact with A- or B-microtubules having differ-ent types of wall lattice (see Section III). The more narrow base of the arm is attached to the A-microtubule (so-called A-type attachment), while the wider head parts are attached to the B-microtubules (B-type attachment). Addition of ATP dissociates B-attachments but not A-attachments.

Dynein attachment to the flagellar microtubules is a cooperative process; that is, usually numerous dynein arms are attached simultaneously. Mutual contacts of neighboring dynein arms attached to the microtubules can be seen in the electron microscopic pictures of flagella. Possibly, dynein arms are bound not only to the microtubule, but to one another. Numerous arms attached to the same microtubule can form a large coordinately working system. Dynein arms can bind not only to flagellar doublets, but also to many other types of microtubules; these attachments can be either of A or of B type.

What are the molecular mechanisms of the force generation by dynein? Triple interactions of dynein, ATP, and microtubule proved to be rather similar to the actin–myosin–ATP interaction (see Chapter 1). The cycle of these interactions (Fig. 2.13) has four main stages:

1. ATP is bound by the dynein arm and detaches this arm from the B-microtubule.
2. Dynein-associated ATP is hydrolyzed.
3. The dynein arm with bound ADP and phosphate is bound to the B-microtubule.
4. ADP and phosphate are released, changing the conformation of microtubule-bound dynein and thus producing the force for sliding.

The first two stages of this cycle have been observed experimentally, but the exact sequence of the last two stages is suggested only by indirect evidence. Dynein ATPase is stimulated by microtubules.

ATPase activity of dynein is characteristically inhibited by two agents: vanadate and erythro-9 [3-(2-hydroxynonyl)] adenine (EHNA). Vanadate acts as an analog of phosphate; EHNA acts as an analog of ATP. Unfortunately, both these agents are not highly specific and may affect a number of other cellular molecules besides dynein.

C. Organization and Regulation of the Movements of Cilia and Flagella

ATP-driven cyclic interactions of dynein arms attached to the A-microtubule with the neighboring B-microtubule lead to mutual directional sliding of these microtubules. This sliding can be observed directly when the axo-

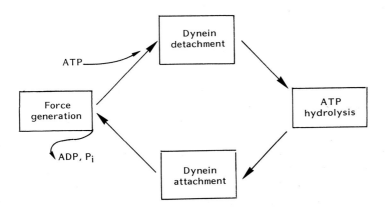

FIG. 2.13. Hypothetical scheme of cycle of dynein–microtubule interactions. (After Johnson, 1983.)

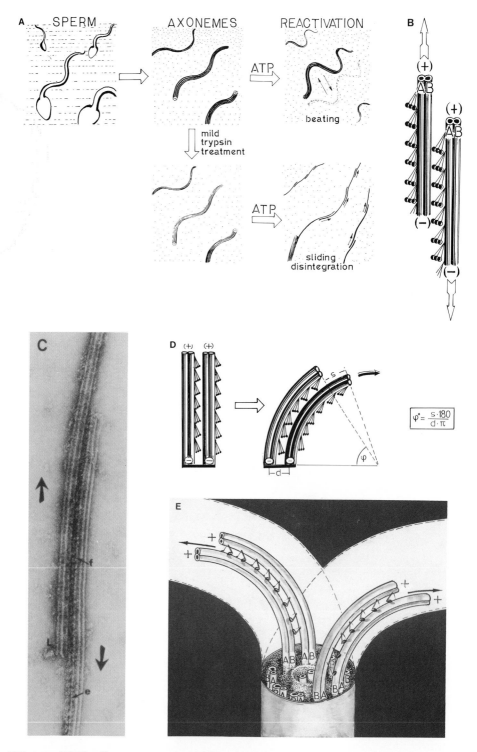

FIG. 2.14. **(A)** Flagellar axonemes can undergo bending or sliding disintegration in the presence of ATP. **(B)** Scheme showing direction of ATP-induced sliding of adjacent microtubule doublets. **(C)** Sliding doublets of *Tetrahymena*. Direction of sliding is shown by arrows. Electron micro-

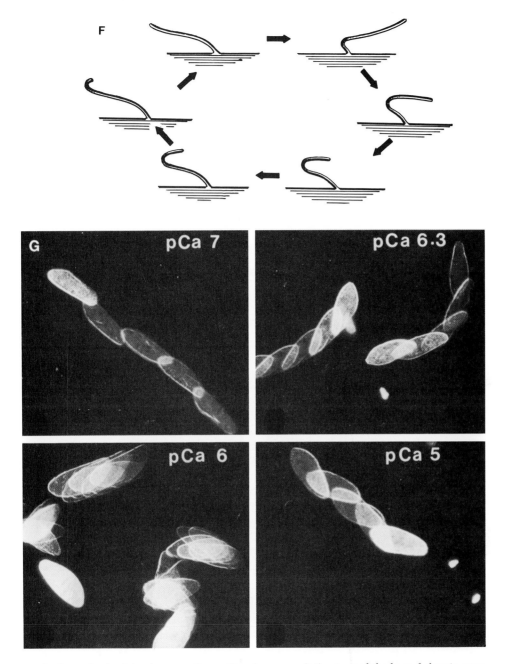

graph of negatively stained preparation. e, Dynein arms pointing toward the base; f, dynein arms between the doublets. (From Sale and Satir, 1977. Courtesy of P. Satir.). **(D)** Transformation of sliding of adjacent doublets into bending within the intact axoneme. The formula shows the dependence of the angle of bending (ϕ) on the length of sliding (s). **(E)** Activity of opposite doublet pairs induces bending in opposite directions. **(F)** The cycle of asymmetrical bendings of cilium in the course of its beating. **(G)** Swimming of Triton-treated models of the ciliate *Paramecium caudatum* in a solution containing ATP and various concentrations of calcium. Each print contains several consecutive images of the dark-field illumination; intervals between photographs 0.9 sec. At pCa 7 (10^{-7} M Ca^{2+}) and pCa 6.3 ($10^{-6.3}$ Ca^{2+}) the models swim forward. At pCa 6.0 the models rock in place. At pCa 5.0 they swim backward. (From Otter *et al.*, 1984. Courtesy of P. Satir.)

neme is first pretreated with proteases and then incubated with ATP. In these conditions the length of the axoneme can increase up to 8–10 times, and the doublets are irreversibly drawn apart. This is so-called sliding disintegration of the axoneme (Fig. 2.14A). As expected, this sliding is directional. The A-microtubule carrying dynein arms always slides toward its minus end relative to the neighboring B-microtubule to which these arms are directed (Fig. 2.14B,C).

Sliding takes place not only in the disintegrated axonemes, but during bending and unbending of the normal moving cilia and flagella. Electron microscopic comparison of the structure of bent and straight cilia has shown that their microtubules do not change their lengths, but slide along one another (Fig. 2.14D).

Mutual sliding of neighboring microtubules will lead to the bending of flagella only if these microtubules are somehow linked to one another. One of these links is probably the basal body. However, axonemes detached from this body bend and unbend in the presence of ATP; thus, other types of links should also exist.

The morphology of the axoneme (see Introduction) suggests that three types of structures can act as intermicrotubular links: nexin interdoublet links, radial spokes, and dynein arms. The crucial role of any of these structures in bending has not yet been proven. Among the normal forms of flagella and cilia there are some variants without a central pair of microtubules, central sheath, and radial spokes: flagella with "9 + 0," "6 + 0," and even "3 + 0" patterns. Some of these deviant flagellar types retain some motility. One particular Chlamydomonas mutant without spokes but with motile flagella has been isolated. These data suggest that radial spokes are not essential for bending. The possible role of interdoublet nexin links also is not clear; any genetic variants of these links have not been isolated. It is possible that some dynein arms attached to the B-walls in the absence of ATP act as temporary links restricting sliding; these links can be broken in the presence of ATP. Some other types of links also can be reversibly formed and broken during bending.

Regular movement of the flagella and cilia requires not only restriction of the sliding of neighboring doublets, but also regulation of distribution of the activity of various doublets. Each pair of outer doublets oriented BA→BA has its symmetrical pair on the other side of the flagellum; that is, there is another pair of the doublet with the opposite orientation AB→AB. Obviously, directional sliding of one or the other pair of doublets will bend the flagellum in the opposite direction. Only when these pairs work by turns will the flagellum or cilium bend forward and back. Therefore, it was suggested that doublet sliding is alternately activated on two sides of the axoneme during the bending and unbending cycle (Fig. 2.14E). This suggestion is supported by the recent experiments of Tamm (1984) who found that selective activation of sliding on the opposite sides of the axoneme occurs at the early stages of sliding desintegration. Some cilia first extruded all the doublets located on one half of the axoneme; some extruded those on the other half. Thus, the mechanism for alternating activation of various doublets lies somewhere in the axoneme itself.

Additional mechanisms are needed to regulate the asymmetry of the bending. Most cilia and flagella do not simply bend symmetrically forward and back. Their bending is highly asymmetrical, so that the cell, propelled by these bendings, swims directionally (Fig. 2.14F). The direction of asymmetrical beating of cilia can be reversed by external factors. For example, *Paramecium* begins to swim backward when it meets an unfavorable environment. Remarkably, reversal of ciliary beatings can be observed not only in live Infusoria, but also in their "models" demembranated by detergent. When activated by ATP, these models swim forward in the solution containing 10^{-7} M Ca; they stop moving at 10^{-6} M Ca and start to move backward at 10^{-5} M (Fig. 2.14G). Certain mutants of *Paramecium* swim forward normally in usual medium but are unable to reverse their movement after external stimulation. The same inability to reverse movement is observed when these mutant cells are demembranated and placed into solution with ATP and high Ca.

These experiments show that the calcium-dependent mechanism regulating asymmetry of cilium movement is localized in the axoneme. This mechanism in living cells can be switched by alterations of calcium concentration within the cytoplasm.

The nature of calcium-dependent regulation remains unknown. Calcium-binding protein, calmodulin, is probably the mediator of this regulation. Calmodulin binds to dynein and can modulate its ATPase activity. Phenothiazines, inhibitors of calmodulin, block the action of Ca^{2+} on ciliary beating. Further study of mutants defective in their responses to calcium may provide information about the nature of these mechanisms.

To summarize, basic mechanisms of dynein–tubulin and of actin–myosin motility have many similarities. Both types of motility are based on cyclic ATP-driven protein interactions, causing the mutual sliding of cytoskeletal fibrils: of actin filaments or of microtubules. However, the structural organization of the motile cytoskeleton in cilia and flagella is more complex than that in the most highly organized actin–myosin system, in the myofibril. The filaments in the myofibrils slide parallel to one another; that is, perform linear movements. In contrast, linear sliding of microtubules in cilia and flagella is transformed into integrated curvilinear movements, which can switch from one pattern to another. It is obvious that the complex organization needed for these movements is built into the cytoskeleton of cilia and flagella. Almost nothing is known about the nature of this organization, which makes cilia and flagella the highly efficient swimming devices of eukaryotic cells.

As indicated at the beginning of this section, microtubule-associated motility is observed not only in cilia and flagella, but also in many other structures. Possibly, some forms of this motility are also based on dynein–tubulin interactions, but as yet this mechanism has not been demonstrated unequivocally in any nonflagellar structure. Certain forms of motility may be based on interactions of the microtubule not with dyneins, but with other ATPases. In particular, special proteins, so-called "translocators," seem to be responsible for the movement of organelles along the microtubules. These topics will be further discussed in Chapters 5 and 8.

IX. Conclusion: Dynamics of Microtubular Systems

Many microtubular arrays are highly organized systems with specified positions, numbers, structures, and lengths of individual microtubules. At the same time, some of these arrays are also highly dynamic. The cell usually contains not only the polymerized microtubules, but also a large pool of non-polymerized tubulin. For instance, about 50% of tubulin in cultured animal cells is nonpolymerized. There is a continuous exchange of tubulin molecules between the soluble pool and the microtubules. As shown by experiments with intracellular injection of fluorochrome-labeled tubulin, the mitotic spindle seems to have the most unstable microtubules. The half-life of tubulin in these arrays can be as short as several seconds. Cytoplasmic microtubules of interphase cells are somewhat more stable; the half-life of their tubulin is of the order of several minutes. The rate of tubulin exchange in cilia and flagella has not yet been measured. It would also be important to compare the rate of exchange of different microtubules with their stability to depolymerizing agents, such as colchicine. As already mentioned, the ciliary microtubules have greater resistance to these agents than microtubules of other structures. Inhibition of drug-induced depolymerization by ATP deprivation (see Section VI) suggests that stability of microtubules can be modulated by some cellular regulatory mechanisms; it would be interesting to compare the rate of tubulin exchange in control and ATP-deprived cells. Results of measurements of tubulin dynamics show that the life of the whole array can be much longer than that of individual microtubules in the array. For instance, the radiating arrays of interphase microtubules can exist for many hours, although their tubulin is exchanged every several minutes. Of course, the structure of the array may also be changed with time reversibly or irreversibly. In particular, the mitotic spindle has a relatively short life span, and its structure is continuously changed during this period.

When the structure remains stable for a long time, this stability is continuously maintained by the cells and restored after external traumas. For instance, the cells strictly control the size of their cilia and flagella. When *Chlamydomonas* is placed into the vessel with vigorously circulating medium, its flagella are amputated by the shearing force. Later, in stationary medium, the cell regrows the flagella to original length. When *Chlamydomonas* loses two flagella, both regenerate synchronously (Fig. 2.15). However,

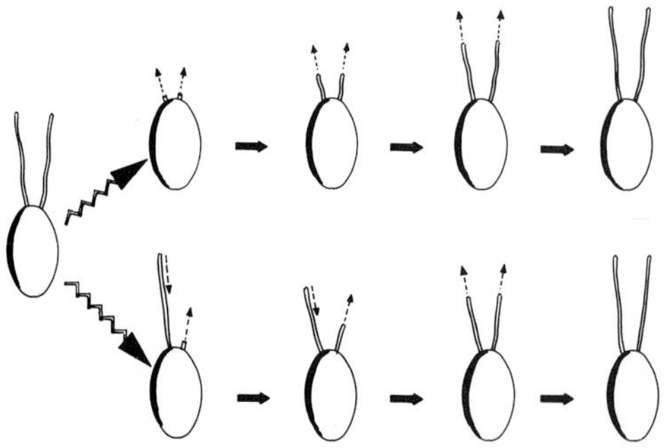

FIG. 2.15. Regeneration of *Chlamydomonas* flagella after mechanical detachment. Detachment of both flagella is followed by their simultaneous growth. When only one flagellum is detached, its regrowth is accompanied by shortening of the second flagellum. When both flagella acquire similar size, they start to grow simultaneously. (Scheme based on Rosenbaum *et al.*, 1969.)

when this cell loses only one flagellum, the remaining one shortens and a new one grows, until both flagella reach the same length. Afterward both flagella grow synchronously to their full length. This remarkable result suggests that the cell has a special mechanism regulating final flagellar length and equalizing the length of two flagella.

All these facts indicate that characteristics of the microtubular arrays are not irreversibly determined at the moment of their formation, but are continuously controlled by the cell.

Experiments *in vitro* described in this chapter have revealed several types of mechanisms that can dynamically control the microtubular system. The first mechanism of control can be based on dynamic instability of microtubule polymerization depending on the formation of GTP caps. Conditions leading to the loss of these caps, e.g., decreased concentration of tubulin or of GTP, can cause rapid depolymerization of microtubules. Instability of microtubules in arrays can be due to random gain and loss of GTP caps.

The second level of control may be based on the attachment of various MAPs to the wall of the microtubule. This attachment usually makes the microtubule more stable and less likely to depolymerize. Phosphorylation of these MAPs can decrease their ability to stabilize microtubules. The cell can reversibly alter the activities of microtubule-associated kinases, which phosphorylate MAPs, and in this way control the properties of microtubules.

MAPs can also control positions of microtubules within the arrays. Mutual sliding of microtubules, mediated by dynein, is the most obvious example of this control. Possibly, certain types of MAPs can also cross-link neighboring microtubules.

Association of the microtubule end with MTOCs of various types can be the factor determining their stability. The centers can nucleate and renucleate microtubules. This renucleation may be the reason why the microtubular arrays associated with MTOC can remain stable for a long time. Association of both ends with MTOC can make the microtubule especially stable.

Unstable microtubules coexisting within the same cell can compete with one another for the common pool of tubulin. Usually the microtubules associated with MTOC win this competition, and free microtubules disappear (Fig. 2.9D–F). The intensity of this competition depends on the stability of the microtubules. When cultured cells are treated with taxol, free microtubules, not connected with any visible MTOC, are formed in their cytoplasm (Fig. 2.9L). Taxol increases the stability of microtubules so that the free microtubules survive even in the absence of MTOC.

Some microtubules in normal cells, e.g., those in the neuronal processes, seem to have no visible attachment to MTOCs. These apparently free microtubules may be stabilized by some MAPs or by unknown microtubule-capping proteins. Recently it had been suggested that free microtubules present in the cell may be nucleated by MTOCs and then detached from them. If this suggestion is correct, even the population of free microtubules is dependent on MTOCs. By nucleating microtubules, the MTOCs determine and stabilize the general structure of arrays with which they are associated. The MTOCs themselves are under the control of some factors regulating their replication and formation *de novo*. MTOC-regulating mechanisms control not only the structure of individual arrays, but also dynamic interrelationships between various arrays. These mechanisms of unknown nature form the highest level of cellular control of microtubular structures.

Thus, the systems of microtubules seem to be regulated by a hierarchy of various mechanisms. At present we can only guess at the possible nature of some of these mechanisms.

Literature Cited

Anderson, K. G. W., and Brenner, R. M. (1971) The formation of basal bodies (centrioles) in the rhesus monkey oviduct, *J. Cell Biol.* **50:**10–34.

Bershadsky, A. D., and Gelfand, V. I. (1981) ATP-dependent regulation of cytoplasmic microtubule disassembly, *Proc. Natl. Acad. Sci. USA* **78:**3610–3613.

Birrell, G. B., Habliston, D. L., Nadakavukaren, K. K., and Griffith, O. H. (1985) Immunophotoelectron microscopy: The electron optical analog of immunofluorescence microscopy, *Proc. Natl. Acad. Sci. USA* **82:**109–113.

Bulinski, J. C., and Borisy, G. G. (1979) Self-assembly of microtubules in extracts of cultured HeLa cells and the identification of HeLa microtubule-associated proteins, *Proc. Natl. Acad. Sci. USA* **76:**293–297.

Burton, P. R., and Paige, J. L. (1981) Polarity of axoplasmic microtubules in the olfactory nerve of the frog, *Proc. Natl. Acad. Sci. USA* **78:**3269–3273.

Carlier, M-F., and Pantaloni, D. (1981) Kinetic analysis of guanosine 5'-triphosphate hydrolysis associated with tubulin polymerization, *Biochemistry* **20:**1918–1924.

Carlier, M-F., Hill, T. L., and Chen, Y. (1984) Interference of GTP hydrolysis in the mechanism of microtubule assembly: An experimental study, *Proc. Natl. Acad. Sci. USA* **81:**771–775.

Cleveland, D. W., Hwo, S-Y., and Kirschner, M. W. (1977) Purification of tau, a microtubule-associated protein that induces assembly of microtubules from purified tubulin, *J. Mol. Biol.* **116:**207–225.

Dirksen, E. R. (1971) Centriole morphogenesis in developing ciliated epithelium of the mouse oviduct, *J. Cell Biol.* **51:**286–302.

Euteneuer, U., and McIntosh, J. R. (1981) Polarity of some motility-related microtubules, *Proc. Natl. Acad. Sci. USA* **78:**372–276.

Field, D. J., Collins, R. A., and Lee, J. C. (1984) Heterogeneity of vertebrate brain tubulins, *Proc. Natl. Acad. Sci. USA* **81:**4041–4045.

Goodenough, U. W., and Heuser, J. E. (1984) Structural comparison of purified dynein proteins with *in situ* dynein arms, *J. Mol. Biol.* **180:**1083–1118.

Haimo, L. T. (1982) Dynein decoration of microtubules—determination of polarity, in *Methods in Cell Biology*, Vol. 24 (L. Wilson, ed.), Academic Press, New York, pp. 189–206.

Haimo, L. T., Telzer, B. R., and Rosenbaum, J. L. (1979) Dynein binds to and crossbridges cytoplasmic microtubules, *Proc. Natl. Acad. Sci. USA* **76:**5759–5763.

Hirokawa, N. (1982) Cross-linker system between neurofilaments, microtubules, and membranous organelles in frog axons revealed by the quick-freeze, deep-etching method, *J. Cell Biol.* **94:**129–1242.

Huber, G., Pehling, G., and Matus, A. (1986) The novel microtubule-associated protein MAP3 contributes to the *in vitro* assembly of brain microtubules, *J. Biol. Chem.* **261:**2270–2273.

Johnson, K. (1983) The pathway of ATP hydrolysis by dynein, *J. Biol. Chem.* **258:**13825–13832.

Jones, J. C. R., and Tucker, J. B. (1981) Microtubule-organizing centers and assembly of the double-spiral microtubule pattern in certain heliozoan axonemes, *J. Cell Sci.* **50:**259–280.

Krauhs, E., Little, M., Kempf, T., Hofer-Warbinek, R., Ade, W., and Ponstingl, H. (1981) Complete amino acid sequence of β-tubulin from porcine brain, *Proc. Natl. Acad. Sci. USA* **78:**4156–4160.

Mitchison, T., and Kirschner, M. (1984) Microtubule dynamics and cellular morphogenesis, in *Molecular Biology of the Cytoskeleton* (G. G. Borisy, D. W. Cleveland, and D. B. Murphy, eds.), Cold Spring Harbor Laboratory, Cold Spring Harbor, New York, pp. 27–44.

Murray, J. M. (1984) Three-dimensional structure of a membrane–microtubule complex, *J. Cell*

Ochs, R. L., and Stearns, M. E. (1981) Colloidal gold immunolabeling of whole-mount erythrophore cytoskeletons: Localization of tubulin and HMW-MAPs, *Biol. Cell* **42**:19–28.

Otter, T., Satir, B. H., and Satir, P. (1984) Trifluorazine-induced changes in swimming behavior of *Paramecium*: Evidence for two sites of drug action, *Cell Motil.* **4**:249–267.

Parysek, L. M., Asnes, C. F,. and Olmsted, J. B. (1984) MAP4: Occurrence in mouse tissues, *J. Cell Biol.* **99**:1309–1315.

Rosenbaum, J. L., Moulder, J. E., and Ringo, D. L. (1969) Flagellar elongation and shortening in *Chlamydomonas*: The use of cycloheximide and colchicine to study the synthesis and assembly of flagellar proteins, *J. Cell Biol.* **41**:600–619.

Sale, W. S., and Satir, P. (1977) The direction of active sliding of microtubules in *Tetrahymena* cilia, *Proc. Natl. Acad. Sci. USA* **74**:2045–2049.

Schultheiss, R., and Mandelkow, E. (1983) Three-dimensional reconstruction of tubulin sheets and re-investigation of microtubule surface lattice, *J. Mol. Biol.* **170**:471–496.

Tamm, S. L. (1984) Alternate patterns of doublet microtubule sliding in ATP-disintegrated macrocilia of the ctenophore *Beroë*, *J. Cell Biol.* **99**:1364–1371.

Tucker, J. B. (1970) Morphogenesis of a large microtubular organelle and its association with basal bodies in the ciliate *Nassula*, *J. Cell Sci.* **6**:385–429.

Tucker, J. B. (1971) Spatial discrimination in the cytoplasm during microtubule morphogenesis, *Nature* **232**:387–389.

Tucker, J. B. (1979) Spatial organization of microtubules, in *Microtubules* (K. Roberts and J. S. Hyams, eds.), Academic Press, New York, pp. 315–357.

Tucker, J. B. (1984) Spatial organization of microtubule-organizing centres and microtubules, *J. Cell Biol.* **99**:55s–62s.

Tucker, J. B., Mathews, S. A., Hendry, K. A. K., Mackie, J. B., and Roche, D. L. J. (1985) Spindle microtubule differentiation and deployment during micronuclear mitosis in Paramecium, *J. Cell Biol.* **101**:1966–1976.

Valdivia, M., and Brinkley, B. R. (1984) Biochemical studies of the kinetochore/centromere of mammalian chromosomes, in *Molecular Biology of the Cytoskeleton* (G. G. Borisy, D. W. Cleveland, and D. B. Murphy, eds.), Cold Spring Harbor Laboratory, Cold Spring Harbor, New York, pp. 79–86.

Vallee, R. B. (1984) MAP2 (microtubule-associated protein 2), in *Cell and Muscle Motility*, Vol. 5, *The Cytoskeleton* (J. W. Shay, ed.), Plenum Press, New York, pp. 289–311.

Vallee, R. B., and Bloom, G. S. (1984) High molecular weight microtubule-associated proteins (MAPs), in *Modern Cell Biology*, Vol. 3 (B. H. Satir, ed.), Liss, New York, pp. 21–75.

Vallee, R. B., Bloom, G. S., and Theurkauf, W. E. (1984) Microtubule-associated proteins: Subunits of the cytomatrix, *J. Cell Biol.* **99**:38s–44s.

Voter, W. A., and Erikson, H. P. (1982) Electron microscopy of MAP2 (microtubule-associated protein 2), *J. Ultrastruct. Res.* **80**:374–382.

Additional Readings

General reviews

Dustin, P. (1970) *Microtubules*, Springer-Verlag, New York.

McKeithan, T. W. and Rosenbaum, J. L. (1984) The biochemistry of microtubules: A review, in *Cell and Muscle Motility*, Vol. 5, *The Cytoskeleton* (J. W. Shay, ed.), Plenum Press, New York, pp. 255–288.

Oakley, B. R. (1985) Microtubule mutants, *Can. J. Biochem. Cell. Biol.* **63**:479–488.

Raff, E. C. (1984) Genetics of microtubule system, *J. Cell Biol.* **99**:1–10.

Roberts, K., and Hyams, I. S., eds. (1979) *Microtubules*, Academic Press, New York.

Tubulin; structure of microtubules; polymerization *in vitro*

Amos, L. A. (1982) Tubulin and associated proteins, in *Electron Microscopy of Proteins*, Vol. 3 (I. R. Harris, ed), Academic Press, London, New York, pp. 207–250.

Bergen, L. G., and Borisy, G. G. (1980) Head-to-tail polymerization of microtubules *in vitro*. Electron microscope analysis of seeded assembly, *J. Cell Biol.* **84**:141–150.

Cleveland, D. W., and Sullivan, K. F. (1985) Molecular biology and genetics of tubulin, *Ann. Rev. Biochem.* **54**:331–365.

Dentler, W. L., Granett, S., Witman, G. B., and Rosenbaum, J. L. (1974) Directionality of brain microtubule assembly *in vitro*, *Proc. Natl. Acad. Sci. USA* **71**:1710–1740.

Detrich, H. W., III, Jordan, M. A., Wilson, L., and Williams, R. C., Jr. (1985) Mechanism of microtubule assembly. Changes in polymer structure and organization during assembly of sea urchin egg tubulin, *J. Biol. Chem.* **260**:9479–9490.

Eichenlaub-Ritter, U., and Tucker, J. B. (1984) Microtubules with more than 13 profilaments in the dividing nuclei of ciliates, *Nature* **307**:60–62.

Hill, T. L. (1985) Phase-changing kinetics for a microtubule with two free ends, *Proc. Natl. Acad. Sci. USA* **82**:431–435.

Kirchner, K., and Mandelkow, E-M. (1985) Tubulin domains responsible for assembly of dimers and protofilaments, *Eur. Mol. Biol. Organ. (EMBO) J.* **4**:2397–2402.

L'Hernault, S. W., and Rosenbaum, J. L. (1985) *Chlamydomonas* α-tubulin is posttranslationally modified by acetylation on the ε-amino group of a lysine, *Biochemistry* **24**:473–478.

Mandelkow, E-M., and Mandelkow, E. (1985) Unstained microtubules studied by cryo-electron microscopy. Substructure, supertwist and disassembly, *J. Mol. Biol.* **181**:123–135.

Margolis, R. L., and Wilson, L. (1978) Opposite end assembly and disassembly of microtubules at steady state *in vitro*, *Cell* **13**:1–8.

McEwen, B., and Edelstein, S. J. (1980) Evidence for a mixed lattice in microtubules reassembled *in vitro*, *J. Mol. Biol.* **139**:123–145.

McIntosh, J. R., and Euteneuer, U. (1984) Tubulin hooks as probes for microtubule polarity: An analysis of the method and an evaluation of data on microtubule polarity in the mitotic spindle, *J. Cell Biol.* **98**:525–533.

Mitchison, T., and Kirschner, M. (1984) Dynamic instability of microtubule growth, *Nature* **312**:237–242.

Rothwell, S. W., Grasser, W. A., and Murphy, D. B. (1985) Direct observation of microtubule treadmilling by electron microscopy, *J. Cell Biol.* **101**:1637–1642.

Thompson, W. C. (1982) The cyclic tyrosination/detyrosination of alpha tubulin, in *Methods in Cell Biology*, Vol. 24 (L. Wilson, ed.), Academic Press, New York, pp. 235–255.

Tilney, L. G., Bryan, J., Bush, D. J., Fujiwara, K., Mooseker, M. S., Murphy, D. B., and Snyder, D. H. (1973) Microtubules: Evidence for thirteen protofilaments, *J. Cell. Biol.* **59**:267–275.

Voter, W. A., and Erikson, H. P. (1984) The kinetics of microtubule assembly. Evidence for a two-stage nucleation mechanism, *J. Biol. Chem.* **259**:10430–10438.

Weisenberg, R. C. (1972) Microtubule formation *in vitro* in solutions containing low calcium concentrations, *Science* **177**:1104–1105.

Weisenberg, R. C., and Deery, W. I. (1976) Role of nucleotide hydrolysis in microtubule assembly, *Nature* **263**:792–793.

Microtubule-associated proteins (MAPs); drugs affecting microtubule assembly

Bergen, L. G., and Borisy, G. G. (1983) Tubulin–colchicine complex inhibits microtubule elongation at both plus and minus ends. *J. Biol. Chem.* **258**:4190–4194.

Böhm, J. J., Vater, W., Fenske, H., and Unger, E. (1984) Effect of microtubule-associated proteins on the protofilament number of microtubules assembled *in vitro*, *Biochim. Biophys. Acta* **800**:119–126.

Bulinski, J. C., and Borisy, G. G. (1980) Widespread distribution of 210,000-mol-wt microtubule-associated protein in cells and tissues of primates, *J. Cell Biol.* **87**:802–808.

Hoebeke, J., Van Nijen, G., and De Brabander, M. (1976) Interaction of oncodazole (R 17934), a new antitumoral drug, with rat brain tubulin, *Biochem. Biophys. Res. Commun.* **69**:319–324.

Huber, G., Alaimo-Beuret, D., and Matus, A. (1985) MAP3: Characterization of a novel microtubule-associated protein, *J. Cell Biol.* **100**:496–507.

Job, D., Pabion, M., and Margolis, R. L. (1985) Generation of microtubule stability subclasses by microtubule-associated proteins: implications for the microtubule "dynamic instability" model, *J. Cell Biol.* **101:**1680–1689.

Jordan, M. A., Margolis, R. L., Himes, R. H., and Wilson, L. (1986) Identification of a distinct class of vinblastine binding sites on microtubules, *J. Mol. Biol.* **187:**61–73.

Kuznetsov, S. A. Rodionov, V. I., Gelfand, V. I., and Rosenblat, V. A. (1981) Microtubule-associated protein MAP1 promotes microtubule assembly *in vitro*, *FEBS Lett.* **135:**241–244.

Kuznetsov, S. A., Rodionov, V. I., Gelfand, V. I., and Rosenblat, V. A. (1984) MAP2 competes with MAP1 for binding to microtubules, *Biochem. Biophys. Res. Commun.* **119:**173–178.

Lindwall, G., and Cole, R. D. (1984) Phosphorylation affects the ability of tau protein to promote microtubule assembly, *J. Biol. Chem.* **259:**5301–5305.

Manfredi, J. J., Parness, J., and Horwitz, S. B. (1982) Taxol binds to cellular microtubules, *J. Cell Biol.* **94:**688–696.

Margolis, R. L., and Wilson, L. (1977) Addition of colchicine–tubulin complex to microtubule ends: The mechanism of substoichiometric colchicine poisoning, *Proc. Natl. Acad. Sci. USA* **74:**3466–3470.

Murphy, D. B., Johnson, K. A., and Borisy, G. G. (1977) Role of tubulin-associated proteins in microtubule nucleation and elongation, *J. Mol. Biol.* **177:**33–52.

Vallee, R. B. (1982) A taxol-dependent procedure for the isolation of microtubules and microtubule-associated proteins (MAPs), *J. Cell Biol.* **92:**435–442.

Williams, R. F., Mumford, C. L., Williams, C. A. Floyd, L. J., Aivaliotis, M. J., Martinez, R. A., Robinson, A. K., and Barnes, L. D. (1985) A photaffinity derivative of colchicine: 6′-(4′-Azido-2′-nitrophenylamino) hexanoyldeacetylcolchicine. Photolabeling and location of the colchicine-binding site on the α-subunit of tubulin, *J. Biol. Chem.* **260:**13794–13802.

Zieve, G., and Solomon, F. (1982) Proteins specifically associated with the microtubules of the mammalian mitotic spindle, *Cell* **28:**233–242.

Zingsheim, H. P., Herzog, W., and Weber, K. (1979) Differences in surface morphology of microtubules reconstituted from pure brain tubulin using two different microtubule associated proteins—High molecular weight MAP2 proteins and tau proteins, *Eur. J. Cell Biol.* **19:**175–182.

Dynein and motility of cilia and flagella

Blum, J. J., Hayes, A., Jamieson, G. A., and Vanaman, T. C. (1980) Calmodulin confers calcium sensitivity on ciliary dynein ATPase, *J. Cell Biol.* **87:**386–397.

Brokaw, C. J., Luck, D. J. L., and Huang, B. (1982) Analysis of the movement of *Chlamydomonas* flagella: The function of the radial spoke system is revealed by comparison of wild type and mutant flagella, *J. Cell Biol.* **92:**722–732.

Gibbons, B. H. (1982) Reactivation of sperm flagella: Properties of microtubule-mediated motility, in *Methods in Cell Biology*, Vol. 25 (L. Wilson, ed.), Academic Press, New York, London, pp. 253–271.

Gibbons, B. H., Bacceti, B., and Gibbons, I. R. (1985) Live and reactivated motility in the 9 + 0 flagellum of *Anguilla* sperm, *Cell Motil.* **5:**333–350.

Gibbons, I. R. (1981) Cilia and flagella of eucaryotes, *J. Cell Biol.* **91** (3, pt. 2):107s–124s.

Goodenough, U. W., and Heuser, I. E. (1985) Substructure of inner dynein arms, radial spokes and the central pair/projection complex of cilia and flagella, *J. Cell Biol.* **100:**2008–2018.

Goodenough, U. W., and Heuser, I. E. (1985) Outer and inner dynein arms of cilia and flagella, *Cell* **41:**341–342.

Haimo, L. T., and Fenton, R. D. (1984) Microtubule crossbridging by *Chlamydomonas* dynein, *Cell Motil.* **4:**371–385.

Hollenbeck, P. J., Suprynowicz, F., and Cande, W. Z. (1984) Cytoplasmic dynein-like ATPase cross-links microtubules in an ATP-sensitive manner, *J. Cell Biol.* **99:**1251–1258.

Johnson, K. A., Porter, M. E., and Shimizu, T. (1984) Mechanism of force production for microtubule-dependent movements, *J. Cell Biol.* **99:**132s–136s.

Luck, D. J. L. (1984) Genetic and biochemical dissection of the eucaryotic flagellum, *J. Cell Biol.* **98:**789–794.

Mitchell, D. R., and Rosenbaum, I. L. (1985) A motile *Chlamydomonas* flagellar mutant that lacks outer dynein arms, *J. Cell Biol.* **100:**1228–1234.

Omoto, C. K., and Johnson, K. A. (1986) Activation of the dynein adenosine triphosphatase by microtubules, *Biochemistry* **25**:419–427.

Pfister, K. K., Haley, B. E., and Witman, G. B. (1985) Labelling of *Chlamydomonas* 18S dynein polypeptides by 8-azido adenosine 5'-triphosphate, a photoaffinity analog of ATP, *J. Biol Chem.* **260**:12844–12850.

Piperno, G. (1984) Monoclonal antibodies to dynein subunits reveal the existence of cytoplasmic antigens in sea urchin egg, *J. Cell Biol.* **98**:1842–1850.

Sale, W. S., and Gibbons, I. R. (1979) Study of the mechanism of vanadate inhibition of the dynein crossbridge cycle in sea urchin sperm flagella, *J. Cell Biol.* **82**:291–298.

Satir, P. (1968) Studies on cilia: further studies on the cilium tip and a "sliding filament" model of ciliary motility, *J. Cell Biol.* **39**:77–94.

Summers, K. E., and Gibbons, I. R. (1971) Adenosine triphosphate-induced sliding of tubules in trypsin-treated flagella of sea urchin sperm, *Proc. Natl. Acad. Sci. USA* **68**:3092–3096.

Tsukita, Sh., Tsukita, S., Usukura, J., and Ishikawa, H. (1983) ATP-dependent structural changes of the outer dyneim arm in *Tetrahymena* cilia: A freeze-etch replica study, *J. Cell Biol.* **96**:1480–1485.

Warner, F. D. (1976) Cross-bridge mechanisms in ciliary motility: The sliding–bending conversion, in *Cell Motility*, Book C (R. Goldman, T. Pollard, and J. Rosenbaum, eds.), Cold Spring Harbor Laboratory, Cold Spring Harbor, New York, pp. 891–914.

Microtubule organizing centers (MTOCs); organization and dynamics of the microtubules in the cell

Brinkley, B. R., Cox, S. M., Pepper, D. A., Wible, L., Brenner, S. L., and Pardue, R. L. (1981) Tubulin assembly sites and the organization of cytoplasmic microtubules in cultured mammalian cells, *J. Cell Biol.* **90**:554–562.

Cabral, F., Wible, L., Brenner, S., and Brinkley, B. R. (1983) Taxol-requiring mutant of Chinese hamster ovary cells with impaired mitotic spindle assembly, *J. Cell Biol.* **97**:30–39.

Calarco-Gillam, P. D., Siebert, M. C., Hubble, R., Mitchison, T., and Kirschner, M. (1983) Centrosome development in early mouse embryos as defined by an autoantibody against pericentriolar material, *Cell* **35**:621–629.

Chalfie, M. (1982) Microtubule structure in *Caenorhabditis elegans* neurons, *Cold Spring Harbor Symp. Quant. Biol.* **46**:255–261.

De Brabander, M., Geuens, G., Nuydens, R., Willebords, R., and De May, J. (1981) Taxol induces the assembly of free microtubules in living cells and blocks the organizing capacity of the centrosomes and kinetochores. *Proc. Natl. Acad. Sci. USA* **78**:5608–5612.

Deery, W. J., and Brinkley, B. R. (1983) Cytoplasmic microtubule assembly–disassembly from endogenous tubulin in a Brij-lysed cell model, *J. Cell Biol.* **96**:1631–1641.

Evans, L., Mitchison, T., and Kirschner, M. (1985) Influence of the centrosome on the structure of nucleated microtubules, *J. Cell Biol* **100**:1185–1191.

Gundersen, G. G., Kalnoski, M. H., and Bulinski, J. C. (1984) Distinct populations of microtubules: Tyrosinated and nontyrosinated alpha tubulins are distributed differently *in vivo*, *Cell* **38**:779–789.

Heidemann, S. R., Hamborg, M. A., Thomas, S. J., Song, B., Lindley, S., and Chu, D. (1984) Spatial organization of axonal microtubules, *J. Cell Biol.* **99**:1289–1295.

Karsenti, E., Kobayashi, S., Mitchison, T., and Kirschner, M. (1984) Role of the centrosome in organizing the interphase microtubule array: Properties of cytoplasts containing or lacking centrosomes, *J. Cell Biol.* **98**:1763–1776.

Keith, C., Di Paola, M., Maxfield, F. R., and Shelanski, M. L. (1983) Microinjection of Ca^{2+}-calmodulin causes a localized depolymerization of microtubules, *J. Cell Biol* **97**:1918–1924.

Miller, M., and Solomon, F. (1984) Kinetics and intermediates of marginal band reformation: evidence for peripheral determinants of microtubule organization, *J. Cell Biol.* **99**:70s–75s.

Mitchison, T., and Kirschner, M. (1984) Microtubule assembly nucleated by isolated centrosomes, *Nature* **312**:232–237.

Mitchison, T. J., and Kirschner, M. W. (1985) Properties of the kinetochore *in vitro*. I. Microtubule nucleation and tubulin binding, *J. Cell Biol.* **101**:755–765.

Nadezhdina, E. S., Fais, D., and Chentsov, Yu. S. (1979) On the association of centrioles with the interphase nucleus, *Eur. J. Cell Biol.* **19**:109–115.

Osborn, M., and Weber, K. (1976) Cytoplasmic microtubules in tissue culture cells appear to grow from an organizing structure towards the plasma membrane, *Proc. Natl. Acad. Sci. USA* **76**:867–871.

Salmon, E. D., Leslie, R. J., Saxton, W. M., Karow, M. L., and McIntosh, J. R. (1984) Spindle microtubule dynamics in sea urchin embryos: Analysis using a fluorescein-labelled tubulin and measurement of fluorescence redistribution after laser photobleaching, *J. Cell Biol.* **99**:2165–2174.

Saxton, W. M., Stemple, D. L., Leslie, R. J., Salmon, E. D., Zavortink, M., and McIntosh, J. R. (1984) Tubulin dynamics in cultured mammalian cells, *J. Cell Biol.* **99**:2175–2186.

Soltys, B. J., and Borisy, G. G. (1985) Polymerization of tubulin *in vivo*: Direct evidence for assembly onto microtubule ends and from centrosomes, *J. Cell Biol.* **100**:1682–1689.

Tassin, A-M., Maro, B., and Bornens, M. (1985) Fate of microtubule-organizing centers during myogenesis *in vitro*, *J. Cell Biol.* **100**:35–46.

Valdivia, M. M., and Brinkley, B. R. (1985) Fractionation and initial characterization of the kinetochore from mammalian metaphase chromosomes, *J. Cell Biol.* **101**:1124–1134.

Vorobjev, I. A., and Chentsov, Yu. S. (1982) Centrioles in the cell cycle. I. Epithelial cells, *J. Cell Biol.* **98**:938–949.

Wheatley, D. N. (1982) *The Centriole: A Central Enigma of Cell Biology*, Elsevier, New York.

3

Systems of Intermediate Filaments

I. General Morphology of Structures Formed by Intermediate Filaments

Intermediate filaments (IFs), as indicated by their name, are thicker than actin filaments, but thinner than microtubules. Their diameter in electron microscopic sections is 8–12 nm (Fig. 3.1A).It is possible that IFs, in contrast to actin filaments and microtubules, are not universal components of the cytoskeleton of all eukaryotic cells.

As of now, there is no convincing evidence of IF presence in the cells of Protozoa. There are some exceptional mammal cells (some embryonic neurons and certain exceptional tumor cell lines) in which IF were not revealed. However, IFs are abundant in most cell types of vertebrates.

In cultured fibroblasts, IF proteins make up about 2–4% of the total protein content. This percentage is much higher (about 30%) in cultured epidermal cells.There is growing evidence that cells of all invertebrates and even plants also contain IF.

The most typical structures formed by IF are three-dimensional loose networks distributed throughout the cytoplasm and intermixed with other cell components. The central parts of these networks are concentrated around the nuclei, while the peripheral parts radiate toward the plasma membranes (Fig. 3.1B, C). Some parts of these networks are aligned into more or less parallel bundles. In particular, numerous parallel filaments are present in the elongated processes of neurons.

A characteristic feature of IF is their chemical stability. Harsh extracting procedures (see Section II) that remove most other cell components, including microfilaments and microtubules, leave these filaments intact. In this way purified frameworks of IF can be obtained. These frameworks remain associated with the nucleus. Thus, networks of IF seem to be linked to the nucleus: the nature of this attachment is not clear. IF were also found to converge in some special centers near nuclei (IF organizing centers); the nature of these centers also has not been identified.

The morphology of the networks of IF has distinctive characteristics in the cells of various tissue (Fig. 3.1D). Epithelial cells are characterized by the

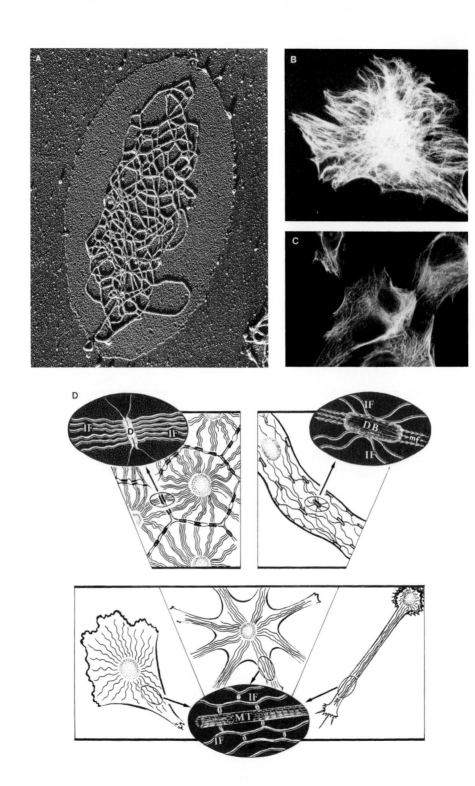

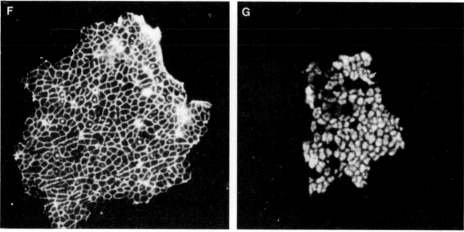

FIG. 3.1. (A) Intermediate filaments of chicken erythrocytes attached to the plasma membrane. Electron micrograph of shadowed preparation of lysed erythrocyte, The preparations were preincubated with antivimentin antibody. ×19,800, reproduced at 35%. (Courtesy of B. L. Granger and E. Lazarides.) **(B)** Intermediate filaments in the cytoplasm of normal cultured mouse fibroblast. Immunofluorescence microscopy using antivimentin antibody. (Courtesy of I. S. Tint.) **(C)** Keratin network in the cytoplasm of cultured rat hepatocytes. Immunofluorescence microscopy using antibody against rat keratin with mol. wt. 49 kD. (Courtesy of G. A. Bannikov and S. M. Troyanovsky.) **(D)** Arrays of intermediate filaments in the cells of various types. Upper left: IF associated with desmosomes (D) in epithelial cells. Upper right: IF associated with dense bodies in smooth muscle. Lower row: IF associated with microtubules in fibroblasts, glial cells, and neurons. **(E)** Bundles of keratin filaments attached to the desmosomes in the multilayered epithelium of human uterine cervix. Electron micrograph of the section. (Courtesy of A. P. Cherny.) **(F,G)** Desmin (F) and α-actinin (G) in the sheets of Z disks isolated from chicken skeletal muscle cell. Immunofluorescence microscopy. α-Actinin is localized within the disks, while desmin filaments form the network between the disks. (Courtesy of E. Lazarides.)

attachment of IF to the cytoplasmic sides of the membranes of specialized cell–cell contacts called spot desmosomes or simple desmosomes (Fig. 3.1E); IF in these cells are also attached to specialized cell substrate contacts, hemidesmosomes. Desmosomes link the networks of IF in adjoining neighboring cells into mechanically integrated multicellular systems, such as sheets and alveoli. The keratinized layers of the epidermis are an extreme case of this integration. Here almost all the cellular organelles are replaced by the networks of IF, strengthened by cross-linking proteins; the networks of individual cells are firmly linked, so that coherent layers of high mechanical stability are formed.

In smooth and striated muscle, IFs link the sites of attachment of actin filaments with one another and with the plasma membrane. More specifically, IFs in smooth muscle cells are linked to the dense bodies present in the cytoplasm. Actin filaments are also attached to the dense bodies (see Chapter 1, Fig. 1.8). IFs were also described as being attached to the focal structures located at the inner sites of the membrane, so-called dense plaques.

In striated muscle, IFs (Figs. 3.1F, G and Chapter 1, Fig. 1.6) are attached to the periphery of the Z disks. Longitudinally oriented filaments go along each myofibril and link its Z disks with one another. Transverse filaments were described to link the Z disks of neighboring myofibrils with one another. Possibly, they also go from the myofibrils to the outer membrane and are attached to it in the costamere regions (see Chapter 1). In addition, in cardiac muscle cells longitudinal IFs are attached to spot desmosomes localized in the intercalated disk region, near the fascia adherens (Chapter 1, Fig. 1.6). IFs in mesenchymal, glial, and neural cells radiate from the nucleus toward the plasma membrane; any special connections of these filaments with the plasma membrane have not yet been identified. IFs in these cells are usually closely associated with radiating cytoplasmic microtubules.

Although the systems of IF in various tissues have different organizations, individual filaments of these tissues are usually morphologically indistinguishable. Nevertheless, the filaments present in the cells of various tissues are composed of different proteins. We shall discuss the properties of these proteins in the next section.

II. Proteins of Intermediate Filaments

A. Isolation

Since IFs are chemically very stable, special harsh procedures are used for their extraction. In experiments with cultured cells, they are first treated with nonionic detergent, e.g., Triton X-100, to obtain insoluble cytoskeleton. Then actin and tubulin are removed by high-salt buffers with KCl or Kl (about 1 M), but the IFs remain insoluble. To extract their proteins, the remaining cytoskeletons are treated with denaturing solvents, e.g., with strong ionic detergents, urea, acetic acid, or salt solutions of very low ionic strength. To extract the least soluble of the IF-proteins, keratins, one has to use such solutions as 8 M urea.

B. Five Classes of IF-Proteins

There are five tissue-specific classes of the proteins of IF: vimentin, glial fibrillary acidic protein, desmin, neurofilament proteins, and keratins (Fig. 3.2). Three groups of tissues contain a single protein each. Vimentin (mol. wt. about 53 kD) is present in all mesenchymal tissues, including connective tissue cells, blood cells, bone, and cartilage cells. Desmin (mol. wt. 52 kD) is present in all types of smooth and striated muscle, except certain types of vascular smooth muscle cells that contain vimentin. Glial fibrillary acidic protein (GFAP) (mol. wt. about 50 kD) is present in astroglial cells. The two remaining classes of IF-proteins contain several polypeptide species each. Neurofilaments contain three proteins (NF-L, NF-M, and NF-H) with molecular weights of about 65, 105, and 135 kD, respectively; sometimes these pro-

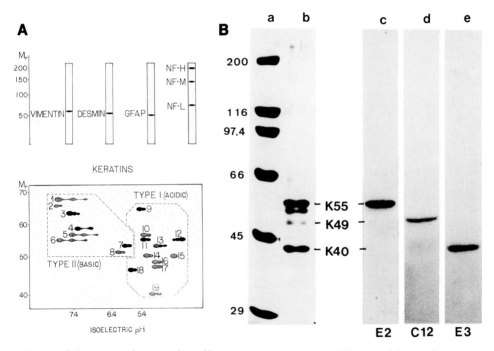

FIG. 3.2. **(A)** Proteins of intermediate filaments in various tissues. Schemes of the results of analysis by sodium dodecyl sulfate–polyacrylamide gel electrophoresis (SDS–PAGE) (top) and by two-dimensional electrophoresis/electrofocusing (bottom). Human cytokeratins are numbered according to the classification of Moll *et al.*, 1982. Epidermal cytokeratins are marked by cross-shadowing. **(B)** Monoclonal antibodies reacting with individual cytokeratins, analyzed by SDS–PAGE. Lanes, from left to right: (a) molecular markers with known molecular weight shown at the left; (b) preparation of IF from rat hepatoma cells; three polypeptides with mol. wt. 55 kD, 49 kD and 40 kD are present; (c,d,e) immune blots of the preparations shown in lane b using monoclonal antibodies against keratin with mol. wt. 55 kD (c), 49 kD (d), and 40 kD (e). Gels in lanes a and b were stained with Coumassie blue. Blots c, d, and e were prepared by transferring the proteins from the gels to nitrocellulose filters; the filters were incubated with antibodies and then with ^{125}I-labeled A-protein reacting with the bound antibodies; autoradiographs were then prepared. Each of the three antibodies used reacts with only one particular keratin. (Courtesy of G. A. Bannikov and S. M. Troyanovsky.)

teins are designated by their apparent molecular weights in sodium dodecyl sulfate electrophoresis (70, 150, and 200 kD, respectively). These proteins are specific for neurons; usually the cells contain all three polypeptides.

Keratins present in most epithelial tissues represent the largest protein family. Each mammalian species has up to 30 distinct keratins with molecular weights varying from 40 to 70 kD. In human epithelia, 19 keratin polypeptides have been identified. The keratins can be divided into two distinct families: type I (acidic) and type II (neutral–basic) (Fig. 3.2). Type I keratins have lower isoelectric points and usually lower molecular weights than type II keratins. Monoclonal antibodies have been raised that are specific for either type I or type II keratins with no significant cross-reactivity. The minimal number of keratins simultaneously present in the same cell is two; the maximal number can be as high as 10. Significantly, simultaneously expressed keratins always contain at least one member of the acidic and neutral–basic subfamilies. Sets of keratin molecules are different in morphologically different epithelia. In particular, the composition of monolayered epithelia in the digestive tract

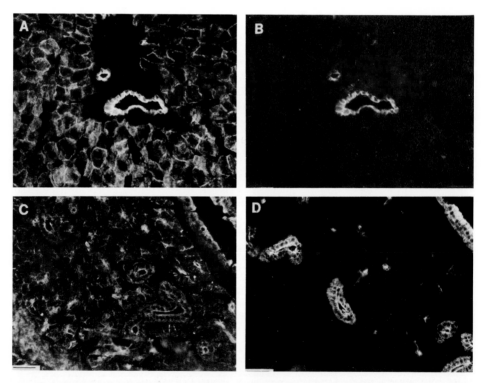

FIG. 3.3. Differential distribution of individual keratins in various epithelial structures of the same organ. Frozen sections of the rat liver **(A, B)** and of the rat submaxillary salivary gland **(C, D)**. Immunofluorescence microscopy using staining with antibodies against keratin with mol. wt. 49 kD (A, C) and keratin with mol. wt. 40 kD (B, D). 49-kD keratin is present in all the epithelial cells of both organs; in contrast, 40-kD keratin is present only in the bile ducts of the liver and in the ducts of salivary glands. (Courtesy of G. A. Bannikov and S. M. Troyanovsky.)

and its associated ducts and glands (liver, pancreas, gall bladder) is different from that of multilayered epidermis. There are additional differences between the various cells of the same tissues, e.g., between ductal and parenchymal cells of liver or between ductal and alveolar cells of salivary glands (Fig. 3.3). Thus, the composition of IF is not only specific for each group of tissues, but is closely correlated with cell morphology in the same group of epithelial tissues.

Various IF-proteins can be distinguished not only by analytical methods of protein chemistry, but also immunologically. There are monoclonal antibodies specific for each molecular species of these five groups of proteins, including each of the keratins. Immunomorphological examination using these antibodies makes it possible to determine the specificity of IF-proteins individually in each cell present in the tissue section (Fig. 3.3) or in culture.

Each species of IF-protein is coded by a special gene. Although certain IF-proteins share only 23–25% amino acid sequence homology with one another (see Section II.C), the structures of all IF genes are remarkably similar. In particular, the positions of intron sequences segmenting the coding portions are highly similar in these genes. These results indicate that all types of IF-protein genes had a common ancestor.

C. Common and Specific Features of Different IF-Proteins

How can the different protein molecules form morphologically similar filaments? The answer is that all these molecules, in spite of all the differences between them, share common structural features (Fig. 3.4A).

All are rod-shaped and have a large central core domain of about 40 kD consisting of several α-helical subdomains interspersed by short nonhelical inclusions. The central core domain is flanked by two nonhelical terminal domains. This secondary structure was determined from amino acid sequence data of IF subunits. Complete sequences of several types of keratins, of vimentin, desmin, and GFAP, and of neurofilament proteins are now known; they were determined both by protein and by cDNA sequencing.

The lengths of corresponding central domains and positions of inclusions are similar in different subunits, whereas the lengths of terminal domains can differ considerably. Differences in the molecular weight of various IF-proteins are largely due to the different sizes of these terminal domains. Various domains of the subunit are likely to have different roles in the formation of filament. Central α-helical domains of each subunit interact laterally with corresponding domains of neighboring subunits, so that two α-helices intertwine around one another, forming a two-stranded coiled coil. Terminal nonhelical domains can be required to link these dimeric rods into higher-order structures. In fact, the subunits from which terminal domains have been removed by proteases become unable to form filaments, but are polymerized only into oligomeric rods of about 2 nm × 50 nm.

The presence of numerous intertwined α-helices (coiled coils) within the different types of intermediate filaments was shown by x-ray diffraction analysis; the axes of these helices are approximately parallel to that of the filament.

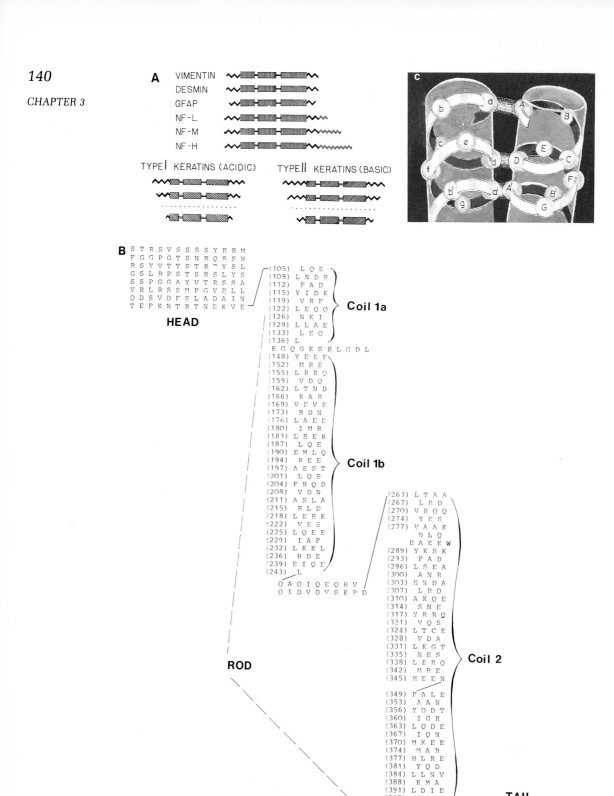

A VIMENTIN
DESMIN
GFAP
NF-L
NF-M
NF-H

TYPE I KERATINS (ACIDIC)   TYPE II KERATINS (BASIC)

C

B
S T R S V S S S S Y R R M
F G G P G T S N R Q S S N
R S Y V T T S T R T Y S L
G S L R P S T S R S L Y S
S S P G G A Y V T R S S A
V R L R S S M P G V R L L
Q D S V D F S L A D A I N
T E F K N T R T N E K V E

HEAD

(105) L Q E
(108) L N D R
(112) F A D
(115) Y I D K
(119) V R F
(122) L E Q Q
(126) N K I
(129) L L A E
(133) L E Q
(136) L

Coil 1a

K G Q G K S R L G D L

(148) Y E E E
(152) M R E
(155) L R R Q
(159) V D Q
(162) L T N D
(166) K A R
(169) V E V E
(173) R D N
(176) L A E D
(180) I M R
(183) L R E K
(187) L Q E
(190) E M L Q
(194) R E E
(197) A E S T
(201) L Q S
(204) F R Q D
(208) V D N
(211) A S L A
(215) R L D
(218) L E R K
(222) V E S
(225) L Q E E
(229) I A F
(232) L K K L
(236) H D E
(239) E I Q E
(243) L

Coil 1b

Q A Q I Q E Q H V
Q I D V D V S K P D

(263) L T A A
(267) L R D
(270) V R Q Q
(274) Y E S
(277) V A A K
        N L Q
        E A E E W
(289) Y K S K
(293) F A D
(296) L S E A
(300) A N R
(303) N N D A
(307) L R Q
(310) A K Q E
(314) S N E
(317) Y R R Q
(321) V Q S
(324) L T C E
(328) V D A
(331) L K G T
(335) N E S
(338) L E R Q
(342) M R E
(345) M E E N

Coil 2

(349) F A L E
(353) A A N
(356) Y Q D T
(360) I G R
(363) L Q D E
(367) I Q N
(370) M K E E
(374) M A R
(377) H L R E
(381) Y Q D
(384) L L N V
(388) K M A
(391) L D I E
(395) I A T
(398) Y R K L
(402) L E G
(405) E E S R
(409) I

ROD

TAIL

S L P L P N F S S L N L R
E T N L E S L P L V D T H S
K R T L L I K T V E T P D G
Q V I N E T S Q H H D D L E

α-Helical domains of various IF-proteins have certain periodicities of amino acid residues required for the formation of coiled coils. In each consecutive group of seven residues (heptad), the first and fourth positions are occupied by hydrophobic amino acids (Fig. 3.4B,C). These two residues are located at the same side of the α-helix strands. These apolar residues on the contacting surfaces of two strands of the coiled coil interact hydrophobically, stabilizing the structure of the coil. Other residues of each heptad, occupying superficial positions on the coil, are usually polar or charged; possibly, they are important for the electrostatic interactions between the neighboring coiled coils in the filament.

Thus, all IF-protein subunits have important structural similarities. Their relationship is confirmed by immunological data. Certain types of monoclonal antibodies are able to recognize all IF regardless of their tissue type specificity. Analysis of amino acid sequences also reveals homology of primary structures of all these proteins. However, the degree of homology between various types of subunits is different (Fig. 3.4A). Desmin, vimentin, and GFAP have more than 60% homologous sequences. Central domains of these proteins have higher homology than terminal domains. The three neurofilament proteins (200, 150, and 70 kD) have similar structures except for the different length of the C-terminal domain; this domain is usually longer in 150- to 200-kD proteins.

Keratins have a lower homology with the four other types of proteins (about 30%). Analysis of the sequences of mRNAs of keratins and of corresponding amino acid sequences had shown that type I and II keratins share a low homology, less than 30% at the nucleic acid level. In contrast, the homology between the individual members of the same keratin group is high: from 50 to 90%. Analyzed individual keratins of the same type have similar central domains; only their terminal domains are different from one another.

Thus, there can be major sequence divergence among distinct IF-proteins, although they all retain certain features required for formation of the common secondary structure.

←

FIG. 3.4. **(A)** Comparison of the structures of various IF-proteins. Shaded boxes, central helical domains (from left to right: 1A, 1B, 2); zigzag lines, nonhelical terminal domains. The longer α-helical domain 2 can be further subdivided into parts 2A and 2B. **(B)** Amino acid sequence of hamster vimentin. (Data from Quax-Jenken *et al.*, 1983.) One-letter symbols for amino acids are used (see legend to Fig. 2.3A). The sequence is to be read line by line from the left upper corner (N-end) to the right bottom corner (C-end). α-Helical domains of the central (rod) portion of the sequence are subdivided into heptads. Numbers of the amino acid residues occupying the "a" and "d" positions of the heptads are indicated in parenthesis to the left of the corresponding letters. The nonpolar amino acids leucine (L), isoleucine (I), valine (V), tyrosine (Y), methionine (M), and phenylalanine (F) occur predominantly in these positions. **(C)** Interaction between α-helices in a coiled coil of IF. The scheme shows contacting α-helical domains of two protein subunits forming the coiled coil; end-on view from amino to carboxyl ends. a, b . . . g and A, B . . . G are two heptads of amino acid residues. Analysis of amino acid sequences has shown that the first and fourth positions of each heptad (a, d and A, D) are usually occupied by hydrophobic amino acids. It is suggested that hydrophobic interactions of these residues in two contacting helices stabilize the coiled coil structures.

III. Formation and Structure of Intermediate Filaments

A. Polymerization of Filaments

Protein subunits, solubilized by the denaturing agents, are readily repolymerized after removal of these agents from solution.

Polymerization of keratin filaments occurs even in very diluted salt solutions (< 0.01 M); that is, in almost pure water. Proteins of other types, e.g., desmin and vimentin, were observed to remain soluble in these low-salt solutions; in these cases filament formation could be induced by raising the ionic strength to physiological level. Thus, assembly of IF occurs in less stringent conditions than that of certain other cytoskeleton structures, especially microtubules.

Distinct types of subunits interact specifically with one another during polymerization. Vimentin, desmin, and GFAP readily copolymerize. Each of these types of subunit alone is also able to form filaments. In contrast, keratin filaments cannot be formed by any single type of subunit, but only by pairs of keratin polypeptides of type I and II (so-called keratin pairing). *In vitro* all the diverse type I keratin polypeptides are able to form IF when allowed to react with equimolar amounts of any of the type II keratins. Examples of successful IF formation *in vitro* include combinations of polypeptides that have never been found in the same cell *in vivo*, e.g., complexes of epidermal keratins with those from simple epithelia or complexes of keratins from different species. *In vitro* copolymerization of certain keratins with desmin or vimentin has been also observed by some investigators; however, *in vivo* keratins are always excluded from vimentin filaments, and vice versa.

Light neurofilament subunit NF-L does readily polymerize on its own. It easily copolymerizes with the other neurofilament proteins (NF-M and NF-H). NF-M or NF-H alone, without NF-L, seems to be unable to form 10-nm filaments under the same conditions.

Thus, some IF are homopolymers made up of a single type of subunit. Other filaments are heteropolymers composed of two or three types of subunits.

B. The Structure of 10-nm Filaments

The building blocks of all types of IF are pairs of subunits forming two-chain coiled coils. Experiments in which the subunits were covalently cross-linked to one another by special reagents confirmed that coiled coils consist of two polypeptide chains aligned in parallel and in register. These dimers are made by two identical subunits in vimentin, desmin, and glial filaments. Keratin filaments contain heterodimers formed by keratin molecules of type I and type II. Two identical dimers further associate to form tetramers. Tetramers were isolated from keratin filaments by solublization in 4 M urea. It is not yet clear how dimers are packed into tetramers. There is some evidence in favor of antiparallel orientation of the two dimeric coiled coils in the tetrameric unit.

Tetrameric units are threads of 2–3 nm diameter with a length of about

48 nm. In the absence of denaturant they associate into typical 8–12nm intermediate-sized filaments as well as into thinner fibrils of 2–3 and about 4.5 nm in diameter.

On the other hand, when complete keratin filaments are incubated in low concentrations of phosphate buffers, they unravel into several protofibrils with an apparent diameter of 4.5 nm. Each protofibril can be further unraveled into protofilaments with a diameter about 2–3 nm. Similar unraveling into protofibrils and protofilaments was observed with most other IFs under a variety of condition. A special technique, scanning transmission electron microscopy, was used to obtain high-contrast images of unstained IFs; analysis of these images made it possible to measure mass-per-unit length of individual IF. Several classes of IF were found in each preparation; the major class had 32–33 molecules of IF protein per cross-section; some "lightweight" filaments (about 22 subunits/cross-section) as well as "heavyweight" (45 subunits/cross-section) filaments were also found.

All these data are consistent with the following model of IF (Fig. 3.5). Protofilament may consist of tetramers packed end to end; two protofilaments may form one protofibril. The "typical" IF is made by four protofibrils, which correspond to $4 \times 2 \times 4 = 32$ protein subunits per cross-section. The lightweight and heavyweight filaments can have less or more protofibrils, respectively. Electron microscopy indicates that protofibrils are helically twisted in

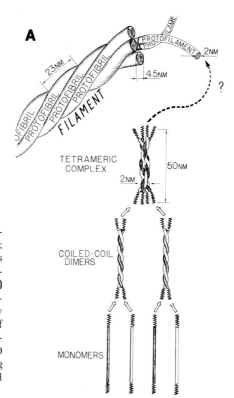

FIG. 3.5. (A) A model of the structure of IF. Protein subunits form coiled coil (hetero) dimers; antiparellel (?) dimers form tetramers; tetramers form protofilaments; protofilaments form protofibrils and protofibrils form 10-nm filament. **(B)** Unraveling of IF revealing their protofibril substructure. Partial unraveling was achieved by treating the keratin filaments with low levels of Na-phosphate. Electron micrograph of freeze-dried and shadowed preparation. Scale bar: 100 nm. **(C)** Keratin filament after glycerol-spraying and shadowing. 23-nm periodicity is seen. (B and C from Aebi *et al.*, 1983. Courtesy of U. Aebi.)

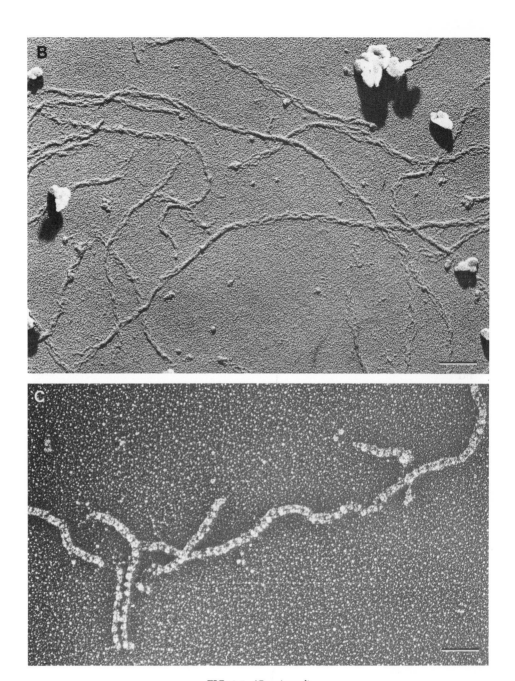

FIG. 3.5. (*Continued*)

the filament with the 92-nm pitch. This leads to $92 \div 4 = 23$ nm periodicity along the filament length (Fig. 3.5).

Thus, IF can be regarded as compact cables composed of numerous intertwined wires. We do not yet understand the construction of these cables well enough to determine whether they have any inherent polarity. Unlike other

IF, neurofilaments exhibit radial projections on their walls. These projections are formed by the radially projecting end domains of NF-H proteins; they bind monoclonal antibodies specific for these domains. Possibly, these periodic projections participate in the formation of bridges connecting neurofilaments with one another and with other structures.

C. Intermediate Filament-Associated Proteins (IFAPs)

Several such proteins are present in preparations of IF. Filaggrin was isolated from keratinizing epidermal cells. This is a basic protein (mol. wt. 30 kD) that can aggregate *in vitro* all types of IF, so that large bundles of these filaments are formed. *In vivo* filaggrin may stabilize bundles of keratin filaments in epidermal cells. Another group of proteins, cysteine-rich proteins, are present in the fully keratinized skin appendages, such as hair. They form numerous disulfide bonds with cysteine-rich terminal domains of certain keratins, producing highly rigid and insoluble keratinized tissues.

Two distinct high-molecular-weight proteins, synemin (230 kD) and paranemin (280 kD), are associated with desmin filaments in muscle cells and also with some, but not all, vimentin filaments in muscle and nonmuscle cells.

Frameworks of IF prepared from glioma cell cultures and from other cultured cells contain considerable amounts of another IFAP, plectin, of about 300 kD; this protein seems to be immunologically related to some MAPs; possibly, it is involved in the linkage between IF and microtubules.

Another 300-kD protein associated with IF was recently isolated from baby hamster kidney fibroblastic cell line (BHK-21); this protein, designated IFAP-300K, seems to be unrelated to MAPs and to plectin. IFAP-300K is associated with IF in a discontinuous manner, forming "bridges" between neighboring filaments.

Thus, IFAPs are tissue-specific proteins involved in the binding of IF to one another and, possibly, to other cellular structures. Many types of IFAPs may still remain unstudied.

IV. Dynamics of Intermediate Filaments in the Cells

A. Rearrangements of Filaments

Most IF protein subunits within the cell are usually assembled in filaments. Nonpolymerized, newly synthesized subunits are rapidly incorporated into the filaments. For instance, the half-life of labeled newly synthesized vimentin in the soluble fraction of chicken erythroid cell was found to be only 7 min; disappearance of the label from the nonsoluble pool was accompanied by a corresponding increase in the label in the soluble fraction. As already mentioned, 10-nm filaments often seem to be associated with the nuclear envelope or with some special near-nuclear centers, but any special role of these structures in the formation of filaments has not yet been demonstrated.

The IF, once formed, are almost insoluble under physiological conditions.

The IF networks in fully keratinized cells are transformed into hard and apparently inert structures. However, in the living nucleated cells, these networks usually remain dynamic and can undergo reversible reorganization. For instance, the meshworks of vimentin and of keratin filaments present in the cytoplasm of many cultured cell lines disappear partially at the beginning of cell division. Keratin and vimentin molecules are collected within the spheroidal bodies scattered throughout the cytoplasm of mitotic cells. After the end of mitosis the new filaments are formed; often they seem to sprout out of the spheres. This morphological reorganization is accompanied by a change in the antigenic properties of IF-proteins. Spheroidal bodies and residual filaments in the mitotic cells are stained by certain monoclonal antibodies that do not react with interphase IF but do react with purified IF-protein (keratin or vimentin). These changes may be due to unwinding of the filaments into constituent protofilaments; partial unwinding can unmask the new antigenic sites on the surface of protofilaments; completely dissociated protofilaments can then reaggregate into a spheroid mass. These suggestions remain hypothetical.

Destruction of microtubules by colchicine or some other disturbance, e.g., heat shock (incubation at 42–43°) or acrylamide treatment, induces retraction of vimentin network from the periphery of cultured fibroblast and

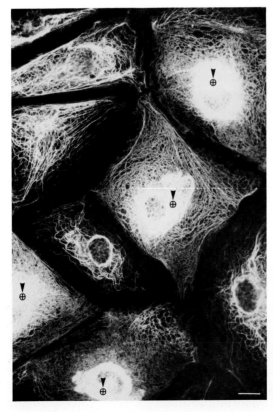

FIG. 3.6. Keratin networks in the epithelial cultured cells of PTK$_2$ line fixed 24 hr after injection of antikeratin antibody. Injected cells are marked by arrowheads and crossed circles. Partial perinuclear collapse of the keratin network within the injected cells. Scale bar: 10 μm. (From Klymkowsky et al., 1983. Micrograph courtesy of M. W. Klymkowsky.)

accumulation of filaments around the nucleus. This "collapse" of IF will be discussed later (see Chapter 5). Similar redistribution of IF of different types can be also induced by injection of antibodies to IF-proteins into the cell (Fig. 3.6).

Enzymatic phosphorylation of IF-proteins can be one of the mechanisms regulating the state of filament assembly. All types of subunits were found to be phosphorylated by several kinases both *in vivo* and *in vitro*. Increased phosphorylation of vimentin was observed after incubation of cells with biologically active agents, such as hormones or growth factors (see also Chapter 9), and in mitotic cells, but the exact effects of these alterations on the filaments are as yet unknown. Strong phosphorylation of NF-M neurofilament protein seems to be correlated with its association with the filament: the less-phosphorylated variant of NF-M does not coassemble with NF-L core protein.

Proteolysis is another mechanism that probably plays an important role in regulation of IF. Certain calcium-activated proteases were found to be specific for particular types of IF proteins. These proteases are present only in the cells expressing corresponding types of filaments. For example, mesenchymal cells contain protease specific for vimentin and desmin, while neural cells contain a protease specific for neurofilament proteins. Obviously, regulated specific proteolysis seems to be the most logical way to remove insoluble IF from the cytoplasm. Future studies are needed to find out in which situations this mechanism actually comes into play.

B. Alterations of Protein Composition of Filaments

In contrast to reversible rearrangements of the morphology of networks, replacement of one type of network by another occurs rarely and only in cases where the cell type changes. Classification of cells according to the type of IF-proteins they contain corresponds almost exactly to the standard histological classification of major types of tissues, e.g., epithelial, mesenchymal, muscular, neural, and glial.

Naturally, expression of various IF proteins during embryonic development is correlated with differentiation of various tissue types. For instance, egg cells and all cells of early mouse embryo contain keratins. However, when the primary mesenchymal cells are detached from the embryonic ectoderm, their cytoskeleton is abruptly changed; keratin filaments disappear, and vimentin filaments appear. In this way each major tissue formed during embryogenesis acquires its specific type of IF-proteins. Only a few cells express simultaneously two types of these proteins. For example, certain astrocytes contain vimentin and GFA; certain smooth muscle cells contain vimentin and desmin; maturing neuroblasts can transiently contain vimentin and neurofilament proteins.

Within the group of epithelial tissues, alterations of cell differentiation are usually accompanied by changes of the spectra of keratin subunits. Keratinization of epidermal cells is the best-studied case of such change. Cultured human basal epidermal cells contain two major type I keratins (46 kD and 50 kD) and two major type II keratins (56 kD and 58 kD). When keratinization of these cells begins, two additional larger species of each type are pro-

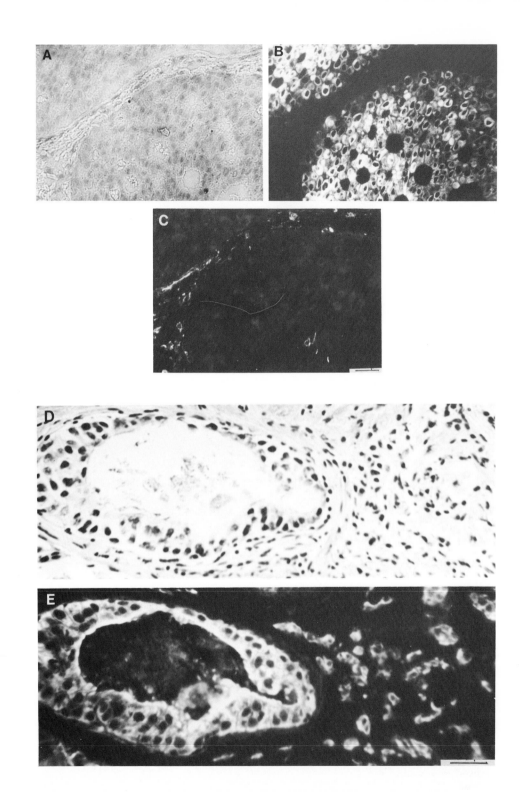

duced: 56.5 kD keratin of type I and 67 kD keratin of type II. Inhibition of keratinization by vitamin A prevents expression of these two proteins.

Drastic alterations of IF-protein specificity occur when epithelial cells of various types are cultured; these cells, in addition to keratin filaments, begin to form vimentin filaments. The simultaneous presence of these two networks is characteristic of many cultured epithelial lines (see also Chapter 6).

What is the mechanism of alteration of the specificity of IF-proteins during differentiation? These proteins are usually present in the cell in the polymerized state. Therefore, to form new filaments of the same or another type, the cell usually must activate the transcription of mRNA from the corresponding gene, and must synthesize new subunits coded by this mRNA. In fact, many developmental alterations of IF were found to be immediately preceded by transcription of mRNA from the genes coding their proteins. For instance, increase of the amount of vimentin filaments during erythropoiesis in chicken is preceded by induction of the synthesis of vimentin mRNA. The appearance of desmin during myogenesis is preceded by the induction of synthesis of desmin RNA, and so forth.

When mRNA from bovine epidermis was injected into cultured human fibroblasts, these cells began to form a network of keratin filaments, in addition to the previously existing vimentin network. Microinjection of epidermal keratin mRNA into epithelial cells of different type (e.g., cultured kidney cells) leads to formation of hybrid keratin filaments containing a combination of keratins that does not occur in normal cells. Thus, differential expression of the subunit-coding genes is both necessary and sufficient to produce cell-specific differences in composition of IF.

Tissue-specific proteins of IF can be used as reliable and stable markers of the cell type in normal and pathological conditions. Of special importance is the fact that the type of IF is largely retained during neoplastic transformation (Fig. 3.7). Therefore, the tissue origin of neoplastic cells can often be diagnosed with the aid of specific antibodies distinguishing various types of IF-proteins. For instance, when the secondary nodule from an unknown primary tumor is examined, immunomorphological analysis can help to determine whether this tumor is of epithelial, neural, or other origin. This differential diagnosis can be important in the choice of the further strategy of the physician.

To summarize, many properties of IF set them apart from the two other

---

FIG. 3.7. Staining of various cells in human breast carcinoma by antibodies to different proteins of IF. **(A–C)** Adjacent serial sections of frozen tumor tissue stained by hematoxylin (A), by antibody to 49-kD keratin (B), and by antibody to vimentin (C). Two aggregates of neoplastic epithelial cells are separated by the thin layer of connective tissue stroma with numerous collagen fibers. Keratin is present only in epithelial cells; vimentin, only in the stroma cells. ×200, reproduced at 62%. (Courtesy of V. I. Guelstein and T. A. Tchipisheva.) **(D, E)** Keratin-containing cancer cells invading connective tissue stroma in the human breast. The same frozen section stained with hematoxylin and examined by immunofluorescence using antibodies to 49-kD keratin. The field contains the epithelial alveoli surrounded by the stroma; immunofluorescence reveals invading epithelial cells in this stroma. ×300. (Courtesy of V. I. Guelstein and T. A. Tchipisheva.)

main groups of cytoskeletal fibrils. IFs are much more stable than actin filaments and microtubules; their polymerization usually seems to be irreversible. In contrast to actin–myosin and dynein–tubulin systems, the systems of IF seem to be unable to move actively and independently of other structures. Formation of networks mechanically connecting peripheral and central cell parts may be one of the main functions of these filaments; these networks may also be essential for the formation of mechanically integrated multicellular systems from single cells. The most intriguing special feature of the systems of IF is the tissue specificity of their protein subunits. It is possible that morphologically similar filaments made of distinct proteins have different mechanical properties or differing ability to link other structures. These subtle differences may be essential for the formation of diverse multicellular structures.

Literature Cited

Aebi, U., Fowler, W. E., Rew, P., and Sun, T-T. (1983) The fibrillar substructure of keratin filaments unraveled, *J. Cell Biol.* **97**:1131.–1141.

Klymkowsky, M. W., Miller, R. H., and Lane, E. B. (1983) Morphology, behavior, and interaction of cultured epithelial cells after the antibody-induced disruption of keratin filament organization, *J. Cell Biol.* **96**:494–509.

Moll, R., Franke, W. W., Schiller, D. L., Geiger, B., and Krepler, R. (1982) The catalog of human cytokeratins: Pattern of expression in normal epithelia, tumors, and cultured cells, *Cell* **31**:11–24.

Quax-Jenken, Y., Quax, W., and Bloemendal, H. (1983) Primary and secondary structure of hamster vimentin predicted from the nucleotide sequence, *Proc. Natl. Acad. Sci. USA* **80**:3548–3552.

Additional Readings

General

Steinert, P. M., Jones, J. C. R., and Goldman, R. D. (1984) Intermediate filaments, *J. Cell Biol.* **99**(1 pt2):22s–27s.

Steinert, P. M., Steven, A. C., and Roop, D. R. (1985) The molecular biology of intermediate filaments, *Cell* **42**:411–419.

Traub, P. (1985) *Intermediate Filaments*, Springer-Verlag, Berlin, Heidelberg, New York, Tokyo, pp. 1–256.

Wang, E., Fishman, D., Liem, R. K. H., and Sun T-T., eds. (1985) Intermediate filaments, *Ann. NY Acad. Sci.* **455**:1–832.

IF-proteins; structure of the filaments; assembly *in vitro*

Engel, A., Eichner, R., and Aebi, U. (1985) Polymorphism of reconstituted human epidermal keratin filaments: Determination of their mass-per-length and width by scanning transmission electron microscopy (STEM), *J. Ultrastruct. Res.* **90**:323–335.

Fuchs, E., Kim, K. H., Hanukoglu, I., Marchuk, D., Tyner, A., and McCrohon, S. (1984) Influence of differential keratin gene expression on the structure and properties of the resulting 8-nm filaments, in *Molecular Biology of the Cytoskeleton* (G. G. Borisy, D. W. Cleveland, D. B. Murphy, eds.), Cold Spring Harbor Laboratory, Cold Spring Harbor, New York, pp. 381–393.

Geisler, N., and Weber, K. (1982) The amino acid sequence of chicken muscle desmin provides a common structural model for intermediate filament proteins, *EMBO J.* **1**:1649–1656.

Geisler, N., Fischer, S., Vandekerckhove, J., VanDamme, J., Flessmann, U., and Weber, K. (1985) Protein–chemical characterization of NF-H, the largest mammalian neurofilament component; intermediate filament-type sequences followed by a unique carboxy-terminal extension, *EMBO J.* **4**:57–63.

Geisler, N., Kaufman, E., and Weber, K. (1985) Antiparallel orientation of the two double-stranded coiled-coils in the tetrameric protofilament unit of intermediate filaments, *J. Mol. Biol.* **182**:173–177.

Hanukoglu, I., and Fuchs, E. (1983) The cDNA sequence of a type II cytoskeletal keratin reveals constant and variable structural domains among keratins, *Cell* **33**:915–224.

Hatzfeld, M., and Franke, W. W. (1985) Pair formation and promiscuity of cytokeratins: Formation *in vitro* of heterotypic complexes and intermediate-sized filaments by homologous and heterologous recombinations of purified polypeptides, *J. Cell Biol.* **101**:1826–1841.

Henderson, D., Geisler, N., and Weber, K. (1982) A periodic ultrastructure in intermediate filaments, *J. Mol. Biol.* **155**:173–176.

Ip, W., Hartzer, M. K., Pang, Y-Y.S., and Robson, R. M. (1985) Assembly of vimentin *in vitro* and its implications concerning the structure of intermediate filaments, *J. Mol. Biol.* **183**:365–375.

Jorcano, J. L., Franz, K. K., Rieger, M., Magim, T. M., and Franke, W. W. (1984) Identification of different types of keratin polypeptides distinguished within the acidic cytokeratins subfamily: Corresponding cytokeratins in diverse species, in *Molecular Biology of the Cytoskeleton* (G. G. Borisy, D. W. Cleveland, and D. B., Murphy, eds.), Cold Spring Harbor Laboratory, Cold Spring Harbor, New York, pp. 395–408.

Kaufmann, E., Geisler, N., and Weber, K. (1984) SDS–PAGE strongly overestimates the molecular masses of the neurofilament proteins, *FEBS Lett.* **170**:81–84.

Krieg, T. M., Schafer, M. P., Cheng, C. K., Filpula, D., Flaherty, P., Steinert, P. M., and Roop., D. R. (1985) Organization of a type I keratin gene. Evidence for evolution of intermediate filaments from a common ancestral gene, *J. Biol. Chem.* **260**:5867–5870.

Lawson, D. (1983) Epinemin: A new protein associated with vimentin in non-neural cells, *J. Cell Biol.* **97**:1891–1905.

Lewis, S. A., Balcarek, J. M., and Cowan, N. J. (1984) Structure of mouse glial fibrillary acidic protein and its expression by *in situ* hybridization using a cloned cDNA probe, in *Molecular Biology of the Cytoskeleton* (G. G. Borisy, D. W. Cleveland, and D. B. Murphy, eds.), Cold Spring Harbor Laboratory, Cold Spring Harbor, New York, pp. 455–463.

Lieska, N., Yang, H-Y., and Goldman, R. D. (1985) Purification of the 300 K intermediate filament-associated protein and its *in vitro* recombination with intermediate filaments, *J. Cell Biol.* **101**:802–813.

Marchuk, D., McCrohon, S., and Fuchs, S. (1984) Remarkable conservation of structure among intermediate filament genes, *Cell* **39**:491–498.

Parry, D. A. D., Steven, A. C., and Steinert, P. M. (1985) The coiled-coil molecules of intermediate filaments consist of two parallel chains in exact axial register, *Biochem. Biophys. Res. Commun.* **127**:1012–1018.

Pruss, R. M., Mirsky, R., Raff, M. C., Thorpe, R., Dowding A. J., and Anderton, B., H. (1981) All classes of intermediate filaments share a common antigenic determinant defined by a monoclonal antibody, *Cell* **27**:419–428.

Quax, W., Quax-Jenken, I., van den Hevel, T., Ergerts, W. V., and Bloomendal, H. (1984) Organization and sequence of the genes for desmin and vimentin, in *Molecular Biology of the Cytoskeleton*, Cold Spring Harbor Laboratory, Cold Spring Harbor, New York, pp. 445–454.

Quinlan, R. A., Golberg, J. A., Schiller, D. L., Hatzfeld, M., and Franke, W. W. (1984) Heterotypic tetramer (A_2D_2) complexes of non-epidermal keratins isolated from cytoskeletons of rat hepatocytes and hepatoma cells, *J. Mol. Biol.* **178**:365–388.

Sandoval, I. V., Colaco, C. A. L. S., and Lazarides, E. (1983) Purification of the intermediate filament-associated protein, synemin, from chicken smooth muscle, *J. Biol. Chem.* **258**:2568–2576.

Sauk, J. J., Krumweide, M., Cocking-Johnson, D., and White, J. G. (1984) Reconstitution of cytokeratin filaments *in vitro*: Further evidence for the role of nonhelical peptides in filament assembly, *J. Cell Biol.* **99**:1590–1597.

Steinert, P. M., Cantieri, J. S., Teller, D. C., Lonsdale-Eccles, J. D., and Dale, D. A. (1981) Characterization of a class of cationic proteins that specifically interact with intermediate filaments, *Proc. Natl. Acad. Sci. USA* **78**:4097–4101.

Steinert, P. M., Rice, P. H., Roop, D. R., Trus, B. I., and Steven, A. C. (1983) Complete amino acid sequence of a mouse epidermal keratin subunit and implications for the structure of intermediate filaments, *Nature* **302**:794–800.

Steinert, P. M., Parry, D. A. D., Raccosin, E. L., Idler. W. W., Steven, A. C., Trus, B. Lq, and Roop, D. R. (1984) The complete cDNA and deduced amino acid sequence of a type II mouse epidermal keratin of 60,000 Da: Analysis of sequence differences between type I and II keratins, *Proc. Natl. Acad. Sci. USA* **81**:5709–5713.

Steven, A. C., Wall, J., Hainfeld, J., and Steinert, P. M. (1982) Structure of fibroblastic intermediate filaments: Analysis of scanning transmission electron microscopy, *Proc. Natl. Acad. Sci. USA* **79**:3101–3105.

Traub, P., and Vorgias, C. E. (1983) Involvement of the N-terminal polypeptide of vimentin in the formation of intermediate filaments, *J. Cell Sci.* **63**:43–67.

Wang, E., Cairncross, J. G., Yung, W. K. A., Garber, E. A., and Liem, R. K. H. (1983) An intermediate filament-associated protein, p50, recognized by monoclonal antibodies, *J. Cell Biol.* **97**:1507–1514.

Weber, K., and Heister, N. (1985) Intermediate filaments: Structural conservation and divergence, *Ann. NY Acad. Sci.* **455**:126–143.

Wiche, G., Herrmann, H., Leichtfried, F., and Pytela, R. (1982) Plectin: A high molecular weight cytoskeletal polypeptide component that copurifies with intermediate filaments of the vimentin type, *Cold Spring Harbor Symp. Quant. Biol.* **46**:475–482.

Wong, J., Hutchison, S. B., and Liem, R. K. H. (1984) An isoelectric variant of the 150,000-dalton neurofilament polypeptide. Evidence that phosphorylation state affects its association with the filament, *J. Biol. Chem.* **259**:10867–10874.

Intermediate filaments in the cells

Capetanaki, Y. C., Ngai, J., and Lazarides, E. (1984) Regulation of the expression of genes coding for the intermediate filament subunits vimentin, desmin, and glial fibrillary acidic protein, in *Molecular Biology of the Cytoskeleton* (G. G. Borisy, D. W. Cleveland, and D. B. Murphy, eds.), Cold Spring Harbor Laboratory, Cold Spring Harbor, New York, pp.415–434.

Cooper, D., Schermer, A., and Sun, T-T (1985) Classification of human epithelia and their neoplasms using monoclonal antibodies to keratins: Strategies, applications, and limitations, *Lab. Invest.* **52**:243–256.

Dahl, D., and Bignami, A. (1986) Neurofilament phosphorylation in development. A sign of axonal maturation? *Exp. Cell Res.* **162**:220–230.

Dawson, P. J., Hulms, J. S., and C. W. Lloyd (1985) Monoclonal antibody to intermediate filament antigen cross-reacts with higher plant cells, *J. Cell Biol.* **100**:1793–1798.

Eckert, B. S. (1985) Alteration of intermediate filament distribution in PtK cells by acrylamide, *Eur. J. Cell. Biol.* **37**:169–174.

Eckert, B. S., Caputi, S. E., and Warren, R. H. (1984) Dynamics of keratin filaments and the intermediate filament distribution center during shape change in PtK$_1$ cells, *Cell Motil.* **4**:169–182.

Eichner, R., Bonitz, P., and Sun, T-T. (1964) Classification of epidermal keratins according to their immunoreactivity, isoelectric point, and mode of expression, *J. Cell Biol.* **98**:1388–1396.

Franke, W. W., Schmid, E., Grund, C., and Geiger, B. (1982) Intermediate filament proteins in nonfilamentous structures: Transient disintegration in granular aggregates, *Cell* **30**:103–113.

Franke, W. W., Schmid, E., Wellsteed, J., Grund, C., Gigi, O., and Geiger, B. (1983) Change of cytokeratin filament organization during the cell cycle: Selective masking of an immunologic determinant in interphase PtK$_2$ cells, *J. Cell Biol.* **97**:1255–1260.

Franke, W. W., Grund, C., Kuhn, C., Lehto, V-F., and Virtanen, I. (1984) Transient change of organization of vimentin filament during mitosis as demonstrated by a monoclonal antibody, *Exp. Cell Res.* **154**:567–580.

Franke, W. W., Schmid, E., Mittnacht, S., Grund, C., and Jorcano, J. L. (1984) Intergration of different keratins into the same filament system after microinjection of mRNA for epidermal keratins into kidney epithelial cells, *Cell* **36**:813–825.

Georgatos, S. D., Weaver, D. C., and Marchesi, V. T. (1985) Site specificity in vimentin–membrane interactions: Intermediate filament subunits associate with the plasma membrane via their head domains, *J. Cell Biol.* **100**:1962–1967.

Gilfix, B. M., and Eckert, R. L. (1985) Coordinate control by vitamin A of keratin gene expression in human keratinocytes, *J. Biol. Chem.* **260**:14026–14029.

Granger, B. L., and Lazarides, E. (1982) Structural associations of synemin and vimentin filaments in avian erythrocytes revealed by immunoelectron microscopy, *Cell* **30**:263–275.

Hirokawa, N., Glicksman, A., and Willard, M. B. (1984) Organization of mammalian neurofilament polypeptides within the neural cytoskeleton, *J. Cell Biol.* **98**:1523–1536.

Ip, W., Danto, S. I., and Fischman, D. A. (1983) Detection of desmin-containing intermediate filaments in cultured muscle and nonmuscle cells by immunoelectron microscopy, *J. Cell Biol.* **96**:401–408.

Jones J. C. R., Goldman, A. E., Yang, H-Y., and Goldman, R. D. (1985) The organizational fate of intermediate filament networks in two epithelial cell types during mitosis, *J. Cell Biol.* **100**:93–102.

Klymkowsky, M. W. (1981) Intermediate filaments in 3T3 cells collapse after intracellular injection of monoclonal anti-intermediate filament antibody, *Nature* **291**:249–251.

Kreis, T., Geiger, B., Schmidt, E., Jorcano, J. L., and Franke, W. W. (1983) *De novo* synthesis and specific assembly of keratin filaments in nonepithelial cells after microinjection of mRNA for epidermal keratin, *Cell* **32**:1125–1137.

Lazarides, E. (1982) Biochemical and immunocytological characterization of intermediate filaments in muscle cells, in *Methods in Cell Biology*, Vol. 25, *The Cytoskeleton*, Part B (L. Wilson, ed.), Academic Press, London, pp. 333–357.

Leterrier, J-F., Liem, R. K. H., and Shelanski, M. L. (1981) Preferential phosphorylation of the 150,000 molecular weight component of neurofilaments by a cyclic AMP-dependent, microtubule-associated protein kinase, *J. Cell Biol.* **90**:755–760.

Murphy, D. B., and Grasser, W. A. (1984) Intermediate filaments in the cytoskeleton of fish chromotophores, *J. Cell Sci.* **66**:353–366.

Osborn, M., and Weber, K. (1982) Intermediate filaments: Cell-type-specific markers in differentiation and pathology, *Cell* **31**:303–306.

Osborn, M., and Weber, K. (1983) Tumor diagnosis by intermediate filament typing: A novel tool for surgical pathology, *Lab. Invest.* **48**:372–394.

Paulin, D., Babinet, C., Weber, K., and Osborn, M. (1980) Antibodies as probes of cellular differentiation and cytoskeletal organization in the mouse blastocyst, *Exp. Cell Res.* **130**:297–304.

Price, M. G., and Sanger, J. W. (1982) Intermediate filaments in striated muscle: A review of structural studies in embryonic and adult skeletal and cardiac muscle, in *Cell and Muscle Motility*, Vol. 3 (R. W. Dowben and J. W. Shay, eds.), Plenum Press, New York, pp. 1–40.

Schmid, E., Schiller, D. L., Grund, C., Stadler, J., and Franke, W. W. (1983) Tissue type-specific expression of intermediate filament proteins in a cultured epithelial cell line from bovine mammary gland, *J. Cell Biol.* **96**:37–50.

Sun, T. T., Eichner, R., Schermer, A., Cooper, D., Nelson, W. G., and Weiss, R. A. (1984) Classification, expression and possible mechanisms of evolution of mammalian epithelial keratins: A unifying model, in *Cancer Cells*, Vol. 1, *The Transformed Phenotype*, Cold Spring Harbor Laboratory, Cold Spring Harbor, New York, pp. 169–176.

Tokuyasu, K. T., Dutton, A. H., and Singer, S. J. (1983) Immunoelectron microscopic studies of desmin (skeletin) localization and intermediate filament organization in chicken cardiac muscle, *J. Cell Biol.* **96**:1736–1742.

Tölle, H-G., Weber, K., and Osborn, M. (1986) Microinjection of monoclonal antibodies to vimentin, desmin, and GFA in cells which contain more than one IF type, *Exp. Cell Res.* **162**:462–474.

Traub, P., and Nelson, W. J. (1981) Occurrence in various mammalian cells and tissues of the Ca^{2+}-activated protease specific for the intermediate-sized filament proteins vimentin and desmin, *Eur. J. Cell Biol.* **26**:61–67.

Tsukita, S., Tsukita, S., and Ishikawa, H. (1983) Association of actin and 10-nm filaments with the dense body in smooth muscle cells of the chicken gizzard, *Cell Tissue Res.* **229**:233–242.

Walter, M. F., and Biessmann, H. (1984) Intermediate-sized filaments in *Drosophila* tissue culture cells, *J. Cell Biol.* **99**:1468–1477.

Wang, K., and Ramirez-Mitchel, R. (1983) A network of transverse and longitudinal intermediate filaments is associated with sarcomeres of adult vertebrate skeletal muscle, *J. Cell Biol.* **96:**562–570.

Weber, K., Shaw, G., Osborn, M., Debus, E., and Geisler, N. (1983) Neurofilaments, a subclass of intermediate filaments: Structure and expression, *Cold Spring Harbor Symp. Quant. Biol.* **48:**(pt 2) 717–729.

Wiche, G., Drepler, R., Artlieb, U., Pytela, R., and Denk, H. (1983) Occurrence and immunolocalization of plectin in tissues, *J. Cell Biol.* **97:**887–901.

4

Unconventional Fibrillar Structures in the Cytoplasm

I. Thin Nonactin Filaments

Besides the classical three groups of cytoskeletal structures, a number of other fibrillar elements have been described in the cytoplasm. These structures have not yet been well explored. Some of them have been seen only in the cells of a few types of lower eukaryotes. Even the existence of some of these structures is still controversial. Nevertheless, it is possible that future studies will show that at least some of these fibrils are not artifacts or exotic rarities, but are important and common components of the cytoskeleton. Therefore, the data on these structures deserve attention.

Several types of unusual structures seen in different cells have similar morphology; they look like very thin filaments about 2–4 nm in diameter.

A. The Spasmoneme of Vorticellid Ciliates

These sessile protozoa, e.g., *Vorticella convallaria*, *Carchesium polybium*, and *Zoothamnium geniculatum*, are attached to the solid substrate by the stalk, which can be as long as 1 mm (Fig. 4.1). The stalk contracts very rapidly in response to various stimuli; contraction to 60% of the extended length takes only 2–10 msec. The contractile part of the stalk is called the spasmoneme. It consists of the parallel 2 to 4-nm filaments and of membranous tubules. Glycerol-extracted demembranated spasmoneme also contracts when the calcium concentration in the incubating solution is changed from 10^{-8} to 10^{-6} M. In contrast to the movements of other extracted structures, e.g., of muscle or of flagella, contraction of a spasmoneme requires only an increase in calcium; ATP or Mg need not be present. Contraction is accompanied by binding of calcium to the spasmoneme. Spasmonemes do not contain significant amounts of actin or tubulin. Instead, two major proteins, spasmin A and B, were revealed, both with mol. wt. about 20 kD. Both these proteins bind calcium, which suggests that the mechanism of contraction of the spasmoneme is profoundly different from the usual mechanism of movement based on the actin–myosin or tubulin–dynein interaction. This contraction seems to be the result of some alteration of conformation of spasmin-containing filaments caused by calcium binding.

B. Striated Rootlets of Cilia and Flagella

Striated rootlets are the bundles of 2-nm filaments attached to the basal bodies of many types of cilia and flagella of lower and higher eukaryotes. These bundles have periodic electron-dense cross-striations (Fig. 4.2). The rootlets isolated from the green alga *Tetraselmis striata* were found to contain one major protein of about 20 kD; two-dimensional electrophoresis resolved this protein into two components. There are some preliminary data on the immunological crossreactivity of this protein and the major protein of the spasmonemes of vorticella. Addition of 2 mM of Ca to living *Tetraselmis* causes contraction of rootlets accompanied by displacement of the basal bodies (Fig. 4.2). A mutant strain of *Chlamydomonas* lacking the rootlets has been obtained. The mode of flagellar beating in this mutant was similar to that of the wild-type organism, with one difference: the direction of beat becomes highly variable. These data suggest that striated rootlets regulate the position of the basal body.

Nonactin 2 to 4-nm filaments, which may be responsible for contractile movements, were also found in a number of other protists. In particular, the bundles of these filaments, called **myonemes,** are present under the plasma membrane of certain marine dinoflagellates. These filaments are straight in the extended state, but appear helically coiled when contracted. Their chemical composition is unknown.

C. Nematode Sperm Movements

Spermatozoa of a free-living nematode, *Caenorhabditis elegans*, are ameboid cells that can crawl on glass or other substrates by extending pseudopods. Unlike many other cells, nematode spermatozoa contain very little actin (less than 0.02% of total cellular protein). Their pseudopods are filled

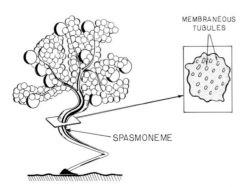

MEMBRANEOUS
TUBULES

SPASMONEME

FIG. 4.1. Contractile stalk (spasmoneme) of a colony of *Zoothamnium geniculatum*. Cross-sectioned 2-nm filaments are present between the membranous tubules. (Scheme based on Routledge *et al.*, 1976.)

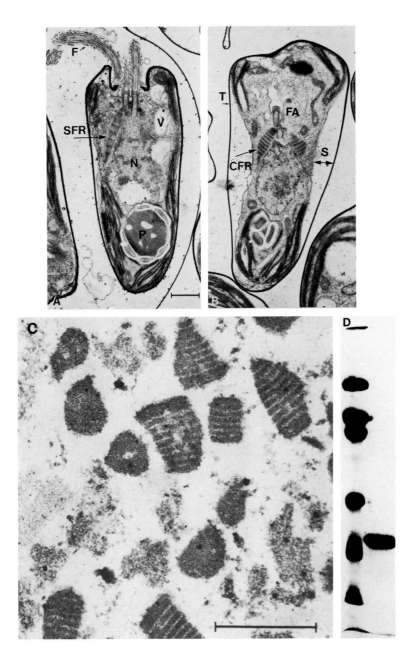

FIG. 4.2. Striated flagellar roots of the green alga *Tetraselmis striata.* **(A)** Thin-sectioned cell fixed in the absence of added calcium. One of the two fully extended striated flagellar roots (SFR) is visible. **(B)** Thin-sectioned cell fixed within 30 sec after addition of 2 mM $CaCl_2$. Both flagellar roots have contracted, resulting in deflagellation, displacement of the flagellar apparatus (FA) and inpocketing of the plasma membrane (S). V, vacuole; N, nucleus; P, pyrenoid; T, theca; F, flagella. **(C)** Fraction of cell homogenate sedimented in sucrose density gradient. Striated flagellar roots isolated from cell homogenate by sucrose density gradient. Scale bars: 1 μm. **(D)** SDS–PAGE of flagellar root preparations. Left lane: Molecular weight markers; top to bottom: 68,000, 45,000, 34,700, 24,000, 18,400, and 14,300. Right lane: Sample of the purified 20,000-D protein. (From Salisbury *et al.*, 1984. Courtesy of J. L. Salisbury.)

with thin filaments about 2 nm in diameter. The major cellular protein of the sperm is a basic 16-kD polypeptide; it constitutes about 10–15% of total cell protein. It is likely, but unproven, that this protein makes up the 2-nm filaments (see Fig. 4.3).

Nothing is known about the interaction of this protein with calcium.

D. 2-nm Filaments in Muscle Sarcomere

When the major muscle proteins are extracted from the sarcomere, numerous thin threads about 2 nm in diameter become apparent at electron microscopy (Fig. 4.4A). These threads connect the neighboring Z disks of the same myofibril with one another and with myosin filaments. In contrast to longitudinal desmin filaments (see Chapter 3), these fibrils go not on the surface of the sarcomere, but inside it. Electrophoresis of these extracted sarco-

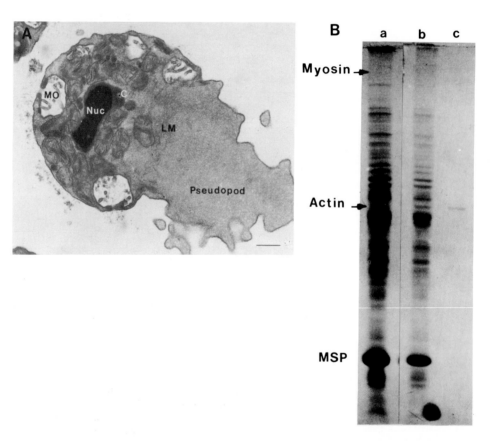

FIG. 4.3. Pseudopods of nematode sperm. **(A)** Electron micrograph of a section of the spermatozoon of the nematode *Caenorhabditis elegans* with extended pseudopod. Nuc, nucleus; LM, laminar membranes; MO, membranous vesicles. Scale bar: 0.5 μm. **(B)** SDS–PAGE of 320 μg of total sperm protein (a), of 80 μg sperm protein plus 0.2 μg rabbit actin (b), and of 0.2 μg rabbit actin (c). The positions of worm actin and myosin from parallel gels of worm actinomyosin are shown by arrows. Actin content in the sperm cell is very low; the major sperm protein (MSP) is about 16 kD. (From Nelson *et al.*, 1982. Micrographs courtesy of S. Ward.)

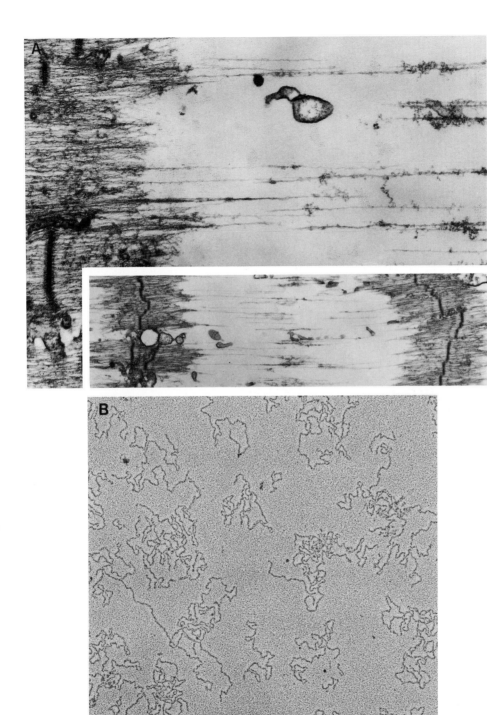

FIG. 4.4. (A) Titin (connectin) filaments in frog skeletal muscle. Myosin filaments have been extracted, leaving actin filaments and a few titin filaments running through a sarcomere. Compare with Fig. 1.6B. ×52,000, inset ×25,000; reproduced at 78%. (From Maruyama *et al.*, 1984. Micrograph courtesy of K. Maruyama.) **(B)** Purified native titin. Electron micrograph of shadowed preparation. ×66,000. (From Trinick *et al.*, 1984. Courtesy of J. Trinick.)

meres reveal two superheavy polypeptides with mol. wt. more than 1000 kD. These proteins were named nebulin (from nebula) and titin (from Greek *Titan*, member of a gigantic race). Titin, also called connectin, was isolated and shown to polymerize into 2-nm filaments *in vitro* (Fig. 4.4B). 2-nm Filaments are thought to form a matrix partly responsible for elastic properties of the muscle.

E. Connecting Filaments

Short filaments about 2–6 nm in diameter are often seen in preparations of cytoskeleton of the vertebrate cells of various types. Usually, these filaments form transverse connections between the two other cytoskeletal fibers, e.g., between the two actin filaments or between actin filaments and the microtubule. These filaments can also link the microtubule or intermediate filament to the surface of vesicular organelles, e.g., of the mitochondria (Fig. 4.5). These filaments are not decorated with myosin fragments; that is, they do not consist of actin. Nonactin 6-nm filaments interconnecting the roots of actin bundles in the brush borders of intestinal cells (see Fig. 1.4) disappear after the extraction of myosin and are stained with antimyosin antibodies;

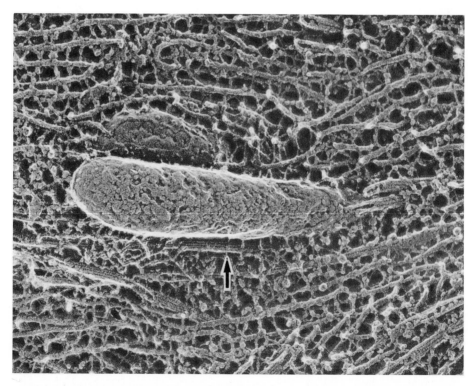

FIG. 4.5. Mitochondrion cross-linked by thin filaments to microtubule (arrow) and intermediate filament in the cytoplasm of the frog axon. Platinum replica of the quick-frozen and freeze-etched axon. ×180,000; reproduced at 88%. (From Hirokawa, 1982. Courtesy of N. Hirokawa.)

they are supposed to consist of the overlapping rod parts of myosin molecules. Another variant of connecting filaments in the brush border was found to consist of spectrinlike protein (terminal web 260/240). The nature of other very thin filaments in nonmuscle cells remains unknown.

F. Conclusion

Descriptions of thin nonactin filaments in various cells are still incomplete because of obvious technical difficulties. The structure of the protein components in most cases remains unstudied. The fragmentary data available at present suggest that there can be several functionally and chemically different groups of these filaments.

Some thin filaments are composed of the known actin-binding proteins, such as myosin or spectrin. Possibly, certain MAPs and IFAPs also participate in the formation of cross-linkers between various cytoskeletal and other structures. Other filaments interconnecting cytoskeletal fibers can be formed by special proteins, such as titin.

Of special interest is the group of thin filaments involved in Ca-promoted contraction. Proteins of about 15–20 kD seem to be present in the three types of structures containing these filaments: in the contractile spasmoneme of sessile Ciliata, in striated rootlets, and in nematode sperm pseudopods. In the first two structures these proteins were shown to have Ca-binding properties. It was suggested that they might belong to the same family of calcium-modulated regulatory proteins as calmodulin and troponin C. However, the nematode sperm protein does not seem to be a member of this family.

At present, however, we do not yet know the structure of any protein of Ca-contractile filaments. Further studies of these nonactin and nontubulin motile systems will be of considerable interest. It is possible that systems of this type are present in most cell types and are essential for important cellular processes, e.g., for organelle movement.

II. The Problem of Microtrabecular Lattice

In the early 1970s Keith R. Porter and his collaborators (see Porter and Tucker, 1981) described the lattice of fine trabeculae filling of cytoplasm of cultured cells and surrounding other organelles. To reveal this lattice they used a special method that makes it possible to examine the three-dimensional internal structure of the cells. Human fibroblasts were grown on the grids used for electron microscopy. Then the cells were fixed with glutaraldehyde, dehydrated with alcohol and acetone, dried, and examined in the high-voltage electron microscope. The potential accelerating electrons in this apparatus is 10 times greater than in the usual electron microscope. Therefore, the kinetic energy of the electrons is so high that they can penetrate specimens up to several micrometers thick. As a result, the internal structure of the thin lamellae of well-spread cells can be examined without previous sectioning.

The slender strands of the lattice revealed by this method varied in diameter from 4 to 20 μm. This structure, named microtrabecular lattice, surrounded all other components of the cytoplasm (Fig. 4.6). A similar lattice was later found by high-voltage electron microscopy in other cells examined. Its morphology is highly dynamic. It is reversibly changed after almost all types of experimental interventions, including altered levels of calcium and magnesium ions and high osmotic pressure. Chilling to 4°C disrupts the lattice and causes clumping of the trabeculae into tiny goblets. Porter and collaborators suggested that the microtrabecular lattice is a dynamic system in which other cell organelles are suspended and which directs their movements.

At present, a decade after the first publications on the microtubular lattice, there is still no agreement about its nature.

Several investigators (e.g., Heuser and Kirschner, 1980) suggested that the microtrabecular lattice is an artifact. They think that various soluble proteins present in the cytoplasm coagulate and form fibrous deposits during fixation. Two groups of facts argue against the artifactual nature of lattice. First, it is found in cells prepared for microscopy by many different methods, including quick freezing. Second, as already mentioned, the lattice undergoes regular alterations in response to various treatments. Both arguments are indirect. Obviously, it would not be easy at present to directly prove or disprove the presence of the lattice in living cells.

If the lattice is a real structure, what is its chemical composition? To answer this question, Schliwa *et al.* (1981) first extracted the cells with a mild detergent, BRIJ 58, and then with the stronger detergent Triton X-100. According to their observations, BRIJ 58 partially destroys the cellular membranes but does not remove the microtrabecular lattice; Triton X-100 destroys all the lattice but leaves intact the more stable cytoskeletal structures. It was suggested that the substances extracted by Triton X-100 from the cell pretreated by BRIJ are probably components of the microtrabecular lattice. These

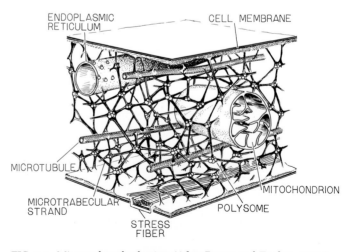

FIG. 4.6. Microtrabecular lattice. (After Porter and Tucker, 1981.)

substances were numerous. More than 400 polypeptides were found in the extracts, with actin being the most prominent component. Thus, these experiments did not reveal any proteins specific for the lattice.

It is likely that the lattice seen at high-voltage electron microscopy is not a special separate system of structures, but a network formed by various labile cytoskeletal components, attached to more stable fibrils. These components may include, for example, labile actin filaments, aggregates of actin-binding proteins, and fibrillar MAP molecules. The very thin filaments described in the previous section may also be part of this dynamic network. Additional proteins from soluble fraction can be reversibly attached to this network.

Investigations of very thin filaments and of the microtrabecular lattice draw attention to the most slender and most unstable cytoskeletal structures. Development of methods for reliable identification of these structures is a challenging problem.

Literature Cited

Heuser. , J. E., and Kirschner, M. (1980) Filament organization revealed in platinum replicas of freeze-dried cytoskeletons, *J. Cell Biol.* **86**:212–234.

Hirokawa, N. (1982) Cross-linker system between neurofilaments, microtubules, and membranous organelles in frog axons revealed by the quick-freeze, deep-etching method, *J. Cell Biol.* **94**:129–142.

Maruyama, K., Sawada, H., Kimura, S., Ohashi, K., Higuchi, H., and Umazume, Y. (1984) Connectin filaments in stretched skinned fibers of from skeletal muscle, *J. Cell Biol.* **99**:1391–1397.

Nelson, G. A., Roberts, T. M., and Ward, S. (1982) *Caenorhabditis elegans* spermatozoan locomotion: Amoeboid movement with almost no actin, *J. Cell Biol.* **92**:121–131.

Porter, K. R., and Tucker, J. B. (1981) The ground substance of the living cell, *Sci. Am.* **244**(3):57–67.

Routledge, L. M., Amos, W. B., Yew, F. F., and Weis-Fohg, T. (1976) New calcium-binding contractile proteins, in *Cell Motility* (R. Goldman, T. Pollard, and J. Rosenbaum, eds.), Cold Spring Harbor Laboratory, Cold Spring Harbor, New York, pp. 93–114.

Salisbury, J. L., Baron, A., Surek, B., and Melkonian, M. (1984) Striated flagellar roots: Isolation and partial characterization of a calcium-modulated contractile organelle, *J. Cell Biol.* **99**:962–970.

Schliwa, M., Van Blerkom, J., and Porter, K. R. (1981) Stabilization of the cytoplasmic ground substance in detergent-opened cells and a structural and biochemical analysis of its composition, *Proc. Natl. Acad. Sci. USA* **78**:4329–4333.

Trinick, J., Knight, P., and Whiting, A. (1984) Purification and properties of native titin, *J. Mol. Biol.* **180**:331–356

Additional Readings

Amos, W. B. (1972) Structure and coiling of the stalk in the Peritrich ciliates *Vorticella* and *Carchesium*, *J. Cell Sci.* **10**:95–122.

Amos, W. B., Routledge, L. M., and Yew, F. F. (1975) Calcium binding proteins in a vorticellid contractile organelle, *J. Cell Sci.* **19**:203–213.

Bridgman, P. C., and Reese, T. S. (1984) The structure of cytoplasm in directly frozen cultured cells. I. Filamentous meshworks and the cytoplasmic ground substance, *J. Cell Biol.* **99**:1655–1668.

Byers, H. R., and Porter, K. R. (1977) Transformations in the structure of the cytoplasmic ground substance in erythrophores during pigment aggregation and dispersion, *J. Cell Biol.* **75**:541–558.

Cachon, J., and Cachon, M. (1981) Movement by non-actin filament mechanisms, *Biosystems* **14**:313–326.

Cachon, J., and Cachon, M. (1984) An unusual mechanism of cell contraction: *Leptodiscinae* dinoflagellates, *Cell Motil.* **4**:41–55.

Cachon, J., and Cachon, M. (1985) Non-actin filaments and cell contraction in *Kofoidinium* and other dinoflagellates, *Cell Motil.* **5**:1–15.

Cachon, J., Cachon, M., Tilney, L. G., and Tilney, M. (1976) Movement generated by interaction between dense material at the ends of microtubules and non-actin containing microfilaments in *Sticholonche zanclea*, *J. Cell Biol.* **72**:213–338.

Ellisman, M. H., and Porter, K. R. (1980) Microtrabecular structure of the axoplasmic matrix: Visualization of cross-linking structures and their distribution, *J. Cell Biol.* **87**:464–479.

Gershon, N. D., Porter, K. R., and Trus, B. L. (1985) The cytoplasmic matrix: Its volume and surface area and the diffusion of molecules through it, *Proc Natl. Acad. Sci. USA* **82**:5030–5034.

Hirokawa, N., Cheney, R. E., and Willard, M. (1983) Location of a protein of the fodrin–spectrin–TW 260/240 family in the mouse intestinal brush border, *Cell* **32**:953–965.

Ip, W., Murphy, D. B., and Heuser, J. E. (1984) Arrest of pigment granule motion in erythrophores by quick-freezing, *J. Ultrastruc. Res.* **86**:162–175.

Porter, K. B., Beckerle, M., and McNiven, M. (1983) The cytoplasmic matrix, in *Modern Cell Biology*, Vol. 2, *Spatial Organization of Eukaryotic Cells* (J. R. McIntosh, ed.), Alan R. Liss, New York, pp. 259–262.

Roberts, T. M. (1983) Crawling *Caenorhabditis elegans* spermatozoa contact the substrate only by their pseudopods and contian 2-nm filaments, *Cell Motil.* **3**:333–347.

Roberts, T. M., and Ward, S. (1982) Centripetal flow of pseudopodial surface components could propel the amoeboid movement of *Caenorhabditis elegans* spermatozoa, *J. Cell Biol.* **92**:132–138.

Schliwa, M., and Van Blerkom, J. (1981) Spatial interaction of cytoskeletal components, *J. Cell Biol.* **90**:222–235.

Wang, K., Ramirez-Mitchell, R., and Palter, D. (1984) Titin is an extraordinarily long, flexible, and slender myofibrillar protein, *Proc. Natl. Acad. Sci. USA* **81**:3685–3689.

Wolosewick, J. J., and Porter, K. R. (1979) Microtrabecular lattice of the cytoplasmic ground substance. Artifact or reality, *J. Cell Biol.* **82**:114–139.

II

General Organization and Function of
Cytoskeleton

Cytoskeleton and Internal Organization of the Cell

I. Introduction

Cytoskeleton is the structure-forming component of the cell. It determines to a large degree the cell shape and distribution of other intracellular organelles and of the components of the plasma membrane. Cytoskeleton integrates various parts of the single cell and various cells into united structures. There are several levels of this integration.

1. Cytoskeletal fibrils of the same type interact with one another forming systems of these fibrils.
2. Various systems of fibrils interact with one another, forming an integrated cytoskeleton.
3. Cytoskeleton interacts with other cell organelles; these interactions determine the positions of these organelles within the cell.
4. Cytoskeletal systems of individual cells interact with one another via cell–cell contacts and contacts with extracellular matrix; these interactions determine the cell shapes and positions in organized multicellular systems, such as tissue or organ.

In previous chapters we have described the main supramolecular structures formed by each type of cytoskeletal fibril: bundles and networks of actin filaments, radiating and parallel arrays of microtubules, and networks of intermediate filaments. In this chapter we shall discuss interactions of various cytoskeletal structures with one another and with other cell components.

II. Integrated Structure of Cytoskeleton

A. General Organization of Cytoskeleton in Various Types of Cells

All cytoskeletal structures present in the same cell interact with one another, so that an integrated cytoskeleton is formed. Characteristic distribution of various cytoskeletal structures within the cell is the most obvious manifestation of this integration. This distribution is always highly anisotropic. Differentiation of central and peripheral cytoskeletal structures is the most general type of this anisotropy observed in almost all cells. Many cells also have other variants of anisotropy, such as differentiation of anterior and posterior structures and of dorsal and ventral structures.

The morphological organization of cytoskeleton has specific features in each cell type. At the same time, cells of very different morphology and origin may have many similar features of cytoskeleton architecture. The most common pattern of cytoskeleton organization has two characteristic features: preferential localization of actin structures at the cell periphery under the plasma membrane and a radiating array of microtubules growing from the perinuclear cell center to the periphery (Fig. 5.1). Intermediate filaments form the network connecting central parts of the cell with the submembranous periphery. This general pattern is characteristic of most interphase cells of higher animals. Most cells with this cytoskeleton pattern can form pseudopods and attach themselves to other cells and to noncellular surfaces. Therefore, we shall designate all cells with this pattern as the pseudopod-forming cells. There are, of course, many variations of this general pattern. For instance, animal pseudopod-forming cells are divided into many classes: epitheliocytes, fibroblastlike cells, neurons, leukocytes. These classes can be distinguished from one another by a number of characteristics, including the type of intermediate filament, the degree of development of the microtubular system and actin cortex, and formation of specialized bundles within the cortex. In the course of differentiation, certain animal pseudopodial cells undergo reduction and specialization of cytoskeleton, so that they lose the ability to form pseudopods and to crawl. During certain specializations most cytoskeletal components disappear, while one of these components is overdeveloped. For instance, the mammal erythrocytes retain only the specialized actin–spectrin network; keratinized cells retain only the specialized keratin network.

In some cases specialization of animal cytoskeleton involves the appearance of microtubular structures at the cell periphery. A single primary cilium is present in many pseudopod-forming cells, but numerous submembranous cilia with basal bodies located in the cortex are characteristic only of ciliated

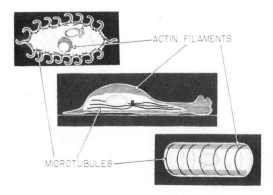

FIG. 5.1. Highly generalized schemes of several types of cytoskeleton patterns: pseudopod-forming cell with peripheral actin and central radiating microtubular system, plant cell with submembranous microtubules and subcortical actin bundles, and ciliate cell with a cortical system of cilia.

epithelia. These epithelia also retain the structures of pseudopod-forming cells: peripheral actin cortex and bundles, central radiating microtubular array, networks of intermediate filaments. Formation of spermatozoa in the vertebrates involves more radical rearrangements of the cell periphery. Here development of the flagellum is accompanied by reduction of the actin cortex and radiating microtubular array.

A great variety of cytoskeleton patterns is characteristic of Protozoa and of other lower eukaryotes. The cytoskeleton architecture of various ameboid cells is similar to that of animal pseudopod-forming cells. Many other unicellular organisms, e.g., various flagellates, have "microtubular" periphery, while actin structures in their cortex seem to be reduced or absent altogether. The ciliates are characterized by an extreme degree of development or cortical microtubular structures.

A special variant of the cytoskeleton pattern is characteristic of the cells of higher plants. Here submembranous parallel arrays of microtubules are present in the peripheral parts of the cytoplasm; actin bundles, if any, are located more to the interior.

Obviously, the organization of cytoskeleton is closely associated with the cell shape and with the mode of cell locomotion. While many cells with submembranous cortices can crawl on solid surfaces, cells with cortical cilia or flagella can swim in fluids. Plant cells with submembranous parallel microtubules are immobile. The cytoskeletons of many lower eukaryotes have not yet been studied in detail, so any comprehensive classification of the organization of cytoskeletons is premature.

Animal pseudopod-forming cells, such as fibroblasts, epitheliocytes, neurons, and others, are at present most widely used in the study of cytoskeletal organization and reorganization. These cells are convenient for these studies because they have both dynamic and well-developed cytoskeletons without excessive specialization. These cells will be the main subject of this and the next chapters.

B. Interdependence of Various Cytoskeletal Systems

Maintenance of the organized pattern of the cytoskeleton depends on the dynamic interactions between the three main types of cytoskeletal systems. The interrelationship between the intermediate filament network and the microtubular system in cultured fibroblastic cells is one of the best-studied examples of these interactions. Double staining of fibroblasts by antibodies against tubulin and vimentin had shown that most intermediate filaments are coaligned with the radiating microtubules (Fig. 5.2A–D). When microtubules are depolymerized by colcemid or by another specific drug, these filaments collapse; that is, they move from the cell periphery toward the central part of the cell and form the bundle encircling the nucleus. This collapse is fully reversible; when the drug is removed from the medium, intermediate filaments move back toward the periphery, following growing microtubules. Dependence of the position of intermediate filaments on the microtubules is not reciprocal. When antivimentin antibodies are injected into the living cell, the network of intermediate filaments collapses around the nucleus, but distribution of microtubules is not visibly changed.

Perinuclear collapse of desmin-containing intermediate filaments after destruction of microtubules can also be observed in certain cultured myoblastic cells. However, desmin filaments of striated muscle are not collapsed by microtubule-dependent drugs. Keratin filaments are the least dependent on microtubules; their collapse is not observed in colchicine-treated cells.

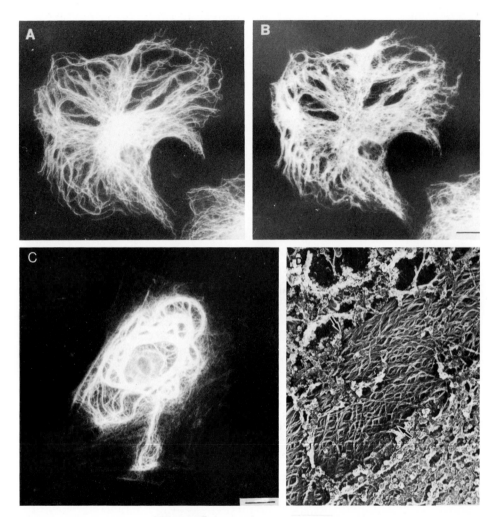

FIG. 5.2. Colocalization of microtubules **(A)** and of intermediate filaments **(B)** in the cultured mouse fibroblast. The cell is doubly stained with antitubulin and antivimentin antibodies. Fluorescence microscopy. Scale bar (B): 10 μm. (Courtesy of I. S. Tint.) **(C,D)** Perinuclear collapse of intermediate filaments in the cytoplasm of colchicine-treated fibroblasts. (C) Cables of intermediate filaments around the nucleus. Immunofluorescence microscopy using antivimentin antibody. Compare with control cells (Fig. 3.1B). Scale bar: 10 μm. (Courtesy of I. S. Tint.) (D) Electron micrograph of replica of the perinuclear cable of intermediate filaments (arrows). Scale bar: 0.5 μm. (Courtesy of T. M. Svitkina.) **(E,F)** Effect of colchicine on the distribution of actin filaments within cultured mouse fibroblast. Immunofluorescence microscopy using antiactin antibodies. (E) Control cell with nearly parallel filament bundles. Scale bar: 10μm. (F) Colchicine-treated cell. The bundles are localized at the cell periphery and are not oriented in parallel to one another. (Courtesy of L. A. Lyass.) **(G)** Multifunctional proteins reacting with various components of the cytoskeleton.

Interrelationships of microtubules with actin structures are complex. Actin bundles, in contrast to vimentin filaments, are not coaligned with microtubules in cultured cells. However, drug-induced depolymerization of microtubules causes considerable contraction of actin cortex and loss of parallel orientation of actin bundles (Fig. 5.2E,F). Reciprocally, formation and attachment of pseudopods, leading to reorganization of actin structures, is followed by alteration of the distribution of microtubules and of the position of the microtubule-organizing center. These interactions of actin cortex and microtubules will be discussed in more detail in Chapter 7.

Thus, three systems of cytoskeletal fibrils interact with one another. Alteration of one system can lead to structural reorganization of other systems.

C. The Links between Various Types of Fibrils

The functional integration of three groups of cytoskeletal fibrils is likely to be due to the formation of dynamic links between these fibrils. Electron microscopy often reveals transverse thin filaments between various fibrils, e.g., between neurofilaments and microtubules in axons (Fig. 2.2L) or between keratin filaments and actin bundles in the intestinal brush borders (Fig. 1.4). The possible nature of these filaments was discussed in Chapter 4.

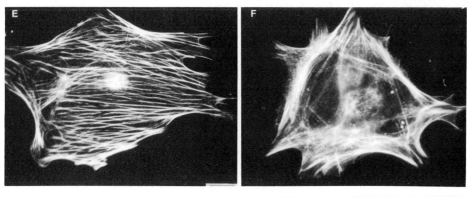

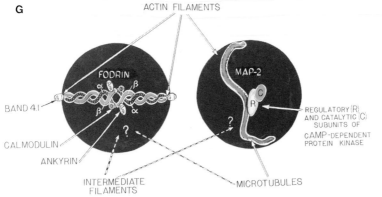

FIG. 5.2. (*Continued*)

The direct end-to-side connections of vimentin and actin filaments were also described. Dense bodies in smooth muscle cells and Z disks in striated muscle cells are examples of specialized structures to which both actin and desmin filaments are attached. In addition to the morphological evidence, there are also some biochemical data suggesting that fibrils of various types can interact.

Certain proteins can bind *in vitro* and *in vivo* to two or even three types of fibrils (Fig. 5.2G). For instance, MAP-2 protein was shown to cross-link actin filaments *in vitro*. These data suggest that MAP-2 can bind not only to microtubules, but also to actin filaments. MAP-2 was also reported to bind vimentin filaments *in vivo* and *in vitro*. Possibly, MAP-2 molecules can link microtubules with actin and vimentin filaments in the cells.

Possibly, proteins of the spectrin group can bind not only to actin and ankyrin (see Chapter 1), but also to vimentin filaments. This suggestion is supported by data showing colocalization of spectrin and vimentin in fibroblasts treated with colchicine or with antispectrin antibodies; this indirect evidence is yet to be confirmed by *in vitro* experiments. Ankyrin was reported to bind *in vitro* not only to spectrin and to certain membrane proteins (see Chapter 1), but also to tubulin. Thus, spectrin and ankyrin can be involved in the formation of cross-links between the microtubules and actin, microtubules and plasma membrane, and so forth.

The data on polyfunctionality of certain cytoskeletal proteins are still scarce and fragmentary. Further work in this field is, obviously, essential for understanding the molecular mechanisms that direct and maintain coordinated localization of various cytoskeletal systems.

III. Cytoskeleton and Distribution of Cellular Organelles

Not only the cytoskeletal structures, but all the other cell organelles have regular anistropic distribution within the cells. For instance, each type of organelle in cultured substrate-attached cells occupies a certain position with regard to the center and periphery. Two zones of cytoplasm can be distinguished in these cells. Their nucleus is surrounded by endoplasm filled with densely packed vesicular organelles, such as cisterns of endoplasmic reticulum, lysosomes, and elements of Golgi apparatus. The zones with lower concentrations of vesicular organelles, but with a high content of cytoskeletal elements, are located externally to the endoplasm; the "ectoplasmic" zones have the shape of thin lamellae or of elongated narrow processes (Fig. 5.3A,B).

The positions and orientation of various organelles are correlated with the organization of cytoskeletal structures; indeed, they are controlled by this organization. For instance, elongated cultured fibroblast has an ellipsoid nucleus with the long axis parallel to that of the cell body. Fibroblasts treated with colchicine acquire less-elongated shapes; the degree of elongation of their nuclei also decreases, and these nuclei lose any definite orientation. Mitochondria of cultured cells are often aligned along the cytoplasmic microtubules, especially in the peripheral parts of the cytoplasm (Fig. 5.3C). Lysosomes are also distributed along the microtubules. Both lysosomes and mitochondria disappear from the cell periphery and are collected only in the endoplasm after destruction of microtubules by colchicine.

Integrity of the microtubular system is also essential for organization of the Golgi apparatus. Normally, the numerous stacks of membrane-bound cisternae forming this apparatus are compactly grouped near the centrosome.

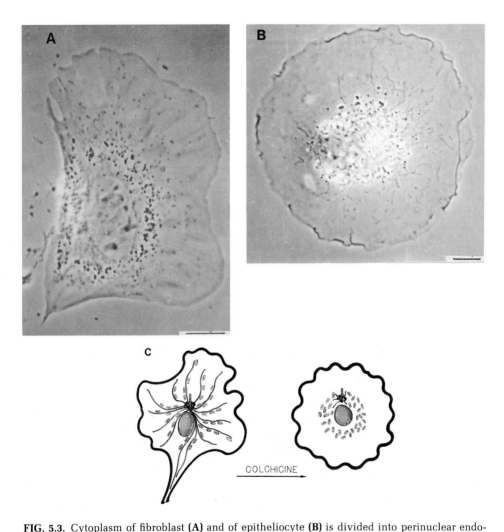

FIG. 5.3. Cytoplasm of fibroblast **(A)** and of epitheliocyte **(B)** is divided into perinuclear endoplasm containing numerous vesicular organelles and lamellar ectoplasm with low numbers of these organelles. Phase-contrast photographs of cultured normal mouse embryo fibroblast and of the cell of IAR-2 line derived from rat liver. Scale bar (A,B): 10μm. (Courtesy of L. V. Domnina.) **(C)** Coalignment of microtubules and of mitochondria in the cultured fibroblast. Accumulation of mitochondria in the perinuclear zone after depolymerization of microtubules. **(D)** Thin fibrils cross-linking the pigment granule with the microtubule. Electron micrograph of platinum replicas of erythrophore of seawater squirrel fish *Holocentrus ascensionis*; the cells were quick-frozen and freeze-etched. × 160,000. (From Ip *et al.*, 1984. Courtesy of W. Ip.) **(E)** Intermediate filaments associated with the surface of pigment granules. Replica of freeze-dried cytoskeleton of the fish melanophore. × 75,000; reproduced at 57%. (From Murphy and Grasser, 1984. Courtesy of D. B. Murphy.) **(F)** T-junction of an actin filament with an intermediate filament (arrow) and of an actin filament with a microtubule (arrowhead). × 120,000; reproduced at 57%. (From Schliwa *et al.*, 1982. Courtesy of M. Schliwa.)

Drug-induced depolymerization of microtubules leads to dispersal of these stacks over the entire cytoplasm.

All these data indicate that the radiating system of microtubules in interphase cells directly or indirectly determines the exact localization of various cytoplasmic organelles, as well as the orientation and shape of the nuclei. As already indicated, thin filaments connecting various vesicular organelles with the microtubules or with accompanying intermediate filaments were seen in electron microscopic pictures (Figs. 5.3D–F and 4.5). Intermediate filament networks also seem to be connected with the nucleus. The role of these links, if any, in the distribution of organelles is not clear.

In all probability, control of the distribution of cytoplasmic organelles by cytoskeleton is dynamic. In other words, this control can be a result of the intracellular movements of organelles mediated by cytoskeletal structures. Characteristics and mechanisms of these movements will be discussed in the following section.

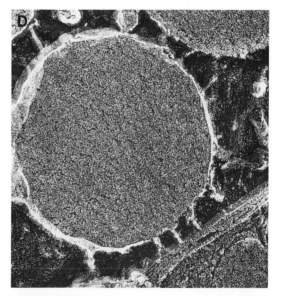

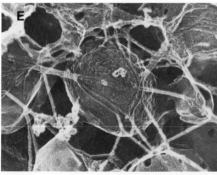
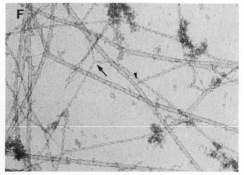

FIG. 5.3. (Continued)

IV. Cytoskeleton and Intracellular Movement of Organelles

175

*CYTOSKELETON
AND INTERNAL
ORGANIZATION OF
THE CELL*

A. Various Forms of Intracellular Motility

Several broad groups of intracellular motile phenomena can be distinguished: mutual translocations of cytoskeletal elements, rotational movements of organelles, and radial movements of organelles.

Movement of cytoskeletal components, such as muscle contraction, or movement of the cilia and flagella is usually associated with visible alterations of the cell shape.

We have already discussed the mechanisms of these phenomena in Chapters 1 and 2. Some other variants of mutual translocations of cytoskeletal structures during mitosis and cell locomotion will be described in Chapters 7 and 8. Here we shall discuss the types of intracellular movements which, in contrast to the first type, are not usually associated with gross cell deformations. They involve the movement of organelles and, sometimes, the flow of the fluid cytoplasm. Depending on their direction, rotational and radial movements of organelles can be distinguished.

B. Rotational Movements of Organelles and Cytoplasm

These movements are conspicuous within plant cells surrounded by immobile walls. Streaming of endoplasm in the giant cells of *Chara* or *Nitella* is the most striking and the best-studied example of this type of movement (Fig. 5.4A,B). In these cells endoplasm flows in the narrow zone between the stationary chloroplast-containing cortical layer and the central vacuole. The flow is rotational; cytoplasm flows along the long axis of the cell, turns around at the end, and flows back on the other side. The flow carries with it the nuclei and many cytoplasmic organelles; all particles flowing in the same cell have a similar velocity, varying from 40 to 100 μm/sec. Bundles of actin filaments with uniform polarity are found at the interface between the cortex and flowing endoplasm; these bundles are believed to be anchored somewhere in the cortex. Microtubules were not found in the area where streaming occurs. Cytochalasin B disorganizes the actin bundles and inhibits streaming. These data suggest that filament bundles directly participate in generation of motile forces for the streaming of endoplasm. However, the exact mechanism of the generation of this force is not yet clear. In particular, it is not known where the myosin molecules interacting with actin are located. One hypothesis is that myosin is attached to some movable network present in the endoplasm; sliding of this network along the bundles pulls the fluid endoplasm and organelles forward. However, there is no direct evidence confirming the existence of this network. Another suggestion is that myosin molecules are attached to the surface of organelles present in the endoplasm; these organelles can then slide along the bundles. In other words, the mechanisms of these movements can be similar to those of myosin-coated polymer beads sliding along the actin filaments in model experiments (see Chapter 1). However, it is not yet clear whether the organelles normally present in the endoplasm of these cells are covered with myosin.

Rotational movements of organelles in animal cells are not conspicuous.

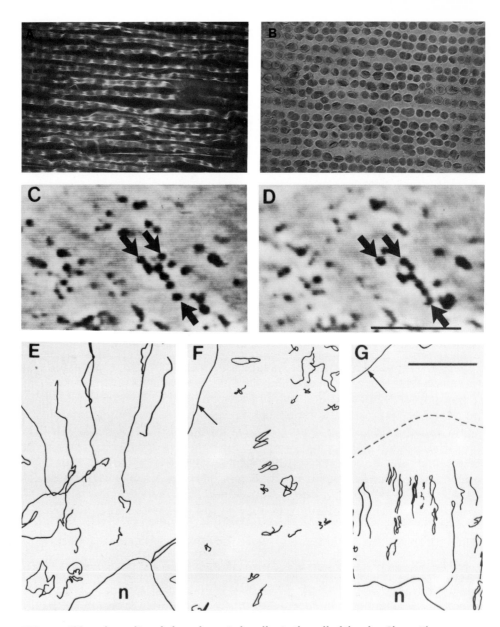

FIG. 5.4. Chloroplasts aligned along the actin bundles in the cell of the alga *Chara*. Fluorescence **(A)** and bright-field **(B)** micrographs of a cell perfused with antiactin antibody. Groups of actin bundles are found beneath each line of chloroplasts. The chloroplasts attenuate the fluorescence of bundles. × 480; reproduced at 57%. (From Tiwari *et al.*, 1984. Courtesy of R. E. Williamson.) **(C–G)** Effects of alterations of microtubules on the translocations of lysosomes in cultured ovarian granulosa cells. (C,D) An example of the actual time-lapse sequence of lysosome travel in cells. Interval 132 sec. (E–G) Tracing of time-lapse sequences of video-enhanced images of individual lysosomes in the cytoplasm (E). Control cell. Long linear excursions of lysosomes. (F) Nocodazole-treated cell with depolymerized microtubules. Limited local movements of lysosomes. (G) Taxol-treated cell with stabilized microtubules. Short-range saltations confined to a perinuclear region, indicated by a dotted line, n, nucleus; arrows indicate cell border. Scale bar: 10 μm. (From Her-

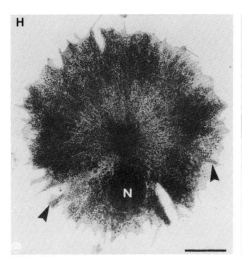

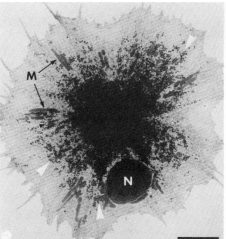

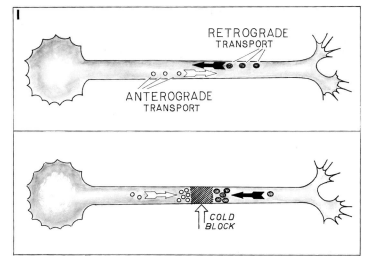

man and Albertini, 1984. Courtesy of B. Herman.) **(H)** Dispersal and aggregation of granules in pigment cells. Cultured squirrel fish erythrophores viewed under high-voltage electron microscope. Files of granules (arrowheads) along the microtubules. N, nucleus; M, mitochondria. (From McNiven and Porter, 1984. Courtesy of K. R. Porter.) **(I)** Scheme of bidirectional fast axonal transport and of its inhibition by local application of cold. **(J,K)** Inhibition of axonal transport in the nerve by local cold block. Electron micrographs of the mouse saphenous nerve in the regions just proximal (J) and just distal (K) to the site of block. The material accumulated proximally by anterograde transport consists mainly of vesiculotubular membranous structures (V). The material accumulated distally by retrograde transport consists of large organelles, mostly multivesicular bodies (Mv) and lamellar bodies (Lb). T, microtubules. Nf, neurofilaments. (From Tsukita and Ishikava, 1984. Micrographs courtesy of S. Tsukita.) **(L)** Time-lapse sequence of the movement of organelles (marked by arrowheads) within the ATP-treated axon permeabilized with saponin. Allen video-enhanced contrast, differential interference contrast (AVEC–DIC) microscopy. (From Forman, 1984. Courtesy of D. S. Forman.) **(M)** Movements of intracellular particles along the lamellipodium formed by the foraminiferum *Allogromia laticollaris*. The arrow points to a particle that moves along a fibril (microtubule). This particle dissociates from the fibril (e) and remains stationary until contact is reinitiated (f). × 2580. (From Travis *et al.*, 1983. Courtesy of R. D. Allen.)

Rotation of nuclei is occasionally seen in time-lapse movies of cultured cells; its mechanisms remain quite obscure.

C. Radial Movements of Organelles

1. Characteristics

These movements transporting the organelles toward and from the perinuclear area are observed in many types of animal cells and of the *Protista*. The three most actively studied types of radial movement are the movement of endosomes and other organelles in cultured cells, the fast transport of vesicular organelles in axons, and the movement of granules in pigment cells.

Movement of the vesicles containing endocytosized material, endosomes, can be observed in cultured fibroblasts, epitheliocytes, and macrophages with

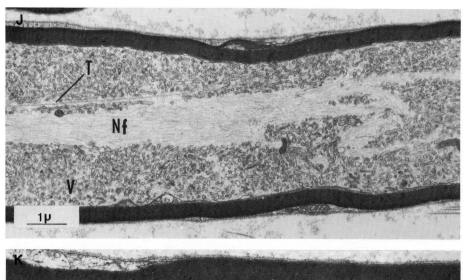

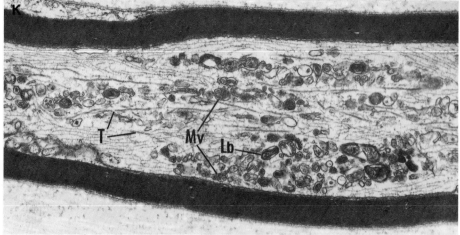

FIG. 5.4. (*Continued*)

the aid of usual phase-contrast microscope or of fluorescent microscopy revealing labeled endocytosized protein molecules. Recently developed methods of video enhancement considerably increase the resolution of the light microscopic image. Endosomes are often formed at the periphery of the substrate-spread cells, and then they move through the cytoplasm into the endoplasm, where they undergo fusion with lysosomes. Endosomes can move in two ways: they either migrate smoothly and continuously or undergo sal-

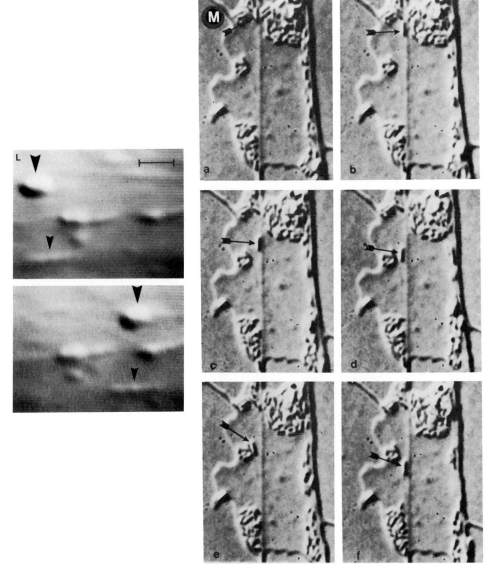

FIG. 5.4. (*Continued*)

tation; that is, a series of jerklike, discontinuous displacements interrupted by stationary periods. Other organelles, such as mitochondria and lysosomes, can also undergo saltation; particles injected into the cytoplasm can saltate. Usually displacement of the particle during each saltation does not exceed several micrometers, the average speed of saltation being about 1–2 μm/sec; consecutive saltations lead to long-range movements of organelles toward the nucleus or away from it. Movements of individual organelles do not seem to be correlated with one another (Fig. 5.4C–G).

Migration of granules in the pigment cells of fish and amphibia is another variant of radial movement. In these flattened cells, numerous pigment granules fill all the cytoplasm. In response to certain neurotransmitters and hormones, all the granules begin to move toward the cell center and are aggregated in this area (Fig. 5.4H). Other stimuli reverse this reaction, so the granules move back and become dispersed over the whole cytoplasm. Aggregation and dispersal of granules lead to reversible protective changes of the skin color of animals. All granules move with similar velocities; this velocity during aggregation can reach 20 μm/sec in certain cells, so aggregation is accomplished in 3–5 sec. Dispersal usually takes several times longer than aggregation. Dynamic observation of the movements of individual granules has shown that they may return after dispersal to exactly the same places they occupied before aggregation. This remarkable fact shows that the particle movements are exactly targeted.

Intense bidirectional movements of organelles over long distances take place within the axons of neurons. These movements are called **fast axonal transport,** to distinguish them from "slow transport," by which not the vesicular organelles but cytoskeletal components are transported toward the periphery with low velocity (see Chapter 7). Organelles traveling away from the cell body (fast anterograde transport) form two subgroups moving with different average velocities: 4–5 μm/sec and 0.6 μm/sec. Anterograde transport carries vesicles of small size. Retrograde transport carries toward the cell body pinocytic vesicles formed at the nerve terminal and multivesicular bodies containing the material destined for degradation; velocity of this transport can reach 3 μm/sec. Until recently, axonal transport was usually studied by indirect methods, especially by following the dynamics of radioactively labeled material and by morphological examination of materials accumulated at the sites of the axon where transport was locally blocked by cold (Fig. 5.4I–K). The previously mentioned new methods of high-resolution light microscopy permitted direct observation of individual organelles within the axon. Vesicles were observed to travel continuously for long distances, but occasionally they underwent saltatory movements with repeated starts and stops (Fig. 5.4L). These vesicles rarely changed anterograde or retrograde direction of movement. Movement of rod-shaped organelles which were probably mitochondria was more variable; they often stopped and changed direction. Foreign inert particles, latex beads, injected into the axon move anterogradely like endogenous organelles. Thus, it seems that the axonal transport system can carry almost any particle present in the cytoplasm. Bidirectional movement of organelles is also observed in the long processes extended by certain Protozoa; these movements are often saltatory (Fig. 5.4M).

Radial movements of organelles can be observed not only in whole cells, but also in simpler experimental systems. Detergent-treated fish pigment cells become open to the external medium, but retain the ability to aggregate pigment granules in response to a specific stimulus, epinephrine. Similarly, axons permeabilized by detergent retain the ability to perform fast transport of organelles (Fig. 5.4L). This transport can also be observed in cytoplasm extruded by compression from giant axons of the squid; cylinders of this cytoplasm can be placed on coverslips and examined microscopically.

Experimental analysis of pigment movement and axonal transport is facilitated in systems in which permeability barriers are absent.

2. Possible Mechanisms

Radial movements of organelles in permeabilized cells and in extruded cytoplasm are activated by addition of ATP. Nonhydrolyzable analogs of ATP are not effective; indeed, some of them inhibit ATP-activated movements. These data suggest that radial movements of organelles are driven by some force-transducing ATPase.

Another important fact established in studies of various types of radial movements is that they are always closely associated with microtubules. Microtubules are always present in the areas where these movements occur, and pathways of movement correspond to the direction of microtubules. This is true for the movement of granules in pigment cells, for the centripetal transport of endosomes, and for saltation of other organelles in cultured cells. Axons and processes of Protista in which particle transport is observed contain numerous microtubules. In contrast, no correlation was observed between the direction of radial movements and localization of actin bundles. Video-enhanced light microscopy revealed fine filaments guiding particle movements in the thin lamellae of living cells. Control staining with antitubulin antibodies had shown that these "filaments" are individual microtubules or small bundles of microtubules. Direct observation had demonstrated that cytoplasmic vesicles contacting these filaments performed saltatory movements exclusively along them. When the particle lost contact with the filaments, its directional movement stopped. Various particles moved along the same filament in both directions.

The role of microtubules in radial movements is also shown by numerous experiments with colchicine and similar drugs. These drugs inhibited endosome migration in cultured cells, dispersal of granules in pigment cells, and fast axonal transport. Although the drugs stopped all the long-range directional transport of organelles, their short-range random wandering within the cytoplasm was often continued.

Microtubule-associated transport of organelles was also observed outside the cells. Native microtubules were prepared from extruded axoplasm of squid axons. Organelles and vesicles, as well as exogenous polystyrene particles, adhered to these microtubules and were transported along them. Presence of ATP was essential for translocation. Vesicles from squid axoplasm also move along the flagellar microtubules placed into solution containing ATP; removal of dynein arms from the flagella did not affect this transport.

What is the exact role of microtubules in the mechanism of radial movement of organelles? By analogy to the movement of cilia, it had been suggested that some microtubule-bound ATPase interacts with the surface of the particle, so the particle slides along the microtubule. Of course, cytoplasmic dynein seems to be the obvious candidate for the role of this ATPase. However, direct evidence of the role of dyneinlike ATPase in organelle movements is still lacking. Another possibility is indicated by the recent experiments of Vale *et al.* (1985a). They have found a new protein, named kinesin, in the cytoplasm of the squid giant axon. Kinesin has an apparent molecular weight of about 600 kD; it consists of 110–120 kD and 60–70 kD subunits. When kinesin was added to a solution containing taxol-stabilized microtubules assembled from purified tubulin and polymer beads or cytoplasmic vesicles, directional sliding of these particles or vesicles along the microtubules was observed. The beads coated with kinesin always moved from the minus to the plus ends of the centrosomal microtubules. This would correspond to the anterograde direction in axons. Kuznetsov and Gelfand (1986) have shown that kinesin is a microtubule-activated ATPase. Thus kinesin is one of the microtubule-associated ATPases ("translocators") responsible for organelle movement. Possibly, distinct translocators are needed for anterograde and for retrograde organelle transport. This possibility is supported by the experiments of Vale *et al.* (1985b), who added a low-speed supernatant of axoplasm to a solution containing oriented microtubules and beads. In these conditions the beads moved along the microtubules in both directions. Removal of kinesin from the supernatant with the aid of antikinesin antibody affinity column changed the direction of bead movement from 70% anterograde to 100% retrograde. Retrograde translocator was recently identified as MAP1C protein. The concept of special microtubule-associated translocators leaves unexplained certain experimental data. In particular, it was found that axon transport is inhibited by proteins severing actin filaments (gelsolin, brevin) and by actin monomer-sequestering protein (DNase I).

To summarize, various types of intracellular movements of organelles are clearly associated with certain cytoskeletal structures. In particular, rotory streaming of the cytoplasm is associated with the bundles of actin filaments. Radial movement of organelles is associated with cytoplasmic microtubules. It is likely that rotory streaming and radial movements are driven by various ATPases attached to actin filaments or to microtubules, respectively. However, the situation may be more complex, and actin may somehow be involved in the radial movement of organelles.

V. Cytoskeleton Control of Distribution and Movement of Components of Plasma Membrane

Interactions of the plasma membrane with the cortical cytoskeleton are of special importance. These two structures together form the surface boundary layer, which shields the cell interior from the external environment and, at the same time, initiates and organizes cell reactions to various components of the environment. Plasma membrane can function properly only in interaction with underlying cytoskeleton. This cytoskeleton is essential for mechanical stabilization of the shape of the plasma membrane. The membrane alone is weak and has a tendency to form vesicles; association with the cortical network of actin filaments prevents deformation of the membrane, so the surface acquires a stable covering. For example, depolymerization of the core actin bundle of brush border microvilli leads to vesiculization and disappearance of these microvilli. Alteration of the surface shape, especially endocytosis and extension of pseudopods, requires active participation of the cortex. Interactions with the cytoskeleton are also necessary for regular nonrandom distribution of the components of the membrane and for directional movements in the plane of the membrane.

A. Immobilization of Membrane Components by Cortical Cytoskeleton

Plasma membrane is a lipid bilayer with incorporated protein molecules. At physiological temperatures most parts of this bilayer are in the liquid-crystalline state, so individual lipid and protein molecules are able to move randomly within this bilayer. This lateral diffusion of the molecules within the membrane can be decreased by two types of peripheral constraints: the external ligands linked to these molecules from the outside and the cytoskeletal components anchoring these molecules from the inside.

At present the most popular method of measurement of the rate of lateral migration of molecules in the membrane is based on determination of fluorescence recovery after local photobleaching. Fluorescent label is attached to certain molecules of the membrane directly or via specific ligands reacting with these molecules. A laser beam is directed on a small spot of the membrane and fluorescence of this spot is measured with a photometer attached to the microscope. The intensity of the light beam is then greatly increased for a short time so that the illuminated spot of the membrane is bleached. Dynamics of fluorescence recovery at the bleached spot is then registered. This recovery is due to lateral diffusion of fluorescent molecules into the bleached area, so the diffusion coefficient of these molecules can be determined from these measurements. Diffusion rates of lipid molecules in contact-free plasma membranes are usually similar to those determined in artificially prepared lipid bilayers ($D \geq 10^{-8}$ cm^2/sec); that is, diffusion of these molecules has no peripheral constraints.

Certain proteins, e.g., rhodopsin in the membranes of retinal cells, also have high diffusion rates ($D > 10^{-8}$). However, other proteins free of external constraints have decreased mobility in the membrane; for instance, acetylcholine receptor molecules in the muscle membrane are immobile ($D = 10^{-12}$). Mobility of this protein is greatly increased when a bleb is made under the plasma membrane ($D = 3 \times 10^{-9}$ cm^2/sec); this bleb is supposed to detach the plasma membrane from the underlying cortical structures. Similarly, the rate of diffusion of band III protein in the membrane of red blood cells of normal mouse is very low ($D = 5 \times 10^{-11}$ cm^2/sec). The rate is much higher ($D = 2 \times 10^{-9}$ cm^2/sec) in mouse cells with a genetic deficiency of spectrin, which is essential for formation of the submembranous cytoskeleton of the erythrocyte. These data suggest that movements of certain membrane proteins within the membrane are restrained by some submembranous components, probably cortical cytoskeleton. Membrane proteins may be temporarily or permanently attached to the cortical actin structures and, less likely, to intermediate filaments or microtubules. As discussed in Chapter 1, this anchorage can be mediated by such proteins as spectrin, ankyrin, possibly talin, and others. However, little is known about the exact degree of anchoring of various proteins and about the mechanisms of this anchoring in any contact-free membrane except that of erythrocytes.

Effects of cytoskeleton-specific drugs on the mobility of various membrane receptors have been measured in a number of experiments. However, these drugs can change not only the lateral mobility of molecules in the membrane, but also the shape and motility of the entire cell surface. For instance, these drugs can induce or inhibit formation of pseudopods, exo- and endocytosis, and so forth. Therefore, alteration of fluorescence recovery after photobleaching caused by these drugs is difficult to interpret.

B. Effects of External Ligands on the Membrane–Cytoskeleton Anchorage

Many data suggest that binding of ligands to the outer parts of membrane molecules alters the interaction of the inner parts of these molecules with the cytoskeleton. More specifically, the degree of anchorage to the cytoskeleton from the inside may be increased when the membrane components are linked to one another by some ligands from the outside. Considerable indirect evidence in favor of this possibility was obtained in experiments with multivalent ligand molecules specifically reacting with certain membrane components, e.g., with antibodies or with lectins (Fig. 5.5A). Lectins are plant proteins specifically binding carbohydrates; therefore they can cross-link membrane molecules containing attached carbohydrate chains. When living cells are treated with these ligands, matching components of the membrane, called receptors, are aggregated into small clusters or large patches. The cells, pretreated with lectins and then extracted with detergent, were found to retain on their cytoskeleton a much larger fraction of corresponding lectin-binding membrane receptors than control cells extracted without this pretreatment. Reciprocally, preparations of plasma membrane detached from cells pretreated with one of the lectins, concanavalin A, contained larger amounts of membrane-linked actin than membranes of nontreated cells. These results indicate that receptors clustered by their ligands become more firmly bound to underlying cortical actin. Another group of observations shows that clustered membrane receptors to concanavalin A are diffusely distributed within the membrane of the fibroblast before cell incubation with this ligand. However, after incubation, ligand-clustered groups of receptors may become collected in the areas of the membrane located over the bundles of actin filaments (stress fibers). Possibly, in this case clustering promotes anchorage of receptors to a specific structure. Almost by definition, certain membrane components of specialized contact structures, such as desmosomes or focal contacts, are bound to the external molecules present on the

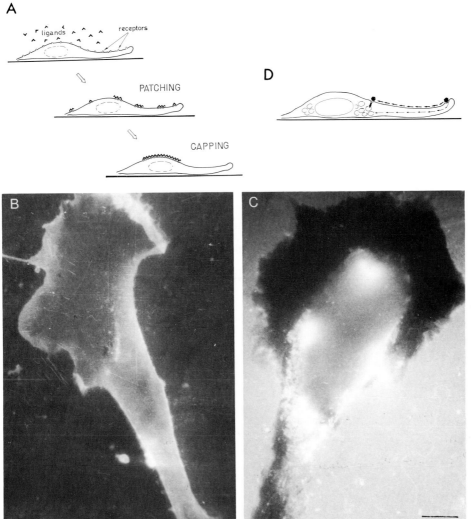

FIG. 5.5. (A) Two stages of redistribution of surface receptors after their cross-linking by corresponding multivalent ligands.

(B,C) Capping of concanavalin A receptors on the surface of cultured fibroblasts. Immunofluorescence microscopy using antibodies against concanavalin A. (B) Fibroblast prefixed before incubation with concanavalin A. Diffuse distribution of concanavalin A receptors; apparent concentration of receptors near the cell edge is due to the presence of surface folds (ruffles). The substrate also gives diffuse fluorescence because concanavalin A reacts with carbohydrate-containing serum proteins absorbed at the surface of the glass. (C) Fibroblast fixed after 30 min of incubation in medium containing concanavalin A at 37°C. Concanavalin A receptors bound to their ligand had been cleared from the leading cell lamellae and concentrated at the surface of the central cell part. Scale bar: 10 μm. (Courtesy of L. V. Domnina and O. Y. Pletyushkina.)

(D) Two possible ways of centripetal movement of the particle attached to the surface of cultured fibroblast: the particle moves at the surface and then is endocytosized, or it is endocytosized and then the endosome moves through the cytoplasm.

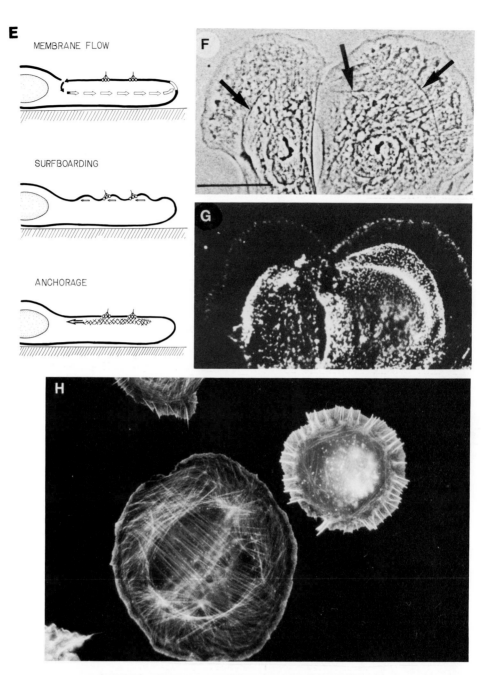

FIG. 5.5 (*Continued*) **(E)** Three hypothetical mechanisms of directional movement of the clusters of membrane receptors cross-linked by ligands: translocation by flow of membrane material, by mechanical waves passing along the surface, and by cortex actin filaments anchored to the cluster.

(**F–H**) Coordinated centripetal movement of surface receptors patched by the corresponding antibody and of the actin-containing arcs in cultured fibroblast. (F,G) Paired phase contrast and immunofluorescence photographs of human fibroblasts showing accumulation of patched receptors over the arcs (arrows). (H) Arc-containing fibroblast stained for actin by phalloidin–rhodamin. Scale bar: 20 μm. (From Heath, 1983. Courtesy of J. P. Heath.) **(I)** Coated vesicle attached to the actin network. Platinum replica of the cytoskeleton preparation of cultured mouse fibroblast. Scale bar: 0.1 μm. (Courtesy of T. M. Svitkina.)

surface of other cells or of the extracellular matrix. Therefore, attachment of actin or keratin filaments to these contact areas can be regarded as yet another example of the anchorage of cytoskeleton fibrils from the inside to clusters of membrane molecules immobilized from the outside.

The mechanism of the effect of external cross-linking on internal anchorage remains obscure. It is possible that anchoring is a cooperative phenomenon, so that a cluster of receptors forms links with actin filaments more effectively than individual receptors not connected to one another. It is also possible that clustering induces some alterations in the local state of the membrane and of submembranous areas; these alterations may induce some changes in the actin cortex. For instance, alterations of the membrane may lead to increased calcium concentration in the cortex; this increase can stimulate the activity of actin-capping proteins in this area and nucleate the formation of new actin filaments, which become attached to the membrane. Membrane-induced local alterations of cytoplasmic pH may also be important. None of these hypotheses is at present supported by any direct experimental evidence.

C. Directional Movements of Membrane Components

Clustered membrane molecules can move directionally along the surface so that they are all eventually collected in one part of the cell, where they form a single large aggregate, called a cap. The direction of capping is usually correlated with the cell shape. For instance, caps are usually formed over the tail pole of elongated moving lymphocytes. Caps in substrate-spread fibroblasts are formed over the endoplasm, while peripheral ruffling lamellae are

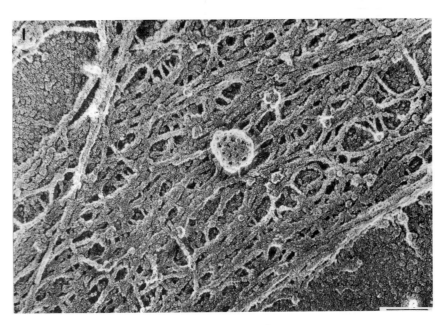

FIG. 5.5. (Continued)

cleared of clustered receptors (Fig. 5.5B,C). Clearing proceeds centripetally from the ruffling active edge toward the endoplasm.

Surface-attached particles can also move directionally along the surface (Fig. 5.5D). Various particles attached to the surface of filopodia and lamellae of fibroblasts move centripetally toward the endoplasm.

Many facts suggest that actin cytoskeleton is involved in capping. Actin, myosin, fodrin, α-actinin, and possibly other cytoskeleton proteins are concentrated under the membrane of the cap in the lymphocytes. Saponin-permeabilized models of the lymphocyte continue to cap the patched concanavalin A receptors after addition of ATP. An inhibitor of actin–myosin interaction, NEM-HMM, prevented capping in these models. Finally, cytochalasins inhibit capping and particle movement in the whole cell, while microtubule-destroying drugs usually do not stop movement.

Thus, the normal function of the actin cortex seems to be essential for capping. However concrete forms of cytoskeleton–membrane interactions leading to directional surface movements are not clear. There are three main groups of hypotheses about the mechanisms of these movements: membrane flow, "surfboarding," and direct anchoring of the clusters (Fig. 5.5E).

According to the membrane flow hypothesis, new material is continuously added to the plasma membrane from the cell interior in certain areas of the surface, e.g., near the anterior edges of cultured fibroblasts. This material then flows back along the surface toward the other sites, where it is interiorized by endocytosis or by some other mechanism. The "sink" may be localized in the central part of the cell. More likely, "coated pits" of the membrane (see next section) diffusely distributed over the surface continuously interiorize the membrane material. The continuous flow from the "spring" to the "sink" may carry with it a receptor cluster, as a flowing river carries a raft. The cortex may be needed for the intracellular transport of membrane material to the surface.

Existence of the "membrane flow" is still controversial, but recent evidence indicates that addition of new membrane material at the outer edge of cultured cells really takes place. However, even if the existence of the membrane flow is finally proven, it remains to be shown that this flow is responsible for capping and particle movement.

Another hypothesis postulates that directional movement of membrane components are driven by mechanical waves passing along the surface: these waves transport the clusters and particles, as waves at the sea surface move surfboards. Waves deforming the surface are, in fact, observed in many cells. For instance, membrane folds, called ruffles, are often formed at the outer edges of fibroblasts; some of these ruffles can move centripetally; that is, in the same direction as patched receptors and particles. Formation of these surface waves can be due to localized coordinated contractions and relaxations of actin cortex.

The third group of hypotheses, in contrast to the other two, postulates a direct role of actin filaments in capping. As already discussed, clustered receptors can become anchored to the underlying filaments before capping. Possibly, these filaments then pull the attached clusters directionally along the surface. According to one particular variant of this hypothesis, new actin filaments are continuously formed near certain areas of the surface and then move centripetally along the surface (see Chapter 7). This "actin flow" may drag with it anchored receptor clusters. Actin filaments can pull not only the receptors directly anchored to them, but also many other receptors laterally attached to the anchored ones.

The existing data are not sufficient to make a final choice between these hypotheses. It seems unlikely that the hypothetical membrane flow is preserved in cell models that are still able to perform capping. Direct association of actin movements with receptor capping was recently shown by Heath (1983). Using video-enhanced fluorescent microscopy of living cells, he had shown that before application of the antibody reacting with certain surface receptors, these receptors were diffusely distributed on the surface of fibroblasts. However, after addition of antibodies to the medium, receptors were clustered. Clusters located near the cell edge were immediately collected over the curved bundle of actin filaments formed under the surface (Fig. 5.5F–H) and began to move centripetally together with this bundle.

Thus, the anchorage hypothesis seems the most probable explanation of capping. However, this hypothesis also has its difficulties. In particular, it cannot easily explain capping of glycolipid molecules cross-linked by corresponding antibodies. These molecules, in contrast to membrane proteins, do not penetrate the lower half bilayer of the membrane, and it is not clear how cortical microfilaments can directly anchor them. It is possible, however, that a lipid cluster becomes attached to some membrane protein, which is then anchored to the filaments.

D. Endocytosis

Receptor clusters and attached solid particles can change their route before, during, or after directional transport along the surface (Fig. 5.5D). These particles and clusters can become surrounded by the plasma membrane and then travel intracellularly as endosomes. Endocytosis of solid particles, phagocytosis, begins with attachment of this particle to the membrane; then the pseudopods are extended near the site of attachment. The membrane of the pseudopod is attached to the free surface of the particle; this attachment–extension cycle is repeated many times until the particle is completely surrounded by the membrane. Local reorganization of actin cortex induced by some membrane-generated signals from the attachment site seems to play a leading role in phagocytosis. The concentration of polymerized actin was observed to increase in the cortex of leukocytes during phagocytosis, and accumulations of actin filaments were often formed around the phagocytized particles. Possibly, newly polymerized filaments within the pseudopods surrounding the particle contract and pull this particle into the cortex. These processes of extension and contraction of pseudopods are similar to those involved in cell spreading and locomotion on solid surfaces (see Chapter 7).

Another variant of submembranous changes is observed during certain instances of receptor-mediated endocytosis; that is, internalization of clustered ligand-bound receptors. This process involves formation of a special membranous lining under the cluster. This lining is a lattice made of the protein clathrin. Clathrin consists of a heavy chain of 180 kD and of two light chains of 33 and 36 kD. The signal inducing self-assembly of this lattice under a particular site of the surface is not known. This assembly is accompanied by local invagination of the surface, so that first the coated pit and then the coated vesicle (Fig. 5.5I) are formed. Preformed coated pits containing receptors to certain proteins also exist at the cell surface. Coated vesicles are formed in these pits after the addition of ligands. Formation of vesicles completes the internalization of the cluster. It is not yet clear whether any cortex actin reorganizations are involved in the formation of these pits and vesicles.

Internalization of phagosomes and of the coated vesicles is usually followed by radial migration of these organelles into the endoplasm. MAPs were found in the preparations of isolated and purified coated vesicles. Possibly, these proteins are involved in the interactions of coated vesicles with the microtubules, along which they migrate.

E. Cytoskeletal Arrays as Sensors of Mechanical Stimuli

Certain submembranous cytoskeletal arrays play a special role in the reception of external mechanical signals. The stereocilium of hair cells of the inner ear cochlea is an example of such an array (Fig. 5.6). The lengths and widths of stereocilia present in each hair cell are strictly controlled. These

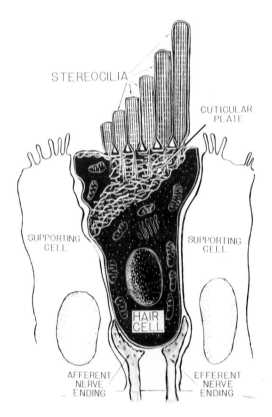

FIG. 5.6. Structure of stereocilium of hair cell of the inner ear.

parameters are different in hair cells that are located in various parts of the cochlea and that are excited by sound waves of different frequencies. The hair cell membrane generates an electric potential in response to the exciting sound. Mechanisms of the induction of this potential by the mechanical sound wave are not clear. Possibly, actin bundles of definite length and width undergo some resonant changes in response to the mechanical sound wave with corresponding parameters; these changes are then somehow transduced to the membrane.

Certain microtubular arrays are also involved in the reception of external stimuli. These arrays were found in certain types of the receptor cells, reacting to mechanical deformations of the environment. For instance, mechano-receptor cells present in the legs of the grasshopper contain numerous cilia without inner doublets (9 + 0 pattern). Of special interest are the touch-receptor neurons of the nematode C. elegans. These cells have long processes filled with numerous parallel microtubules. The structure of these microtubules is unusual; they have 15 protofilaments, while most other cells of the worm have 11 protofilaments (see Chapter 2). The nematodes carrying mutation of one particular gene, mec 7, have touch neurons lacking 15 protofilament microtubules; instead, the processes of these cells are filled with the "usual" 11-protofilament microtubules. Remarkably, this subtle change of microtubular organization is correlated with functional deficiency of the receptor cells; the mutant animals are insensitive to touch.

These data indicate that alterations of the mechanical state of microtubules, e.g., bending induced by external force, can be transduced to the membrane and can alter its components in such a way that specific membrane signals are generated.

Extending these considerations, it is possible that mechanical alterations of cytoskeletal structures can alter the membrane not only in specialized sensory cells, but also in other types of cells. The cells may sense the mechanical state of their environment via their cytoskeleton. For instance, the primary cilium present in many animal cells can have some unknown sensory function. These interesting possibilities need further investigation.

To summarize, organization of the plasma membrane depends on its continuous interactions with the underlying cytoskeleton. Because of these interactions, the structure of the membrane is dynamic and anisotropic; that is, distribution of membrane components may become nonrandom, and these components may move directionally to the surface, along the surface, and from the surface. It is also possible that alteration of the state of certain cytoskeletal structures can induce generation of electric and other signals in the membranes.

VI. Cytoskeleton and Nucleus

Until recently, cytoskeleton had been regarded as purely cytoplasmic structure. However, certain new data suggest that the nucleus may contain structures that are similar to or even identical to the cytoskeletal fibers in the cytoplasm.

One important group of data was obtained in the studies of the nuclear envelope separating the nucleoplasm from the cytoplasm. The major structural component of this envelope is the nuclear lamina, a fibrous layer on the internal surface of the nuclear membrane. The nuclear lamina is composed of three main proteins, lamin A, B, and C. Primary and secondary structure of the lamin A and C shows striking homology with the intermediate filament proteins. In particular, both groups of proteins contain α-helical regions of repeating heptads of amino acids (McKeon *et al.*, 1986). Electron microscopic examination had confirmed that the nuclear lamina is a meshwork of intermediate-type filaments (Aebi *et al.*, 1986).

Polymerized and/or nonpolymerized actin may also be present in the nuclei. This suggestion is supported by a number of experiments. For instance, when antibodies to actin or actin-binding protein, fragmin, were injected into nuclei of living amphibian oocytes, morphology of the lampbrush chromosomes was changed and transcription of protein-coding genes of these chromosomes was inhibited (Scheer *et al.*, 1984).

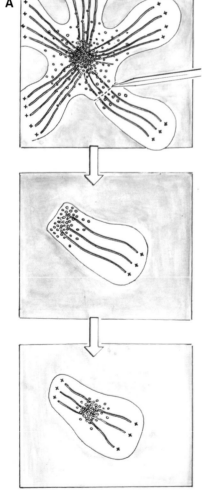

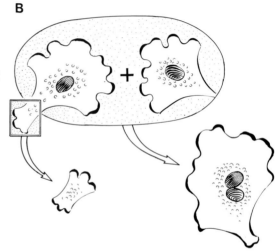

FIG. 5.7. (A) Self-organization of fragment detached from the peripheral cytoplasm of pigment cell. Pigment granules first concentrated near one end of the elongated fragment move toward the new center. A new microtubular system is developed with the microtubules radiating from the new center toward both ends of the fragment. (Scheme based on McNiven *et al.*, 1984.) **(B)** Self-organization of the distribution of cell organelles in cultured fibroblasts. The cytoplasm of mononucleate cells is divided into perinuclear endoplasm with numerous vesicular organelles and lamellar cytoplasm free of these organelles. The cytoplasm of a small fragment cut from the cell periphery is spontaneously divided into the central endoplasm and peripheral lamellar cytoplasm. The multinucleate cell formed from fused mononuclear cells forms united perinuclear endoplasm and peripheral lamellar cytoplasm.

The exact structure and function of "nucleoskeleton" as well as its inter-relationship with cytoplasmic cytoskeletal structures remain to be eluci-dated. The fundamental importance of these problems is obvious.

VII. Conclusion: Self-Integration of Cytoskeleton and of the Cellular Organization

Cytoskeleton is dynamic and anisotropic at several levels of organization: individual fibrils, individual arrays formed by the fibrils, and the integral structure formed by several types of arrays. Anisotropic integral organization of the whole cytoskeleton is manifested by regular distribution of various structures from the cell center to the periphery, between the anterior and posterior parts of the cell, and so forth. As shown in this chapter, positions of cellular organelles and distribution of the plasma membrane components are closely correlated with this anisotropic organization of cytoskeleton.

In particular, many cultured cells have characteristic anisotropic orga-nization, with peripheral actin structures and central radiating microtubular array, with cytoplasm divided into central endoplasm and peripheral ecto-plasmic parts, and with centripetal capping of clustered membrane receptors.

Development of this complex organization is due to intrinsic properties of the system, because even small parts of the cell can restore similar integral organization. In experiments with several types of cells it was shown that each small cytoplasmic fragment, detached mechanically from the whole cell, develops its own center and periphery (Fig. 5.7A,B). In particular, when frag-ments were taken from the peripheral cytoplasm of pigmented cells, mela-nophores, the polarity of remaining distal fragments of microtubules was altered. Immediately after detachment, all the microtubules in the fragment had similar polarity, determined by the position of the previous center. How-ever, several hours later, microtubules located at the two sides of the new center developed opposite polarities. In other words, a new kind of radiating array was developed. Fragments of fibroblasts, like the whole cells, developed actin bundles attached to substrate at the sides of the new center (Fig. 5.7C–E). These reorganizations of cytoskeleton were accompanied by correspond-ing alterations of the position of organelles. Pigment granules in melanophore fragments and cytoplasmic vesicles in the fragments of fibroblasts were col-lected in the new central areas, while new "ectoplasmic" zones free of visible organelles developed at the periphery. Membrane receptors clustered by the corresponding ligand, concanavalin A, were observed to cap toward the new centers of the fragments (Fig. 5.7F).

Another variant of cytoskeleton reorganization is observed when several mononucleate cells are fused into a giant multinucleate structure. This fusion can be induced in culture by certain viruses, e.g., Sendai virus, and by certain chemicals, e.g., polyethylene glycol. All the centrosomes of fused cells migrate toward the new center and aggregate there, forming a new single MTOC, from which large bundles of microtubules radiate toward the periph-ery. Intermediate filaments also radiate from the new center accompanying the microtubules. Actin bundles with the ends attached at the cell periphery

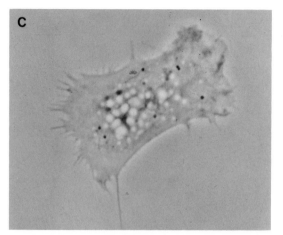

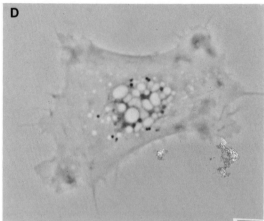

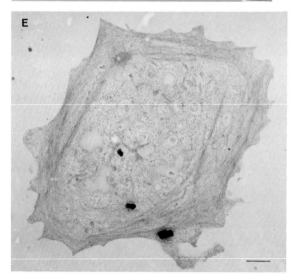

Fig. 5.7. (*Continued*) **(C–E)** Organization of small cytoplasmic fragments cut from the cultured fibroblasts. (C,D) Two phase-contrast photographs of the same fragment made with an interval of 115 min. Accumulation of cytoplasmic vesicles in the center of the fragment: extension and contraction of pseudopods at the edges. Scale bar: 3 μm. (E) Electron micrograph of the fragment sectioned parallel to the substrate. Actin filament bundles at the periphery. Scale bar: 1 μm. (From Albrecht-Buehler, 1980. Courtesy of G. Albrecht-Buehler.)

(F) Capping of concanavalin A receptors on the surface of the cytoplasmic fragment of fibroblast. Methods as in Fig. 5.5C. Notice that receptors are cleared from both ends of the fragment and collected at the surface of its central part. Scale bar: 5 μm. (From Gelfand *et al.*, 1985.)

are formed sometimes; repositioning of cell organelles also occurs. In particular, the nuclei of fused fibroblasts are collected around the new cell center; they are surrounded by a belt of endoplasmic organelles; a circle of thin lamellar cytoplasm is formed at the periphery (Fig. 5.7G–I). Thus, similar reorganization of the cytoskeleton into a single anisotropic system with its own center and periphery occurs in two opposite situations: in miniature fragments cut from the cell and in giant structures obtained by fusion of many single cells. Self-organization of the cytoskeleton in both situations is accompanied by corresponding redistribution of other cell organelles. This remarkable ability for the development of similar patterns of organization in different situations can be designated as self-integration of cytoskeleton. We do not yet know the mechanisms of this self-integration.

It is possible that the contractile properties of actin cortex are important

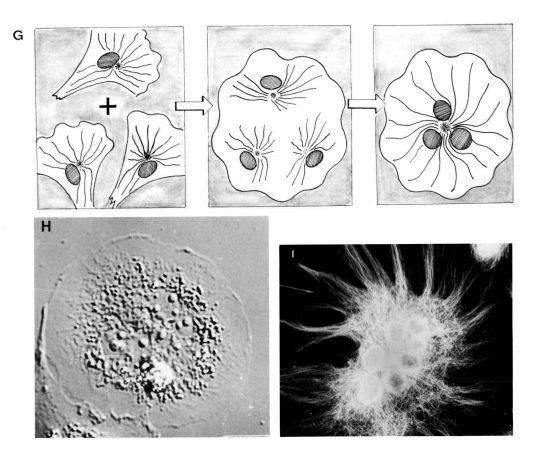

(G–I) Self-organization of multinuclear cells. (G) Formation of a common microtubular system in multinucleate cell obtained by fusion of several mononucleate cells. (H) Three-nuclear cell formed by fusion of three mouse L fibroblasts has united perinuclear endoplasm and lamellar ectoplasm. Phase contrast. Scale-bar: 40μm. (Courtesy of L. A. Lyass.) (I) United system of microtubules radiating to the periphery in multinucleate cell. Immunofluorescence microscopy using antitubulin antibody. Scale bar: 20μm. (From Lyass *et al.*, 1984.)

for the development of anisotropy: each isolated small part of the cortex will contract toward its own center and several fused cortices will also contract toward the single center. These properties of the cortex can be compared with those of a stretched rubber band. Each fragment cut from this band contracts toward its new center, and when several pieces of the band are linked to one another, they have a single common center of contraction. Similar integration can be the result of the presence of any other continuous contractile lattice besides actin cortex, e.g., the hypothetical microtrabecular lattice.

Contraction of actin or of another lattice may determine the position of a new microtubule-organizing center. As suggested by McNiven, Wang, and Porter (1984), contraction leads to accumulation of proteins needed for formation of MTOC at the new center of the fragment. Development of a central radiating microtubular array can then specify the exact positions and pathways of movements of cellular organelles. These suggestions need to be tested.

The phenomenon of self-integration emphasizes the essential role of the cytoskeleton in the structural organization of cells. In this chapter we discussed mainly the "global" aspects of this organization; that is, general distribution of large cellular organelles. However, cytoskeleton can also be essential for the "local" organization of each small area of the cytoplasm. In particular, cytoskeletal fibrils can adsorb at their surfaces certain proteins present in the fluid fraction of the cytoplasm. In this way the organized complexes of these proteins can be formed. In fact, attachment of several glycolytic enzymes to the surface of actin filaments has been observed in various cells. Possibly, a series of glycolytic reactions and of other metabolic processes can proceed in an organized way at the surface of cytoskeletal structures. It was also suggested that diffusing substances as well as cytoplasmic water may be reversibly bound by the surfaces of these structures. In this way the viscosity of the fluid cytoplasm may be controlled. At present, very little is known about these effects, but their role in the regulation of metabolic activities may be significant.

Thus, the data indicate that the cytoskeleton can control all the main aspects of the organization of the cell, especially the cytoplasm. Studies of the mechanisms of this control have just begun.

Literature Cited

Aebi, U., Cohn, J., Buhle, L., and Gerace, L. (1986) The nuclear lamina is a meshwork of inter-mediate-type filaments, *Nature* **323**:560–564.

Albrecht-Buehler, G. (1980) Autonomous movements of cytoplasmic fragments, *Proc. Natl. Acad. Sci. USA* **77**:6639–6644.

Forman, D. S. (1984) Axonal transport of organelles, *Trends Neurosci.* **7**:112–116.

Gelfand, V. I., Glushankova, N. A., Ivanova, O. Y., Mittelman, L. A., Pletyushkina, O. Y., Vasiliev, J. M., and Gelfand, I. M. (1985) Polarization of cytoplasmic fragments microsurgically detached from mouse fibroblasts, *Cell Biol. Intern. Rep.* **9**:883–892.

Heath, J. P. (1983) Direct evidence for microfilament-mediated capping of surface receptors on crawling fibroblasts, *Nature* **302**:532–534.

Herman, B., and Albertini, D. F. (1984) A time-lapse video image intensification analysis of cytoplasmic organelle movements during endosome translocation, *J. Cell Biol.* **98**:565–576.

Ip, W., Murphy, D. B., and Heuser, J. E. (1984) Arrest of pigment granule motion in erythrophores by quick-freezing, *J. Ultrastruct. Res.* **86**:162–175.

Kuznetsov, S. A., and Gelfand, V. I. (1986) Bovine brain kinesin is a microtubule-activated ATPase, *Proc. Natl. Acad. Sci. USA* **83**:8530–8534.

Lyass, L. A., Bershadsky, A. D., Gelfand, V. I., Serpinskaya, A. S., Stavrovskaya, A. A., Vasiliev, J. M., and Gelfand, I. M. (1984) Multinucleation-induced improvement of the spreading of transformed cells on the substratum, *Proc. Natl. Acad. Sci. USA* **81**:3098–3102.

McKeon, F. D., Krischner, M. W., and Caput, D. (1986) Homologies in both primary and secondary structure between nuclear envelope and intermediate filament proteins, *Nature* **319**:463–468.

McNiven, M. A., and Porter, K. R. (1984) Chromatophores—Models for studying cytomatrix translocation, *J. Cell Biol.* **99**:152s–158s.

McNiven, M. A., Wang, M., and Porter, K. R. (1984) Microtubule polarity and the direction of pigment transport reverse simultaneously in surgically severed melanophore arms, *Cell* **37**:753–765.

Murphy, D. B., and Grasser, W. A. (1984) Intermediate filaments in the cytoskeletons of fish chromatophores, *J. Cell Sci.* **66**:353–366.

Scheer, U., Hinssen, H., Franke, W. W., and Jockusch, B. M. (1984) Microinjection of actin-binding proteins and actin antibodies demonstrates involvement of nuclear actin in transcription of lampbrush chromosomes, *Cell* **39**:111–122.

Schliwa, M., Pryzwansky, K. B., and van Blerkom, J. (1982) Implications of cytoskeletal interactions for cellular architecture and behavior, *Phil. Trans. R. Soc. Lond. B.* **299**:199–205.

Tiwari, S. C., Wick, S. M., Williamson, R. E., and Cunning, B. E. S. (1984) Cytoskeleton and integration of cellular function in cells of higher plants, *J. Cell Biol.* **99**:63s–69s.

Travis, J. L., Kenealy, J. F. X., and Allen, R. D. (1983) Studies on the motility of the foraminifera II. The dynamic microtubular cytoskeleton of the reticulopodial network of *Allogromia laticollaris*, *J. Cell Biol.* **97**:1668–1676.

Tsukita, S., and Ishikava, H. (1980) The movement of membranous organelles in axons. Electron microscopic identification of anterogradely and retrogradely transported organelles, *J. Cell Biol.* **84**:513–530.

Vale, R. D., Reese, T. S., and Sheetz, M. P. (1985a) Identification of a novel force-generating protein, kinesin, involved in microtubule-based motility, *Cell* **42**:39–50.

Vale, R. D., Schnapp, B. J., Mitchison, T., Steuer, E., Reese, T. S., and Sheetz, M. P. (1985b) Different axoplasmic proteins generate movement in opposite directions along microtubules *in vitro*, *Cell* **43**:623–632.

Additional Readings

General organization of the cytoskeleton; association between different cytoskeletal elements and between cytoskeleton and cell organelles

Arakawa, T., and Frieden, C. (1984) Interaction of microtubule-associated proteins with actin filaments. Studies using the fluorescence-photobleaching recovery technique, *J. Biol. Chem.* **259**:11730–11734.

Ball, E. H., and Singer, S. J. (1982) Mitochondria are associated with microtubules and not with intermediate filaments in cultured fibroblasts, *Proc. Natl. Acad. Sci USA* **79**:123–126.

Bennet, V. (1984) Brain ankyrin, membrane-associated protein with binding sites for spectrin, tubulin and the cytoplasmic domain of the erythrocyte anion channel, *J. Biol. Chem.* **259**:13550–13559.

Bloom, G. S., and Vallee, R. B. (1983) Association of microtubule-associated protein 2 (MAP2) with microtubules and intermediate filaments in cultured brain cells, *J. Cell Biol.* **96**:1523–1531.

Brinkley, B. R. (1982) Summary: Organization of the cytoplasm, *Cold Spring Harbor Symp. Quant. Biol.* **46**:1029–1040.

Clegg, J. S. (1984) Intracellular water and the cytomatrix: Some methods of study and current views, *J. Cell Biol.* **99**:167s–171s.

Cohen, J., de Loubresse, N. G., and Beisson, J. (1984) Actin microfilaments in *Paramecium*: Localization and role in intracellular movements, *Cell Motil.* **4**:443–468.

Collot, M., Louvard, D., and Singer, S. J. (1984) Association between lysosomes and microtubules in cultured fibroblasts, as studied by double immunofluorescence labelling, *J. Submicrosc. Cytol.* **16**:65–67.

Couchman, J. R., and Rees, D. A. (1982) Organelle–cytoskeleton relationships in fibroblasts: Mitochondria, Golgi apparatus, and endoplasmic reticulum in phases of movement and growth, *Eur. J. Cell Biol.* **27**:47–54.

Geiger, B., and Singer, S. J. (1980) The association of microtubules and intermediate filaments in chicken gizzard cells as detected by double immunofluorescence, *Proc. Natl. Acad. Sci. USA* **77**:4769–4773.

Gershon, N. D., Porter, K. R., and Trus, B. L. (1985) The cytoplasmic matrix: Its volume and surface area and the diffusion of molecules through it, *Proc. Natl. Acad. Sci. USA* **82**:5030–5034.

Geuens, G., De Brabander, M., Nuydens, R., and De Mey, J. (1983) The interaction between microtubules and intermediate filaments in cultured cells treated with taxol and nocodazole, *Cell Biol. Int. Rep.* **7**:35–47.

Heiman, R., Shelanski, M. L., and Liem, R. K. H. (1985) Microtubule-associated proteins bind specifically to the 70-kDa neurofilament protein, *J. Biol. Chem.* **260**:12160–12166.

Hirokawa, N. (1982) Cross-linker system between neurofilaments, microtubules, and membranous organelles in frog axon revealed by quick-freeze, deep-etching method, *J. Cell Biol.* **94**:129–142.

Hirokawa, N., Cheney, R. F., and Willard, M. (1983) Location of a protein of the fodrin-spectrin-TW260/240 family in the mouse intestinal brush border, *Cell* **32**:953–965.

Leterrier, J. F., Liem, R. K. H., and Shelanski, M. L. (1982) Interactions between neurofilaments and microtubule-associated proteins: A possible mechanism for intraorganellar bridging, *J. Cell Biol.* **95**:982–986.

Lloyd, C. W. (1984) Toward a dynamic helical model for the influence of microtubules on wall patterns in plants, *Int. Rev. Cytol.* **86**:1–51.

Mangeat, P. H., and Burridge, K. (1983) Immunoprecipitation of nonerythrocyte spectrin within live cells following microinjection of specific antibodies: Relation to cytoskeletal structures, *J. Cell Biol.* **98**:1363–1377.

Pollard, T. D., Selden, S. C., and Maupin, P. (1984) Interaction of actin filaments with microtubules, *J. Cell Biol.* **99**(1pt2):33s–37s.

Rogalski, A. A., and Singer, S. J. (1984) Association of elements of the Golgi apparatus with microtubules, *J. Cell Biol.* **99**:1092–1100.

Sandoval, I. V., Bonifacino, J. S., Klausner, R. D., Henkart, M., and Wehland, J. (1984) Role of microtubules in the organization and localization of the Golgi apparatus, *J. Cell Biol.* **99**:113s–118s.

Schliwa, M., and van Blerkom, J. (1981) Structural interaction of cytoskeletal components, *J. Cell Biol.* **90**:222–235.

Thyberg, J., and Moskalewski, S. (1985) Microtubules and the organization of the Golgi complex, *Exp. Cell Res.* **159**:1–16.

Tiwari, S. C., Wick, S. M., Williamson, R. E., and Gunning, B. E. S. (1984) The cytoskeleton and integration of cellular function in cells of higher plants, *J. Cell Biol.* **99**(1pt2):63s–69s.

Intracellular transport

Adams, R. J. (1982) Organelle movement in axons depends upon ATP, *Nature* **297**:327–329.

Adams, R. J., and Bray, J. D. (1983) Rapid transport of foreign particles microinjected into crab axons, *Nature* **303**:718–720.

Allen, N. S., and Allen, R. D. (1978) Cytoplasmic streaming in green plants, *Ann. Rev. Biophys. Bioeng.* **7**:497–526.

Allen, R. D. (1985) New observations on cell architecture and dynamics by video-enhanced contrast optical microscopy, *Ann. Rev. Biophys. Biophys. Chem.* **14**:265–290.

Allen, R. D., Weiss, D. G., Hayden, J. H., Brown, D. T., Fujiwake, H., and Simpson, M. (1985) Gliding movement of and bidirectional transport along single native microtubules from squid axoplasm: Evidence for an active role of microtubules in cytoplasmic transport, *J. Cell Biol* **100**:1736–1752.

Bard, F., Bourgeois, C. A., Costagliola, D., and Bouteille, M. (1985) Rotation of the cell nuclei in living cells, a quantitative analysis, *Biol. Cell* **54**:135–142.

Beckerle, M. C. (1984) Microinjected fluorescent polystyrene beads exhibit saltatory motion in tissue culture cells, *J. Cell Biol.* **98**:2126–2132.

Beckerle, M. C., and Porter, K. R. (1983) Analysis of the role of microtubules and actin in erythrophore intracellular motility, *J. Cell Biol.* **96**:354–362.

Brady, S. T., Lasek, R. J., and Allen, R. D. (1982) Fast axonal transport in extruded axoplasm from squid giant axon, *Science* **218**:1129–1131.

Brady, S. T., Lasek, R. J., Allen, R. D., Yin, H. L., and Stossel, T. P. (1984) Gelsolin inhibition of fast axonal transport indicates a requirement for actin microfilaments, *Nature* **310**:56–58.

Brady, S. T., Lasek, R. J., and Allen, R. D. (1985) Video microscopy of fast axonal transport in extruded axoplasm: A new model for study of molecular mechanisms, *Cell Motil.* **5**:81–101.

Clark, T. G., and Rosenbaum, J. L. (1984) Energy requirements for pigment aggregation in *Fundulus* melanophores, *Cell Motil.* **4**:431–441.

De Brabander, M., Geuens, G., Nuydens, R., Moeremans, M., and De Mey, J. (1985) Probing microtubule-dependent intracellular motility with nanometre particle video ultramicroscopy (nanovid ultramicroscopy), *Cytobios* **43**:273–283.

Forman, D. S., Brown, K. J., and Promersberger, M. E. (1983) Selective inhibition of retrograde axonal transport by erythro-9-3(2-hydroxynonyl) adenine, *Brain Res.* **272**:194–197.

Freed, J. J., and Lebowitz, M. M. (1970) The association of a class of saltatory movements with microtubules in cultured cells, *J. Cell Biol.* **45**:334–354.

Gilbert, S. P., Allen, R. D., and Sloboda, R. D. (1985) Translocation of vesicles from squid axoplasm on flagellar microtubules, *Nature* **315**:245–248.

Hayden, J. H., and Allen, R. D. (1984) Detection of single microtubules in living cells: Particle transport can occur in both directions along the same microtubule, *J. Cell Biol.* **99**:1785–1793.

Hayden, J. H., Allen, R. D., and Goldman, R. D. (1983) Cytoplasmic transport in keratocytes: Direct visualization of particle translocation along microtubules, *Cell Motil.* **3**:1–19.

Herman, B., and Albertini, D. F. (1984) A time-lapse video image intensification analysis of cytoplasmic organelle movements during endosome translocation, *J. Cell Biol.* **98**:565–575.

Koonce, M. P., and Schliwa, M. (1985) Bidirectional organelle transport can occur in cell processes that contain single microtubules, *J. Cell Biol.* **100**:322–326.

Lasek, R. J., Garner, J. A., and Brady, S. T. (1984) Axonal transport of the cytoplasmic matrix, *J. Cell Biol.* **99**(1pt2):212s–221s.

Martz, D., Lasek, R. J., Brady, S. T., and Allen, R. D. (1984) Mitochondrial motility in axons: membranous organelles may interact with the force generating system through multiple surface binding sites, *Cell Motil.* **4**:89–101.

Nothnagel, E. A., Barak, L. S., Sanger, J. W., and Webb, W. W. (1981) Fluorescence studies on modes of cytochalasin B and phallotoxin action on cytoplasmic streaming in *Chara*, *J. Cell Biol.* **88**:364–372.

Schliwa, M. (1984) Mechanisms of intracellular organelle transport, in *Cell and Muscle Motility*, Vol. 5 (J. W. Shay, ed.). Plenum Press, New York, pp. 1–80.

Stearns, M. E., and Ochs, R. L. (1982) A functional *in vitro* model for studies of intracellular motility in digitonin-permeabilized erythrophores, *J. Cell Biol.* **94**:727–739.

Cytoskeleton and membrane; capping and related phenomena

Abercrombie, M., Heaysman, J. E. M., and Pegrum, S. M. (1972) The locomotion of fibroblasts in culture. V. Surface marking with concanavalin A, *Exp. Cell Res.* **73**:536–539.

Ben-Ze'ev, A., Duerr, A., Solomon, F., and Penman, S. (1979) The outer boundary of the cytoskeleton: A lamina derived from plasma membrane proteins, *Cell*, **17**:859–865.

Bourguignon, L. Y. W., and Bourguignon, G. L. (1984) Capping and the cytoskeleton, *Int. Rev. Cytol.* **87**:195–224.

Bowser, S. S., and Bloodgood, R. A. (1984) Evidence against surf-riding as a general mechanism for surface motility, *Cell Motil.* **4**:305–314.

Bowser, S. C., and Rieder, C. L. (1985) Evidence that cell surface motility in *Allogromia* is mediated by cytoplasmic microtubules, *Can. J. Biochem. Cell Biol.* **63**:608–620.

Boyles, J., and Baiton, D. F. (1981) Changes in plasma-membrane-associated filaments during endocytosis and exocytosis in polymorphonuclear leukocytes, *Cell* **24**:905–914.

Bretscher, M. S. (1976) Directed lipid flow in cell membranes, *Nature* **260**:21–23.

Carboni, J. M., and Condeelis, J. S. (1985) Ligand-induced changes in the location of actin, myosin, 95K (α-actinin), and 120K protein in amebae of *Dictyostelium discoideum*, *J. Cell Biol.* **100**:1884–1893.

Condeelis, J., (1979) Isolation of concanavalin A caps during various stages of formation and their association with actin and myosin, *J. Cell Biol.* **80**:751–758.

Coudrier, E., Reggio, H., and Louvard, D. (1983) Characterization of an integral membrane glycoprotein associated with the microfilaments of pig intestinal microvilli, *EMBO (Eur. Mol. Biol. Organ.) J.* **2**:469–475.

Dellagi, K., and Brouet, J. C. (1982) Redistribution of intermediate filaments during capping of lymphocyte surface molecules, *Nature* **298**:284–286.

Dembo, M., and Harris, A. K. (1981) Motion of particles adhering to the leading lamella of crawling cells, *J. Cell Biol.* **91**:528–536.

Dentler, W. L., Pratt, M. M., and Stephens, R. E. (1980) Microtubule-membrane interactions in cilia. II. Photochemical cross-linking of bridge structures and the identification of a membrane associated dynein-like ATPase, *J. Cell Biol.* **84**:381–403.

De Petris, S. (1977) Distribution and mobility of plasma membranae components on lymphocytes, in *Dynamic Aspects of Cell Surface Organization* (G. Poste and G. L. Nicolson, eds.), Elsevier/North-Holland Biochemical Press, Amsterdam, pp. 643–728.

Geiger, B. (1983) Membrane-cytoskeleton interaction, *Biochimi. Biophys. Acta* **737**:305–341.

Geiger, B. Z., Avnur, J. E., Kreis, T. E., and Schlessinger, J. (1984) The dynamics of cytoskeletal organization in areas of cell contact, in *Cell and Muscle Motility*, Vol. 5 (J. W. Shag, ed.), Plenum Press, New York, pp. 195–235.

Georgatos, S. D., and Marchesi, V. T. (1985) The binding of vimentin to human erythrocyte membranes: A model system for the study of intermediate filament–membrane interactions, *J. Cell Biol.* **100**:1955–1961.

Goodloe-Holland, C. M., and Luna, E. J. (1984) A membrane cytoskeleton from *Dictyostelium discoideum*. III. Plasma membrane fragments bind predominantly to the sides of actin filaments, *J. Cell Biol.* **99**:71–78.

Harris, A. K. (1976) Recycling of dissolved plasma membrane components as an explanation of the capping phenomenon, *Nature* **263**:781–783.

Heath, J. P. (1983) Direct evidence for microfilament-mediated capping of surface receptors on crawling fibroblasts, *Nature* **302**:532–534.

Hewitt, J. A. (1979) Surf-riding model for cell capping, *J. Theor. Biol.* **80**:115–127.

Jacobson, B. S. (1983) Interaction of the plasma membrane with the cytoskeleton: An overview, *Tissue Cell* **15**:829–852.

Laub, F., Kaplan, M., and Gitler, C. (1981) Actin polymerization accompanies thy-1-capping on mouse thymocytes, *FEBS Lett.* **124**:35–38.

Lehto, B-P., Vartio, T., Badley, R. A., and Virtanen, I., (1983) Characterization of a detergent resistant surface lamina in cultured human fibroblasts, *Exp. Cell Res.* **143**:287–294.

Levine, J., and Willard, M. (1983) Redistribution of fodrin accompanying capping of cell surface molecules, *Proc. Natl. Acad. Sci. USA* **80**:191–195.

Mangeat, P., and Burridge, K. (1984) Actin–membrane interaction in fibroblast: What proteins are involved in this association? *J. Cell Biol.* **99**:95s–103s.

Moran, D. T., Varela, F. J., and Rowley, J. C. (1977) Evidence for active role of cilia in sensory transduction, *Proc. Natl. Acad. Sci. USA* **74**:793–797.

Oliver, J. M., and Berlin, R. D. (1982) Distribution of receptors and functions on cell surfaces: Quantitation of ligand-receptor mobility and a new model for the control of plasma membrane topography, *Phil. Trans. R. Soc. Lond. B* **299**:215–235.

Rogalski, A. A., Bergman, J. E., and Singer, S. J. (1984) Effect of microtubule assembly status on the intracellular processing and surface expression of an integral protein of the plasma membrane, *J. Cell Biol.* **99**:1101–1109.

Rogers, K. A., Khoshbaf, M. A., and Brown, D. L. (1981) Relationship of microtubule organization in lymphocytes to the capping of immunoglobulin, *Eur. J. Cell Mol.* **24**:1–8.

Schlessinger, J. (1983) Mobilities of cell-membrane proteins: How are they modulated by the cytoskeleton? *Trends Neurosci. (TINS)* **6**:360–363.

Singer, S. J., Ash, J. F., Bourguignon, L. Y. W., Heygeness, M. H., and Louvard, D. (1978) Trans-membrane interactions and the mechanisms of transport of proteins across membranes, *J. Supramol. Struct.* **9**:373–389.

Smith, B. A., Clark, W. R., and McConnell, H. M. (1979) Anisotropic molecular motion of cell surfaces, *Proc. Natl. Acad. Sci. USA* **76**:5641–5644.

Spiegel, S., Kassis, S., Wilchek, M., and Fishman, P. H., (1984) Direct visualization of redistribution and capping of fluorescent gangliosides on lymphocytes, *J. Cell Biol.* **99**:1575–1581.

Vasiliev, J. M., Gelfand, I. M., Domnina, L. V., Dorfman, N. A., and Pletyushkina, O. Y. (1976) Active cell edge and movements of concanavalin A receptors of the surface of epithelial and fibroblastic cells, *Proc. Natl. Acad. Sci. USA* **73**:4085–4089.

Multinuclear cells and cell fragments

Albrecht-Buehler, C. (1982) Does blebbing reveal the convulsive flow of liquid and solutes through the cytoplasmic meshwork? *Cold Spring Harbor Symp. Quant. Biol.* **46**:45–50.

Cain, H., Kraus, B., Fringes, B., Osborn, M., and Weber, K. (1981) Centrioles, microtubules and microfilaments in activated mononuclear and multinucleate macrophages from rat peritoneum: Electron-microscopic and immunofluorescence microscopic studies, *J. Pathol.* **133**:301–323.

Euteneuer, U., and Schliwa, M. (1984) Persistent, directional motility of cells and cytoplasmic fragments in the absence of microtubules, *Nature* **310**:58–60.

Holmes, K., and Coppin, P. W. (1968) On the role of microtubules in movement and alignment of nuclei in virus-induced syncytia, *J. Cell Biol.* **51**:752–762.

Keller, H. U., and Bessis, M. (1975) Migration and chemotaxis of anucleate cytoplasmic leukocyte fragments, *Nature* **258**:723–724.

Shaw, J., and Bray, D. (1977) Movement and extension of isolated growth cones, *Exp. Cell Res.* **104**:55–62.

Wang, E., Cross, R. H., and Choppin, P. W. (1979) Involvement of microtubules and 10 nm filaments in the movement and positioning of nuclei in syncytia, *J. Cell Biol.* **83**:320–327.

Wang, E., Roos, D. S., Hegeness, N. H., and Choppin, P. W. (1982) Function of cytoplasmic fibers in syncytia, *Cold Spring Harbor Symposia Quant. Biol.* **46**:997–1012.

6

Regulation of Synthesis of Cytoskeletal Proteins

I. Introduction

As discussed in Chapter 5, the cytoskeleton regulates the organization of cytoplasm. At the same time, the state of the cytoskeleton is continuously regulated by a complex of controlling cellular mechanisms. These regulations are essential both for maintenance of the steady state of the cytoskeleton and for reorganization of the cytoskeleton which can occur in the course of cell life. Many cells, e.g., locomoting pseudopod-forming cells, continuously reorganize their cytoskeleton, so that it is not always possible to draw a sharp boundary between regulations of steady state and of reorganization. Regulation of the state of the cytoskeleton involves two processes: (1) regulation of the expression of genes coding cytoskeletal proteins and of the synthesis of these proteins, and (2) regulations of the state of assembly of the cytoskeletal proteins and of the distribution of cytoskeletal elements.

The first group of regulations will be discussed in this chapter, the second group in Chapters 7 and 8.

Expression of the main cytoskeletal proteins is of vital importance for the eukaryotic cells. This conclusion is strongly supported by recent experiments in which yeast cells with inactivated actin genes were constructed (Fig. 6.1). This construction involved subcloning of a fragment of actin gene on a vector plasmid. This plasmid was then intergrated into the normal actin locus of the yeast genome. The integration caused disruption of this locus. Disruption of a single functioning actin locus was lethal for a haploid yeast. This mutation was recessive; that is, diploid cells containing one normal and one disrupted actin gene were viable. A similar technique was used for disruption of the β-tubulin gene, and the results obtained were also similar. Thus, functioning actin and tubulin genes are essential for the growth of yeast cells. Further studies of cells with mutant actin and tubulin genes may elucidate the exact role of these proteins in various cellular processes. The methods of selective gene alteration may also be very useful for investigation of the function of other cytoskeletal proteins.

Each type and class of cells has not only the characteristic architecture of the cytoskeleton, but also the characteristic spectrum of active cytoskeletal genes and of corresponding proteins. Accordingly, each alteration of cell differentiation is accompanied by alteration of the synthesis of cytoskeletal proteins. These alterations of synthesis are usually due to the changes of expression of corresponding genes. They are manifested by alteration of the quantities of specific mRNAs, which may be detected by various methods, especially hybridization of cellular RNAs with the labeled cloned DNA of corresponding genes. In some cases, alteration of the transcription of genes can be combined with alteration in the processing of transcribed RNA. Posttranslational alterations can further increase the diversity of the molecular forms of synthesized proteins. In the following sections we shall describe several examples of regulated alteration of the synthesis of cytoskeletal proteins in various experimental systems.

II. Flagellar Regeneration in *Chlamydomonas*: Induction of Synthesis of Cytoskeletal Protein

In *Chlamydomonas*, mechanical detachment of the flagella is followed by their rapid regeneration. This process involves a large increase in the synthesis of flagellar proteins preceded by increased accumulation of their mRNAs. Flagella are assembled within 60–90 min after the amputation, and then the synthesis of flagellar-specific proteins returns to the initial level.

Deflagellation causes activation of the expression of the genes of α- and

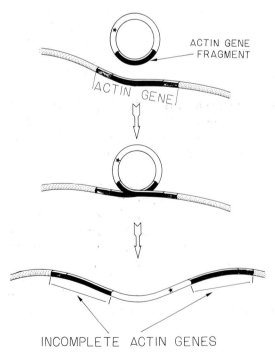

FIG. 6.1. Stages of construction of yeast mutant with disrupted actin gene. From top to bottom: fragment of actin gene subcloned into plasmid vector containing a selectable marker gene (indicated by star); integration of plasmid into the yeast genome by homologous recombination between the actin gene of the yeast and actin sequence found on the plasmid; and disruption of the actin gene by plasmid integration and formation of two incomplete actin sequences in the yeast genome.

β-tubulin. The rate of synthesis of the RNAs coded by these genes reaches its peak within 10–15 min and then declines more slowly to the initial value.

More tubulin mRNA accumulates during regeneration than can be attributed to new synthesis, suggesting that this RNA becomes considerably more stable than in control nonregenerating cells. In contrast, at the later stages the stability of mRNA is decreased, leading to its rapid decay.

The complex process of flagellar regeneration may involve not only activation of tubulin synthesis, but also induction of partial resorption of the remaining flagellum (see Chapter 2). The mechanisms of this resorption are not clear.

As described in Chapter 2, α-tubulin in the flagella of *Chlamydomonas*, in contrast to cytoplasmic tubulin, is acetylated. When resorption of the flagella was chemically induced (by dibucaine), this posttranslational modification was reversed; that is, acetylated α-tubulin was changed back to its unmodified precursor. Possibly, addition and removal of the acetyl group is involved in control of disassembly and assembly of flagella.

Thus, amputation of flagella induces a strictly programed complex of alterations, including transient increase in the expression of many genes, biphasic changes of mRNA stability, and possibly changes of mechanisms performing posttranslational modification.

As yet we do not know what signal from the stump of the amputated flagellum switches on this program and how its course is then controlled.

III. Differentiation of *Naegleria:* Alternative Expression of Actin and Tubulin Genes

Regeneration of flagella is an example of the regulatory program that restores the preexisting structure of cytoskeleton. Other processes lead not to restoration, but to radical reorganization of the cytoskeleton. In particular, certain differentiation events are accompanied by radical alterations in the expression of actin and tubulin genes. A remarkable example of alternative expression of these two groups of genes has been observed in experiments with the protist *Naegleria gruberi*. This organism grows and divides as an ameba in a nutrient-rich environment. When transferred into nutrient-free aqueous medium, amebae undergo reversible conversion into flagellates. Full conversion takes less than 2 hr. This conversion seems to be an adaptive reaction to a "poor" environment. The flagellate can more easily find the new convenient environment: it swims about 100 times as fast as the ameba walks.

Differentiation of cells with actin-based motility into cells with tubulin-based motility is correlated with alteration of synthesis of corresponding proteins and of the abundance of their RNA (Fig. 6.2). For example, actin mRNA decays rapidly during differentiation with a half-life of 20 min, while α-tubulin mRNA increases about 80-fold in less than 50 min and then declines rapidly, with a half-life of 8 min. Transformation from ameba to flagellate, and vice versa, may be accompanied by altered expressions not only of actins and tubulins, but also of many other proteins participating in the formation of the actin cortex or of the flagella; these alterations remain to be studied. Somewhat similar transformations of pseudopodial cells into flagellate cells can occur during many differentiation processes in higher organisms, e.g., during spermatogenesis.

IV. Differential Expressions of Genes in Multigene Families Coding Tubulins, Actins, and Myosins

Each major cytoskeletal protein in higher eukaryotes is coded by a family of genes. As we have just seen, differentiation can radically change the ratio of expression of actin and tubulin gene families. Differentiation can also induce more subtle changes of expression. It can switch different genes of the same family on or off so that various cells will express various isotypes of tubulin, actin, or myosin. For example, several genes coding various isotypes of β-tubulins were found in the genomes of rats, chickens, and *Drosophila* (see Chapter 2). Usually several isoforms of this protein and of corresponding mRNAs are present in each cell, but their relative abundance varies from one cell type to another. In a few cases the switch from one β-tubulin isotype to

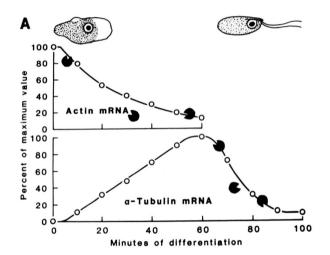

FIG. 6.2. (A) Alterations of the amounts of actin and tubulin mRNAs in *Naegleria* in the course of its differentiation from the ameba into the flagellate. mRNAs were hybridized with ^{32}P-labeled clones of actin or tubulin cDNAs, and the amount of radioactivity was measured. (From Lai *et al.*, 1984. Courtesy of C. Fulton.)

(B) Differences of β-tubulins from the brain and erythrocytes. Peptide maps of α- and β-tubulins from chicken brain (B) and chicken erythrocytes (E). Peptides were generated by cleavage of proteins with α-chymotrypsin. Electrophoresis was performed horizontally with the cathode at the right. Ascending chromatography was performed vertically with butanol pyridine acetic acid. Sketches compare the relative positions of peptides in the maps of α- and β-tubulins from the brain (dotted lines) and the erythrocytes (solid lines). Notice the identity of maps of α-subunits and significant differences in the maps of β-subunits. Only one-third of the 15 major peptides are common to both β-tubulin species. (From Murphy and Wallis, 1983. Courtesy of D. B. Murphy.)

(C) Differential RNA transcription from a single gene and splicing of primary transcript leading to production of two mRNAs for the fast skeletal muscle myosin light chains A1 and A2. (Scheme based on Periasamy *et al.*, 1984.)

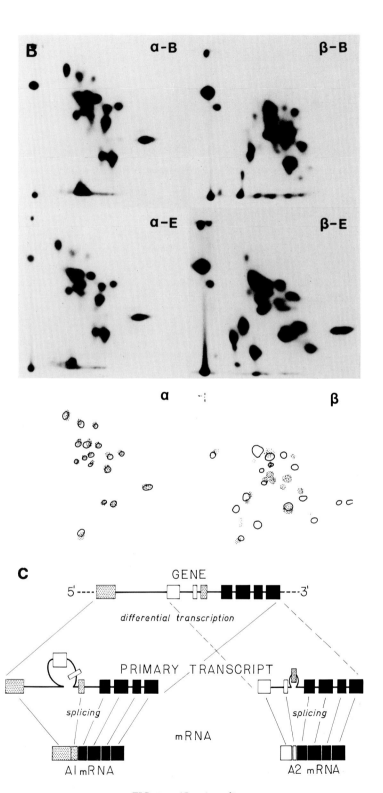

FIG. 6.2. (*Continued*)

another was found to be correlated with specific differentiation. For instance, the unique β-tubulin is contained in the marginal band of microtubules of chick erythrocytes (Fig. 6.2B). This β-tubulin isotype is not present in the erythroblast precursors, which contain another variant of this protein. Thus, the switch in the expression of tubulins occurs during erythroid differentiation.

Elegant evidence showing tissue-specific function of a particular β-tubulin was obtained in experiments with *Drosophila*. One specific isotype of β-2-tubulin is expressed only in the testis. Genetic analysis has shown that mutations of the β-2-tubulin gene are phenotypically manifested by male sterility. Some of these mutations lead to a loss of many functions associated with spermatogenesis: meiotic cell division, alterations of nuclear shape, and assembly of the spermatid axoneme. These data indicate that the mutant β-tubulin is an essential component of at least three types of microtubular structures: the spindle, the flagellum, and the cytoplasmic array. It is possible that the program of differentiation of the testis cells includes the switch to synthesis of a new tubulin. Once expressed, this variant of tubulin is used for many purposes.

Expression of microtubule-associated proteins also undergoes controlled alteration during development. For example, distinct spectra of proteins of the tau group were expressed in the brains of newborn and of 12-day-old rats; correlated alterations of RNAs for these proteins were also observed. Various isoforms of actin and myosins are preferentially expressed in different tissues (see Chapter 1). Close correlation was found between the relative abundance of α- and β-isotypes of myosin heavy chains and of corresponding mRNAs in cardiac ventricular muscle at various stages of development. Here again, the ratio of protein isotypes seems to be determined by the relative expression of corresponding genes. Another mechanism regulating the synthesis of certain isoforms of cytoskeletal proteins is differential processing of RNAs. For example, this type of regulation controls the synthesis of myosin light chains. One class of these chains, the alkali light chain, has two isoforms (A1 and A2) in the fast skeletal muscle. In rabbit skeletal muscle, A1 (190 amino acids long) and A2 (149 amino acids long) have complete sequence homology for the last 141 amino acids at the carboxy terminus. The two proteins differ, however, in length and amino acid sequences at their amino termini. These two proteins were found to be encoded by a single gene, but are translated from different RNAs (Fig. 6.2C). It was shown that transcription of two RNAs starts at two different points of the single gene; later, these two RNAs undergo different forms of splicing; that is, various parts of the primary transcripts are cut out. In this way, two different mature RNAs are generated by the same gene.

A somewhat different variant of alternative splicing leads to generation of multiple isoforms of troponin T in the rat skeletal muscle. The mRNAs for many isoforms of fast skeletal muscle troponin T are identical, with the exception of the small internal regions. The gene of troponin has several miniexons encoding these regions. In the course of splicing, some of these internal regions are removed from the primary transcript, so that many types of mature mRNA are formed.

There is also evidence that various tissue-specific isoforms of tropomyosins can be formed by differential splicing. Thus, this form of regulation seems to play an important role in generation of developmentally determined variability of many cytoskeletal proteins.

V. Regulation of the Synthesis of Intermediate Filament Proteins in Cell Cultures: The Role of Cell Shape and Cell–Cell Interactions

Alterations of the synthesis of proteins of intermediate filaments during animal development are the most obvious shifts of the specificity of cytoskeletal components correlated with differentiation. Several examples of these alterations were described in Chapter 4. To further analyze the mechanisms of these shifts, it is essential to reproduce them in a relatively simple experimental system. The cultures of epithelial cells provide such a system. Many types of epithelial cells brought into culture continue to express keratins but in addition begin to synthesize vimentin. This change of expression of vimentin seems to be induced by attachment of cultured cells to artificial substrates such as glass or plastics. In fact, vimentin synthesis was drastically decreased when the cultured cells were detached from the substrate and grown in suspension; the amount of vimentin in these cells was rapidly decreased (Fig. 6.3). This synthesis was restored again when the cells were allowed to attach to the substrate. The synthesis of keratin in these cultures, in contrast to that of vimentin, was similar in substrate-attached and in suspended cultures. At the same time, the synthesis of keratin increased in the dense culture, where each cell had maximal numbers of cell–cell contacts. When the cells from dense cultures were transferred into a culture dish with large areas of free substrate, the synthesis of keratin rapidly decreased in these sparse cultures. Thus, the synthesis of vimentin seems to be regulated mainly by cell–substrate contacts; the synthesis of keratin, by cell–cell contacts. This control may have important functional significance, because keratin fibrils are involved in the formation of specialized cell–cell contact structures, desmosomes. Exprimentally induced alterations of the ratio between vimentin and keratin synthesis are correlated with alterations of the levels of corresponding RNAs. Thus, shifts of protein synthesis are likely to be due to changes of gene expression. Cell–cell and cell–substrate contacts may change the state of the plasma membrane and of membrane–cytoskeleton relationships. It is important to determine how these changes can induce specific shifts of gene expression and of the synthesis of cytoskeletal proteins (see also Chapter 9).

VI. Unpolymerized Tubulin Modulates the Level of Tubulin Synthesis: Specific Feedback Control of the Synthesis

As shown in the previous sections, alterations of the synthesis of cytoskeletal proteins can be induced by factors external to the cell. However, the cell may also have internal feedback mechanisms adjusting the level of syn-

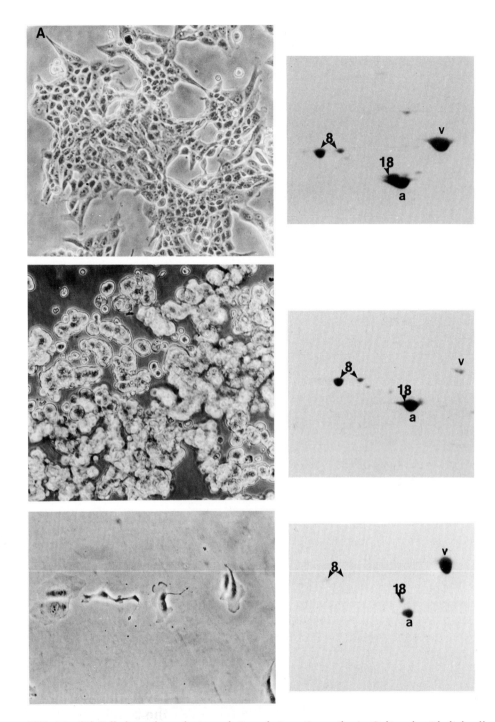

FIG. 6.3. (A) Cell-shape-dependent regulation of vimentin synthesis. Cultured epithelial cells (MDBK line) were grown in monolayer (upper row) or in suspension (middle row); sparse culture of MDBK (lower row). Left: Photomicrographs showing morphologies of the cultures at phase contrast. The cultures were labeled with [^{35}S]methionine; Triton-treated cytoskeletons were prepared and analyzed by two-dimensional gel electrophoresis. Right: Autoradiographs of these gels. a, Actin; v, vimentin; 8, 18, two types of cytokeratins. The amount of labeled vimentin is decreased in suspension culture. (From Ben-Ze'ev, 1984. Courtesy of A. Ben-Ze'ev.)

thesis of specific cytoskeletal protein to the level of unpolymerized molecules of this protein present in the cytoplasm. In fact, the rate of tubulin synthesis was shown to be decreased by an increased level of unpolymerized tubulin. This suppression was demonstrated in two different types of experiments. First, it was shown that colchicine induces rapid cessation of tubulin synthesis in cultured cells. In contrast, vinblastin does not inhibit this synthesis. Colchicine depolymerizes microtubules and raises the level of monomeric

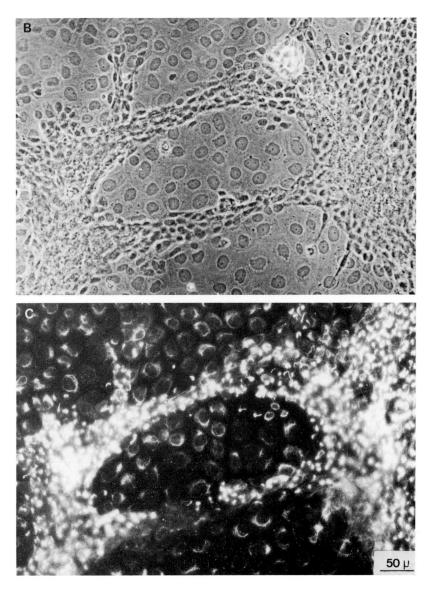

(B,C) Epitheliocytes in densely packed areas of the culture contain more keratin. Fourteen-day-old culture of rat liver-derived cells of IARC-7 line. Phase contrast (B) and immunofluorescence using antibody against keratin with mol. wt. 55 kD. (Courtesy of G. A. Bannikov and S. M. Troyanovsky.)

tubulin in the cultured cells. Vinblastin also depolymerizes microtubules, but in addition it causes aggregation of tubulin into paracrystals (see Chapter 2); therefore, vinblastin, unlike colchicine, does not increase the level of monomeric tubulin in the cell.

Second, it was shown that microinjection of unpolymerized tubulin into the cell rapidly and specifically suppresses tubulin synthesis. The amount of injected tubulin in these experiments was roughly equivalent to 25–50% of the amount initially present in the cell. Suppression of tubulin synthesis is mediated by alterations of the level of tubulin mRNA. Decreased levels of these mRNAs were found in colchicine-treated cells.

Enucleated cells (cytoplasts) respond to drug-induced microtubule depolymerization in the same way as whole cells; that is, by inhibiting tubulin synthesis and by decreasing the levels of tubulin mRNAs. This result eliminates transcription, processing, and transport of these RNAs from the nucleus as sites of regulation. Possibly, nonpolymerized tubulin somehow decreases the stability of its mRNAs, which in turn decreases tubulin synthesis.

Thus, the cell has an autoregulatory cytoplasmic mechanism that determines tubulin content. It is important to determine whether similar mechanisms control the levels of other cytoskeletal proteins, especially actin.

VII. Incorporation of Synthesized Proteins into Cytoskeletal Structures

Newly synthesized cytoskeletal molecules first enter the soluble monomer pools in the cytoplasm and later are incorporated into polymeric structures from these pools. Actin and tubulin pools are equilibrated with corresponding polymers, so the same monomer molecule can participate in many cycles of assembly and disassembly. An alternative suggestion is that some newly synthesized molecules can bypass the soluble pool and be directly incorporated into cytoskeletal fibrils. This suggestion is supported by experiments of Fulton and Wan (1983) in which detergent-extracted cytoskeletons were prepared from cultured cells and translation of proteins from the ribosomes associated with these cytoskeletons was examined. It was found that these ribosomes preferentially translate cytoskeletal proteins. Direct proof of "cotranslational" incorporation of monomers into the polymeric structures at the site of synthesis is still to be obtained. In any case, even the monomers directly incorporated into the fibrils have a good chance of entering the soluble pool later, as many tubulin and actin structures undergo a rapid exchange of molecules with this pool (see Chapters 1 and 2).

Certain specialized cytoskeletal structures are irreversibly assembled from the newly synthesized molecules. This type of assembly may be characteristic of the development of certain finally differentiated cell types. In particular, strictly ordered coupling of the synthesis and irreversible assembly of a series of cytoskeletal proteins was observed by Lazarides (1984) in differentiating mammalian red blood cells. In this system the membrane band III protein anchored to the cytoskeleton is first synthesized and incorporated into the plasma membrane. Then the submembranous cytoskeletal proteins ankyrin and spectrin are formed and bound to this membrane protein. Spectrin and ankyrin are synthesized in excess of the amount assembled. These data suggest that the network is assembled step by step. Band III molecule may serve as the binding site for ankyrin; availability of membrane-bound ankyrin may determine the amount of assembled β-spectrin, and so forth. Excess amounts of unassembled cytoskeletal molecules may be destroyed by proteases, while the assembly stabilizes the constituent proteins against catabolism. In this way the cytoskeleton with precise ratios of components can be formed irreversibly from a soluble pool, in which the ratio of individual components varies considerably. Possibly, similar mechanisms regulate the formation of other permanent cytoskeletal structures, e.g., of keratin network in keratinized cells. Stabilization of proteins against catabolism by the assembly can be a general principle related to many types of cytoskeletal molecules.

VIII. Conclusion

Two large groups of regulatory alterations of cytoskeleton syntheses can be distinguished: alterations that are part of the complex program of reorganization of synthesis and specific alterations of synthesis of individual proteins. Transformations of actin and tubulin in *Naegleria*, specializations of erythroid cells, and development of tissue-specific spectra of proteins are examples of changes of the first type associated with cell differentiation. Each differentiation leads, of course, not only to cytoskeletal changes, but to alterations of synthesis of many other proteins.

Modulations of the synthesis of cytoskeletal protein can be associated not only with differentiations, but also with cell growth. For instance, induction of proliferation in cell cultures, among other protein syntheses, also stimulates the synthesis of major cytoskeletal proteins. In particular, actin synthesis is induced almost immediately after addition of mitogens. Pronounced cyclical alterations of tubulin gene expression were found in the plasmodium of the slime mold *Physarum*. This organism has no cytoplasmic microtubules during the interphase. Tubulins in this organism are synthesized during the premitotic G2 period and used to construct the mitotic spindle. The level of tubulin mRNA rises during that period and mitosis and falls rapidly after metaphase.

Still another group of general reprogrammings of synthesis involves cell reactions to various damaging agents. We have discussed earlier one reaction of this group, regeneration of flagella after mechanical deflagellation. Alterations of cytoskeletal structures and possibly of the synthesis of cytoskeletal proteins may be important components of many other cellular reactions to various stresses. For instance, deflagellation in *Chlamydomonas* can be induced by high and low temperature, by increase or decrease of pH of the medium, by ethyl alcohol, and so forth; restoration of flagella takes place after cessation of the action of these agents. Heat shock can induce the collapse of intermediate filaments and loss of the actin bundles in cultured animal cells. These cytoskeletal alterations in response to injury have not yet been studied in detail.

Mechanisms of all global reprogrammings of the cells are still far from clear. In most cases we know the factors initiating the reprogrammings, e.g., alteration of the composition of medium in *Naegleria*, alteration of the cell–cell and cell–substrate contacts in cultured cells, and action of special regulating molecules, such as the proliferation-inducing hormones. In certain systems we also know something about the final stages of reprogramming, especially about alterations of transcription and processing of RNAs. The molecular mechanisms of most intermediate steps of these alterations are in most cases obscure (see further discussion in Chapter 9).

Besides global reprogramming of synthesis, there may also be special mechanisms regulating the synthesis of individual cytoskeletal proteins and controlling the amount of these proteins in the cell. At present we know well one example of this type of regulation: inhibition of tubulin synthesis by unpolymerized tubulin. Another interesting phenomenon was discovered by Giudice and Fuchs (1987), who transfected epidermal keratin type II gene into fibroblasts. Expression of this foreign gene triggered the expression of endogenous type I epidermal keratin gene. Thus, accumulation of one unpolymerized protein may control the synthesis of another protein. Further studies of these mechanisms are obviously necessary.

The degree of polymerization of various cytoskeletal structures is altered considerably during morphogenetic reorganization, such as cell spreading and locomotion, phagocytosis, and growth of cytoplasmic processes (see Chapter 7). Alteration of the pools of soluble proteins caused by these reorganizations may affect the corresponding synthesis. Feedback regulation of this type may be involved, for instance, in the activation of synthesis of microtubular proteins in neurons during the growth of axons (see Chapter 7). Feedback regulation of the synthesis of individual proteins may be incorporated into the general reorganization of synthesis discussed earlier. To summarize, these regulations may be one method by which cytoskeletal organization controls the synthetic activities of the cell (see Chapter 9). At present, however, all these suggestions remain hypothetical.

Literature Cited

Ben-Ze'ev, A. (1984) Cell–cell interaction and cell–shape-related control of intermediate filament protein syntheses, in *Molecular Biology of Cytoskeleton* (G. G. Borisy, D. W. Cleveland, and D. B. Murphy, eds.), Cold Spring Harbor Laboratory, Cold Spring Harbor, New York, pp. 435–444.

Fulton, A. B., and Wan, K. M. (1983) Many cytoskeletal proteins associated with the HeLa cytoskeleton during translation *in vitro*, *Cell* **32:**619–625.

Giudice, G. J., and Fuchs, E. (1987) The transfection of epidermal keratin genes into fibroblasts and simple epithelial cells: Evidence for inducing a type I keratin by a type II gene, *Cell* **48:**453–463.

Lai, E. I., Remillarg, S. P., and Fulton, C. (1984) Tubulin and actin: Yin–yang gene expression during *Naegleria* differentiation, in *Molecular Biology of the Cytoskeleton* (G. G. Borisy, D. W. Cleveland, and D. B. Murphy, eds.), Cold Spring Harbor Laboratory, Cold Spring Harbor, New York, pp. 257–266.

Lazarides, E. (1984) Assembly and morphogenesis of the avian erythrocyte cytoskeleton, in *Molecular Biology of the Cytoskeleton* (G. G. Borisy, D. W. Cleveland, and D. B. Murphy, eds.), Cold Spring Harbor Laboratory, Cold Spring Harbor, New York, pp. 131–150.

Murphy, D. B., and Wallis, K. T. (1983) Brain and erythrocyte microtubules from chicken contain different β-tubulin polypeptides, *J. Biol. Chem.* **258**:7870–7875.

Periasamy, M., Strehler, E. E., Garfinkel, L. I., Gubits, R. M., Ruiz-Opazo, N., and Nadal-Ginard, B. (1984) Fast skeletal muscle myosin light chains 1 and 3 are produced from a single gene by a combined process of differential RNA transcription and splicing, *J. Biol. Chem.* **259**:13595–13604.

Additional Readings

Alexandrov, V. Y. (1981) Stimulation of flagella recovery in *Chlamydomonas eugametos* after heat injury, *Arch. Protintenk.* **124**:345–352.

Baker, E. J., Schloss, J. A., and Rosendaum, J. L. (1984) Rapid changes in tubulin RNA synthesis and stability induced by deflagellation in *Chlamydomonas*, *J. Cell Biol.* **99**:2074–2081.

Ben-Ze'ev, A., Farmer, S. R., and Penman, S. (1979) Mechanisms regulating tubulin synthesis in cultured mammalian cells, *Cell* **17**:319–325.

Breitbart, R. E., Nguyen, H. T., Medford, R. M., Destree, A. T., Mahdavi, V., and Nadal-Ginard, B. (1985) Intricate combinatorial patterns of exon splicing generate multiple regulated troponin T isoforms from a single gene, *Cell* **41**:67–82.

Brunke, K., Anthony, J., Kalish, F., Stenberg, E., and Weeks, D. (1984) Coordinate expression of the four tubulin genes in *Chlamydomonas*, in *Molecular Biology of the Cytoskeleton* (G. G. Borisy, D. W. Cleveland, and D. B. Murphy, eds.), Cold Spring Harbor Laboratory, Cold Spring Harbor, New York, pp. 367–379.

Caron, J. M., Jones, A. L., and Kirschner, M. W. (1985) Autoregulation of tubulin synthesis in hepatocytes and fibroblasts, *J. Cell. Biol.* **101**:1763–1772.

Caron, J. M., Jones, A. L., Rall, L. B., and Kirschner, M. W. (1985) Autoregulation of tubulin synthesis in enucleated cells, *Nature* **317**:648–651.

Carrino, J. J., and Laffler, T. G. (1985) The effects of heat shock on the cell cycle regulation of tubulin expression in *Physarum polycephalum*, *J. Cell Biol.* **100**:642–647.

Cleveland, D. W., and Havercroft, J. C. (1983) Is apparent autoregulatory control of tubulin synthesis nontranscriptionally regulated? *J. Cell Biol.* **97**:919–924.

Cleveland, D. W., and Sullivan, K. F. (1985) Molecular biology and genetics of tubulin, *Annu. Rev. Biochem.* **54**:331–365.

Cleveland, D. W., Lopata, M. A., Sherline, P., and Kirschner, M. W. (1981) Unpolymerized tubulin modulates the level of tubulin mRNAs, *Cell* **25**:537–546.

Cleveland, D. W., Pittenger, M. F., and Feramisco, J. R. (1983) Elevation of tubulin levels by microinjection suppresses new tubulin synthesis, *Nature* **305**:738–740.

Drubin, D. G., Caput, D., and Kirschner, M. W. (1984) Studies on the expression of the microtubule-associated protein, tau, during mouse brain development, with newly isolated complementary DNA probes, *J. Cell Biol.* **98**:1090–1097.

Farmer, S. R., Wan, K. M., Ben-Ze'ev, A., and Penman, S. (1983) Regulation of actin mRNA levels and translation responds to changes in cell configuration, *Mol. Cell Biol.* **3**:182–190.

Farmer, S. R., Bond, J. F., Robinson, L. S., Mbangkollo, D., Fenton, M. J., and Berkowitz, E. M. (1984) Differential expression of the rat β-tubulin multigene family, in *Molecular Biology of the Cytoskeleton* (G. G. Borisy, D. W. Cleveland, and D. B. Murphy, eds.), Cold Spring Harbor Laboratory, Cold Spring Harbor, New York, pp. 333–342.

Ginzburg, I., and Littauer, U. Z. (1984) Expression and cellular regulation of microtubule proteins, in *Molecular Biology of the Cytoskeleton* (G. G. Borisy, D. W. Cleveland, and D. B. Murphy, eds.), Cold Spring Harbor Laboratory, Cold Spring Harbor, New York, pp. 357–366.

Glass, J. R., De Witt, R. G., and Cress, A. E. (1985) Rapid loss of stress fibers in Chinese hamster ovary cells after hyperthermia, *Cancer Res.* **45**:258–262.

Gulick, J., Kropp, K., and Robbins, J. (1985) The structure of two fast-white myosin heavy chain promoters. A comparative study, *J. Biol. Chem.* **260**:14513–14520.

Guttman, S. D., Glover, C. V. C., Allis, C. D., and Gorovsky, M. A. (1980) Heat shock, deciliation and release from anoxia induce the synthesis of the same set of polypeptides in starved *T. pyriformis*, *Cell* **22**:299–307.

Hastings, K. E. M., Bucher, E. A., and Emerson, C. P., Jr. (1985) Generation of troponin T isoforms by alternative RNA splicing in avian skeletal muscle. Conserved and divergent features in birds and mammals, *J. Biol. Chem.* **260**:13699–13703.

Havercroft, J. C., and Cleveland, D. W. (1984) Programmed expression of β-tubulin genes during development and differentiation of the chicken, *J. Cell Biol.* **99**:1927–1935.

Kemphues, K. J., Raff, E. C., Raff, R. A., and Kauffman, T. C. (1980) Mutation in a testis-specific β-tubulin in *Drosophila*: Analysis of its effects on meiosis and map location of the gene, *Cell* **21**:445–451.

Lewis, S. A., Lee, M. C-S., and Cowan, N. J. (1985) Five mouse tubulin isotypes and their regulated expression during development, *J. Cell Biol.* **101**:852–861.

L'Hernault, S. W., and Rosenbaum, J. L. (1985) Reversal of the posttranslational modification on *Chlamydomonas* flagellar α-tubulin occurs during flagellar resorption, *J. Cell Biol.* **100**:457–462.

Lompre, A-M., Nadal-Ginard, B., and Mahdavi, V. (1984) Expression of cardiac ventricular α- and β-myosin heavy chain is developmentally and hormonally regulated, *J. Biol. Chem.* **259**:6437–6446.

Murphy, D. B., Wallis, K. T., and Grasser, W. A. (1984) Expression of a unique β-tubulin variant in chicken red-cell development, in *Molecular Biology of the Cytoskeleton* (G. G. Borisy, D. W. Cleveland, and D. B. Murphy, eds.), Cold Spring Harbor Laboratory, Cold Spring Harbor, New York, pp. 59–70.

Nabeshima, Y., Fujii-Kuriyama, Y., Muramatsu, M., and Ogata, K. (1984) Alternative transcription and two modes of splicing results in two myosin light chains from one gene, *Nature* **308**:333–338.

Nelson, W. J., and Lazarides, E. (1984) The pattern of expression of two ankyrin isoforms demonstrate distinct steps in the assembly of the membrane skeleton in neuronal morphogenesis, *Cell* **39**:309–320.

Pittenger, M. F., and Cleveland, D. W. (1985) Retention of autoregulatory control of tubulin synthesis in cytoplasts: Demonstration of a cytoplasmic mechanism that regulates the level of tubulin expression, *J. Cell Biol.* **101**:1941–1952.

Raff, E. C. (1984) Genetics of microtubule systems, *J. Cell Biol.* **99**:1–10.

Riederer, B., and Matus, A. (1985) Differential expression of distinct microtubule-associated proteins during brain development, *Proc. Natl. Acad. Sci. USA* **82**:6006–6009.

Rosenbaum, J. L., Moulder, J. E., and Ringo, D. L. (1969) Flagellar elongation and shortening in *Chlamydomonas*: The use of cycloheximide and colchicine to study the synthesis and assembly of flagellar proteins, *J. Cell Biol.* **41**:600–619.

Ruiz-Opazo, N., Weinberger, J., and Nadal-Ginard, B. (1985) Comparison of α-tropomyosin sequences from smooth and striated muscle, *Nature* **315**:67–70.

Schedl, T., Burland, T. G., Gull, K., and Dove, W. F. (1984) Cell cycle regulation of tubulin RNA level, tubulin protein synthesis, and assembly of microtubules in *Physarum*, *J. Cell Biol.* **99**:155–165.

Silflow, C. D., Lefebvre, P. A., McKeithan, T. W., Schloss, J. A. Keller, L. R., and Rosenbaum, J. L. (1982) Expression of flagellar protein genes during flagellar regeneration in *Chlamydomonas*, *Cold Spring Harbor Symp. Quant. Biol.* **46**:157–169.

Sullivan, K. F., Havercroft, J. C., and Cleveland, D. W. (1984) Primary structure and expression of a vertebrate β-tubulin gene family, in *Molecular Biology of the Cytoskeleton* (G. G. Borisy, D. W. Cleveland, and D. B. Murphy, eds.), Cold Spring Harbor Laboratory, Cold Spring Harbor, New York, pp. 321–332.

Tannenbaum, J., and Brett, J. G. (1985) Evidence for regulation of actin synthesis in cytochalasin D-treated Hep-2 cells, *Exp. Cell Res.* **160**:435–448.

Thomas, J. H., Novick, P., and Botstein, D. (1984) Genetics of the yeast cytoskeleton, in *Molecular Biology of the Cytoskeleton* (G. G. Borisy, D. W. Cleveland, and D. B. Murphy, eds.), Cold Spring Harbor Laboratory, Cold Spring Harbor, New York, pp. 153–174.

Welch, W. J., and Suhan, J. P. (1985) Morphological study of the mammalian stress response: Characterization of changes in cytoplasmic organelles, cytoskeleton, and nucleoli, and appearance of intranuclear actin filaments in rat fibroblasts after heat-shock treatment, *J. Cell Biol.* **101**:1198–1211.

Woods, C. M., and Lazirides, E. (1985) Degradation of unassembled α- and β-spectrin by distinct intracellular pathways: Regulation of spectrin topogenesis by β-spectrin degradation, *Cell* **40**:959–969.

7

Reorganization of Cytoskeleton

Morphogenesis and Locomotion of Pseudopod-Forming Cells

I. Introduction: The Main Features of Pseudopodial Reactions

In addition to alteration of protein synthesis, another group of processes leads to reorganization of cytoskeleton, namely, modulations of the assembly and distribution of cytoskeletal elements. These reorganizations are usually reversible; the cell during its life can undergo numerous modulations of the cytoskeleton. Reorganization of the assembly can be combined with alteration of synthesis or occur independently.

Pseudopodial reactions and mitosis are the two best-known and most actively studied groups of modulation of assembly and distribution of cytoskeletal structures. We shall discuss them in this and in the next chapters, respectively.

As already mentioned, pseudopod-forming cells have peripheral actin structures and a central radiating microtubular array. Many classes of these cells can change their shape and move on the noncellular substrates. They can also establish and break specialized contacts with these substrates and with other cells. These morphological alterations are essential for the formation of organized multicellular structures, tissues; they are also essential for many other processes, ranging from the destruction of invading bacteria to blood clotting. There seem to be no common traits between such diverse processes as growth of nerve, formation of the epithelial sheet, chemotatic movement of leukocytes, and activation of platelets. Nevertheless, as we will try to show, cellular mechanisms of these and of many other morphogenetic alterations have significant common features; that is, they all involve numerous and repeated pseudopodial reactions. Many decades ago the Oxford English Dictionary defined pseudopod (pseudopodium) as "a process temporarily formed by protrusion of any part of the protoplasm of the body and serving for locomotion, prehension or ingestion of food." Two comments can be added to this exact definition. First, immediately after protrusion, pseudopods formed by different cells share one common feature; namely, they

217

consist of numerous actin filaments covered by the plasma membrane. Second, temporary protrusions, pseudopods, can sometimes be transformed into permanent processes with more complex structures.

Soon after protrusion pseudopods can attach themselves to various cellular and noncellular surfaces. Extension of the pseudopod is followed sooner or later by its retraction. Retraction can pull the pseudopod back into the cell body. Alternatively, traction of a well-attached pseudopod can pull the cell body forward, toward the site of attachment. Thus, each typical pseudopodial reaction has two stages (extension and retraction) or three stages (extension, attachment, and retraction).

Each pseudopodial reaction involves some local reorganization of cytoskeleton. Integration of the results of many repeated reactions can lead to global reorganization of the whole cytoskeleton. Before discussing the mechanisms of these reorganizations, we shall describe briefly several well-studied types of morphological alteration of various cells that are based on repeated pseudopodial reactions.

II. Morphological Reorganization Based on Pseudopodial Reactions

A. Spreading and Locomotion of Fibroblasts

Primary cultures of embryo fibroblasts of mammals and birds, as well as permanent lines of fibroblastic cells, are easy to cultivate on adhesive substrates, such as glass or plastic. Living cells on these substrates are easy to observe. Therefore, the cultured fibroblast is at present the main model system used in the study of mechanisms of reversible morphological reorganization of the cells of higher eukaryotes. Fibroblasts form pseudopods of various shapes: flattened lamellipods or thin, threadlike filopods. These pseudopodial reactions provide the basis for two main types of closely related processes: spreading and locomotion. Spreading is the transformation of an unattached spherical cell into a substrate-attached cell with a highly flattened shape. When fibroblast, suspended in fluid medium, touches the surface of adhesive substrate, it begins to extend numerous pseudopods from various parts of its spherical body (Fig. 7.1). Few of these pseudopods are attached to the substrate, while others contract and disappear. Attached pseudopods gradually merge into the thin peripheral plate, lamella. The central cell body surrounded by the circular lamella is gradually flattened. The first stage of spreading is usually isotropic; that is, the cell spreads in all directions and acquires the shape of a flattened disk. This stage is followed by anisotropic spreading, also called the polarization stage (Fig. 7.2A). At this stage, formation of pseudopods is stopped in certain zones of the outer edge but continued in other zones. Attachment of pseudopods formed in the active zones leads to elongation of lamellae; in contrast, lamellae near the stable edges retreat. As a result, the cell acquires an elongated shape (Fig. 7.2A–D). Detachment of the cell from the substrate leads to contraction of all the lamellae and return to spherical shape. Thus, spreading is fully reversible.

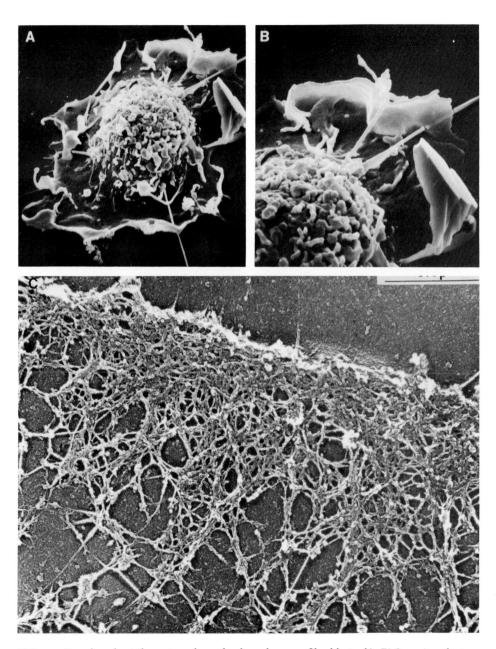

FIG. 7.1. Pseudopods at the outer edges of cultured mouse fibroblasts. **(A, B)** Scanning electron micrographs of spreading fibroblast 30 min after seeding on the glass. Field width: A, 70 μm; B, 30 μm. (A from Vasiliev and Gelfand, 1981. Micrographs courtesy of Y. A. Rovensky.) **(C)** Actin network in pseudopods at the outer edge of the fibroblast. Electron micrograph of platinum replica. Scale bar: 0.5 μm. (Courtesy of T. M. Svitkina.)

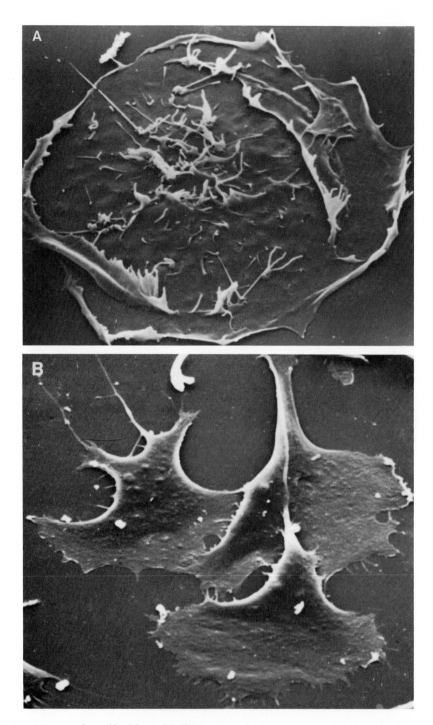

FIG. 7.2. Mouse embryo fibroblasts. **(A, B)** Scanning electron micrographs. (A) Isotropic stage of spreading, 1 hr after seeding. (B) Polarized spreading, 24 hr. Field widths: A, 100 μm; B, 500 μm. (Courtesy of E. V. Leonova.) **(C)** Actin in discoid isotropically spread fibroblast, 1 hr after seeding. Indirect immunofluorescence using antiactin antibody. Scale bar: 20 μm. (Courtesy of T. M. Svit-kina.) **(D)** Scheme illustrating alterations of three main types of cytoskeletal fibers in the course of spreading of fibroblast. Spheroid cell spreads first to discoid shape and then acquires elongated

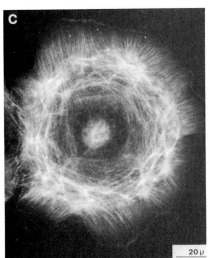

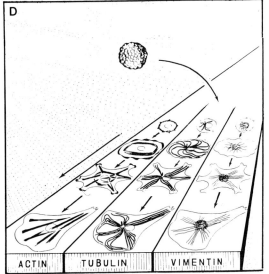

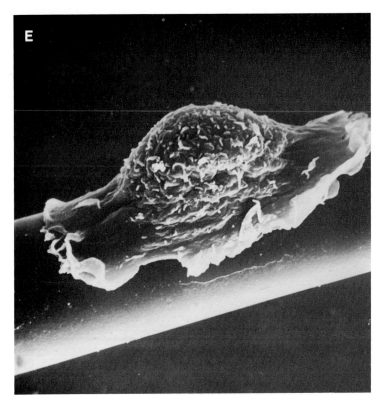

polarized shape. First, circular bundles of actin filaments are formed; polarization is accompanied by formation of the system of straight bundles. Microtubules and intermediate filaments radiate from the perinuclear zone into the spreading lamellae. (From Vasiliev, 1985.) **(E)** Normal mouse fibroblast spreading on the surface of a thin glass cylinder 1 hr after seeding. Pseudopods are formed in all directions, but the attached lamellae are preferentially oriented along the axis of the cylinder. Scanning electron micrograph. Field width: 280 μm. (From Vasiliev, 1985. Micrograph courtesy of Y. A. Rovensky.)

Spreading is closely related to phagocytosis. Sometimes spreading is described as an unsuccessful effort of the cell to phagocytize an infinitely large foreign body, the Petri dish.

Directional locomotion of polarized fibroblast can be regarded as continual anisotropic spreading going in one particular direction. The moving cell extends pseudopods only at one part of its edge, called the leading edge. Attachment and contraction of pseudopods at the leading edge draws the cell body forward. Direction of locomotion can be regulated by several groups of external factors, which change the extension and attachment of pseudopods at the active edge. There are three main types of regulation: regulation by the substrate structure (so-called contact guidance), regulation by contact with another cell (contact paralysis), and regulation by gradient of soluble molecules in the medium (chemotaxis). When the cell moves on the nonhomogenous substrate, its locomotion will always be directed toward the most adhesive area of this substrate; that is, toward the areas where pseudopods are most efficiently attached.

The fibroblast can become oriented along the preferred substrate area, for instance, along the collagen fiber, used as a substrate. This direction of fibroblast orientation and locomotion by the substrate is designated as contact guidance (Fig. 7.2E). When the active edge of fibroblast touches the surface of another cell, formation of pseudopods in the contact area is locally and reversibly stopped; this phenomenon is known as contact paralysis. Because of contact paralysis, fibroblasts usually do not move over the surface of another cell but turn away from it after contact (so-called contact inhibition of locomotion). Many turns of this type can lead to parallel orientation of cells in dense cultures.

Chemotactic regulation of fibroblast movements has also been observed; these cells move directionally along the gradient of soluble molecules of certain proteins, such as fibronectin and elastin. Thus, regulated pseudopodial reactions of individual fibroblasts provide the basis for varied and complex behavior reactions, leading to adaptation of cell shape and positions to all three parts of its microenvironment: the substrate, neighboring cells, and fluid medium.

B. Spreading of Epitheliocytes and Formation of Epithelial Sheets

Cultured cells of epithelia, like fibroblasts, extend pseudopods and spread from spheroid to discoid shape. However, in contrast to fibroblasts, these cells are not polarized, but retain a discoid shape (Fig. 7.3A). Accordingly, the single spread epithelial cells do not move directionally on the substrate. Another characteristic feature of epitheliocytes is the ability of their pseudopods to form efficient specialized intercellular contacts, zonulae adherens and desmosomes. When active lateral edges of two neighboring epitheliocytes touch one another, these cells, in contrast to fibroblasts, do not migrate away from one another, but remain permanently linked by firm contacts. Numerous cells, contacting side by side, form coherent monolayered sheets (Fig. 7.3B,C).

C. Neurons: The Growth of Neurites

The bodies of cultured neurons are usually round or elongated; small pseudopods are continually extended and retracted at all parts of their edges. When one or two larger processes, neurites, emerge from the edges, the pseu-

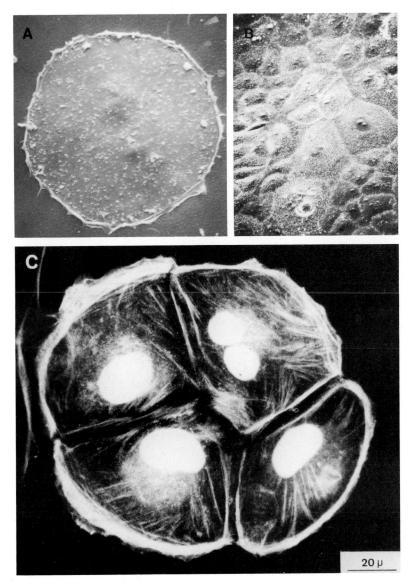

FIG. 7.3. Cultured epitheliocytes (IAR-2 rat line). **(A)** Scanning electron micrographs of the substrate-spread discoid epitheliocyte; ruffles at the edge. ×500. **(B)** Coherent epithelial sheet. Scanning electron micrograph. ×300. **(C)** Peripheral actin bundles near the outer edges of four cells forming a coherent island. Immunofluorescence microscopy using antiactin antibodies. (Courtesy of L. V. Domnina.)

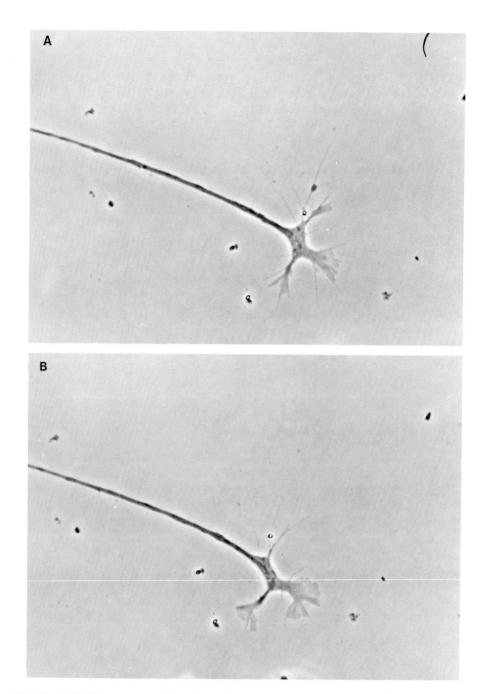

FIG. 7.4. (A–D) The sequence of time-lapse photographs of the advancing growth cone. The end of the neurite of the living chick embryo neuron cultured on a glass coverslip was photographed with intervals of 60 sec using phase-contrast microscopy. Forward movement of the leading edge of the growth cone is easily seen when immobile particles attached to the substrate are used as

C

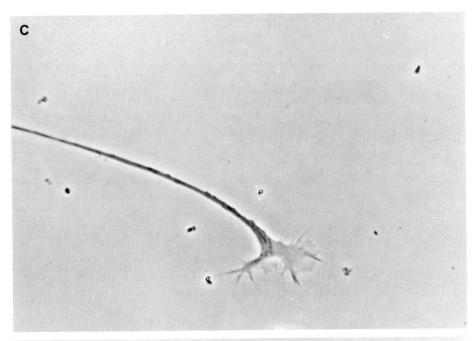

D

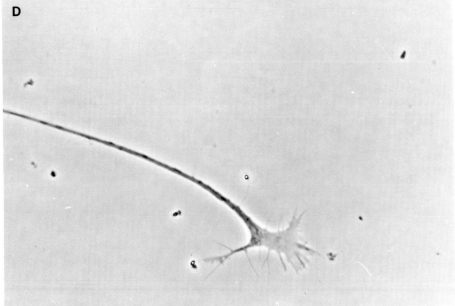

reference points. Extension, attachment, and retraction of thin microspikes and flattened lamellipods at the edges of growth cone. ×1500, reproduced at 85%. (Courtesy of D. Bray and K. Chapman.) **(E)** Organization of the cytoskeleton in the directionally moving polarized fibroblast and in the growing axon of the neuron. Thick black lines, microtubules; thin lines, sites with high concentration of actin filaments. Direction of pseudopodial extensions is shown by arrows.

dopodial activity gradually becomes restricted to the tips of these processes. These tips, called growth cones, are the small lamellae attached to the substrate (Fig. 7.4). Small pseudopods, microspikes, are continually formed, attached, and retracted at the front edges of growth cones. The growth cone is gradually displaced toward the attached pseudopods at the front edge; its opposite edge contracts from time to time; the more central cylindrical part of the process is gradually elongated. The growth of processes can be guided by various environmental cues. For instance, the processes show contact guidance; that is, they can grow toward the most adhesive part of the substrate or along the cylindrical surface of another neurite, which serves as a guiding substrate. The growth of neurites can also be guided by chemotaxis. In particular, the processes of many neurons respond to the gradients of concentration of a special protein called nerve growth factor.

Thus, the growth of neurite is similar in many respects to the locomotion of polarized fibroblast (Fig. 7.4). Both processes involve pseudopodial reactions localized at one particular edge of the cell; the other edges remain stable. There are also significant differences between these two processes. The leading edge of neurons is much narrower and the stable edges much longer than in fibroblast. Attachment and retraction of pseudopods at the leading edge of fibroblast pulls the cell body forward; in contrast, the distance between the leading edge and the body of the neuron is gradually increased, so the neurite is elongated.

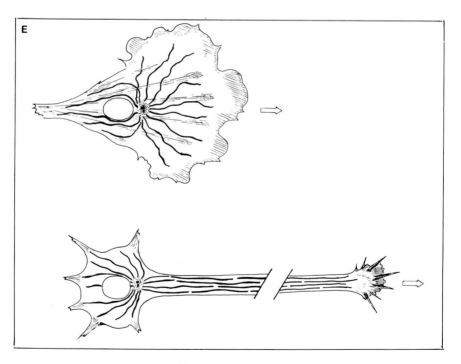

FIG. 7.4. (*Continued*)

D. Movement of Polynuclear Leukocytes and of Amebae

Obviously, there are enormous biological and morphological differences between different classes of amebae and between the amebae and leukocytes. Nevertheless, they all have certain common features. In contrast to fibroblasts, these cells do not spread well on most substrates; their body is elongated but not flattened (Fig. 7.5AB). The cells are able to move very quickly; this movement is accompanied by the formation of pseudopods and by contraction of the tail part of the body. Mammal leukocytes move in culture at a rate almost an order higher than that of fibroblast (10–15 μm compared with 1–2 μm/min). Locomotion of amebae and of leukocytes can be directed by environmental stimuli, especially chemotactic gradients. For instance, myxamebae of the myxomycete *Dictyostelium discoideum* move along the gradients of cAMP (Fig. 7.5C,D). The list of chemotactic attractants for leukocytes includes various bacterial metabolites, denatured proteins, and several synthetic N-formylated peptides.

Leukocytes exhibit high ability to change their shape during locomotion; for instance, leukocytes moving in chemotactic gradient can migrate through filters with pores as small as 5 μm.

E. Spreading of Platelets

Platelets are small anuclear bodies that exist either in nonactivated or in activated state. Activation of platelets is an important part of the complex mechanism of blood coagulation. Nonactivated platelets circulating in the blood have the shape of disks with a smooth surface.

Many soluble molecules (e.g., ADP or serotonin) and many surfaces (e.g., collagen or fibrin) activate platelets; they begin to extend pseudopods (Fig. 7.6). These pseudopods can attach to the surface of other platelets; in this way platelet aggregates are formed. Extended pseudopods can also be attached to noncellular substrates, such as collagen and fibrin fibers and glass. Repeated extension and attachment of pseudopods leads to spreading of platelets on these substrates. The spread platelets degenerate after 1 or 2 hr. Thus, the platelets can be regarded as naturally formed cytoplasmic fragments specially designed for the performance of one set of pseudopodial reactions.

F. Alterations of Cytoskeleton during Various Morphological Reorganizations

In previous sections we briefly described several examples of locomotion and shape alterations in pseudopod-forming cells. Reorganization of the cytoskeleton is central to all these processes. Formation of actin networks and outward growth of microtubules are the most conspicuous of these reorganizations. Actin networks are formed within the extended pseudopods of all cells described above (Figs. 7.1, 7.6, and 1.9). These networks contract and disappear when the pseudopods retract. When pseudopods are attached to other cells or to the substrate, actin networks can be reorganized. This reorganization is most obvious in fibroblasts. Their attached pseudopods form focal cell–substrate contacts that become associated with large, well-ordered bundles of actin filaments, stress fibers (see Chapter 1). Much smaller focal cell–substrate contacts associated with small actin filament bundles are formed in the growth cones of neurons and in platelets.

Another type of reorganization is an outward growth of microtubules and

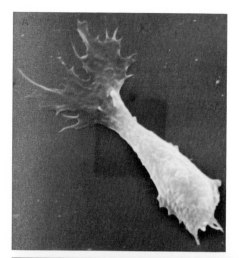

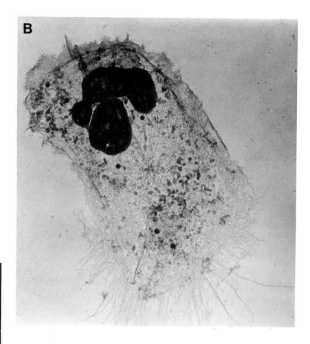

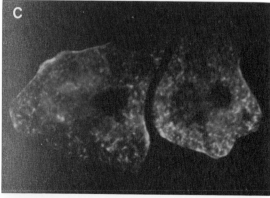

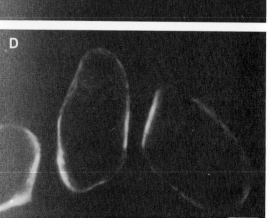

FIG. 7.5. **(A)** Elongated mouse polynuclear leukocyte on the glass surface. Scanning electron micrograph. ×10,000, reproduced at 50%. (Courtesy of I. A. Vorobjev.) **(B)** Overview of the cytoskeleton of extracted leukocyte. (Top) advancing pseudopods with ruffles extending from it; the centrosome is in the center. (Bottom) Tail fibers. ×6000, reproduced at 65%. (Courtesy of M. Schliwa.) **(C, D)** Response of *Dictyostelium amoebae* to added cAMP. Immunofluorescence using antimyosin antibodies. (C) Control cells with numerous myosin rods in the endoplasm. (D) Two minutes after addition of 10^{-6} M cAMP. Notice elongation of cells, disappearance of endoplasmic rods, and accumulation of myosin in the cortex. Scale bar: 5 μm. (From Yumura and Fukui, 1985. Courtesy of S. Yumura.)

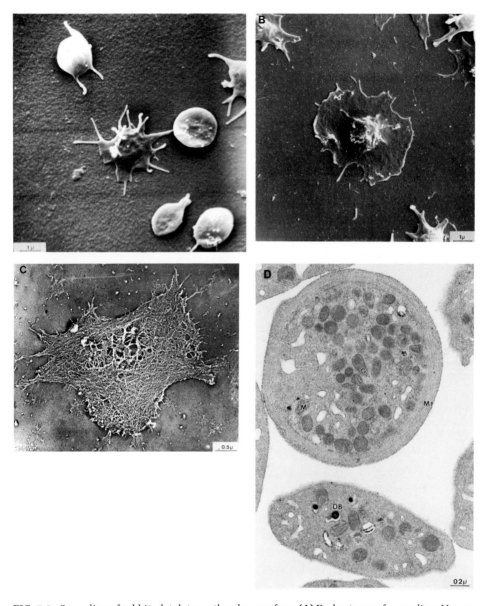

FIG. 7.6. Spreading of rabbit platelets on the glass surface. **(A)** Early stages of spreading. Nonactivated discoid platelet (arrow). Several platelets beginning to form pseudopods. Better-spread platelet with lamellae. Scanning electron micrograph. **(B)** Well-spread platelet with wide lamellae. Scanning electron micrograph. **(C)** Cytoskeleton of well-spread platelet. Dense network of actin filaments at the periphery. Platinum replica. (Courtesy of A. Y. Alexandrova.) **(D)** Submembranous circular bundle of microtubules (Mt) in the nonactivated discoid platelet. Electron micrograph of sectioned cell. G, D, M, various types of internal organelles. (From Lewis, 1984. Courtesy of J. C. Lewis.)

intermediate filaments into the process and lamellae formed from the attached pseudopods. The centrifugal growth of microtubules into lamellae is observed in spreading fibroblasts and epitheliocytes (Fig. 7.2). Intermediate filaments also grow centifugally following the microtubules. The ends of some microtubules and intermediate filaments become associated with the focal contacts at the periphery of fibroblasts. Intermediate filaments of epitheliocytes establish connections with desmosomes at the cell periphery. During neuron growth microtubules and intermediate filaments fill the proximal cylindrical part of the neurite; their growth follows the advancement of the growth cone.

In the following sections we shall discuss possible mechanisms of these cytoskeletal reorganizations and their functional roles. We shall first consider transformations of cortical actin structures directly involved in pseudopodial reactions. Later we shall review alterations of the microtubular system and the role of this system in the general organization of pseudopodial activities of the cell.

III. Reorganization of Actin Cytoskeleton during Pseudopodial Reactions

A. Mechanisms of the Extensions of Pseudopods

Accumulations of actin filaments are present in all types of pseudopods. Usually these filaments form dense networks; more rarely they form small bundles. Pseudopods also contain a characteristic spectrum of actin-binding proteins; the list of proteins enriched in the leading-edge zone of fibroblasts includes α-actinin, filamin, fimbrin, and talin.

What are the mechanisms of formation of new actin structures during pseudopod extension? There is reason to suggest that this extension includes two main groups of processes: (1) external ligand molecules bound locally to the membrane receptors induce some membrane-generated signals near the site of future extension; and (2) these signals induce organization of the network of actin filaments at the extension sites.

Induction of pseudopods by the membrane-bound ligand molecules has been best demonstrated in several cell types: platelets, myxamebae, and leukocytes. The membrane of platelets contains a spectrum of specific receptors to various external activating molecules, e.g., ADP, serotonin, collagen, and many others. The membrane of myxamebae contains receptors to cAMP and the membrane of polynuclear leukocytes contains receptors to chemotactic peptides. Contact of these molecules with the surfaces of corresponding cells induces extension of pseudopods at their surface. The nature of molecules inducing the pseudopods during spreading of fibroblasts and of epitheliocytes is not yet known. By analogy, it seems probable that pseudopods in these cells are induced by various substrate-bound molecules which the surface membrane contacts during spreading and locomotion.

In various systems it was shown that interactions of membrane receptors with corresponding ligands cause alterations of the regulatory membrane components, which eventually lead to generation of signals for intracellular changes. More specifically, ligand–receptor interactions can induce activation of the membrane-bound enzymes which catalyze the formation of molecules, such as cAMP, diacylglycerol, and inositol triphosphate. These molecules serve as "secondary messengers" inducing a cascade of intracellular changes. In particular, inositol triphosphate can stimulate the release of calcium from intracellular stores (see Chapter 9 for more details). Each of these signal changes, in turn, can induce a cascade of various alterations of cytoplasmic components. It is not known which membrane-generated signal is essential for the extension of pseudopods. Certain observations suggest that local increase of the calcium concentration in submembranous cytoplasm can be important. This increase was registered experimentally in the anterior zone of the cytoplasm of moving amebae. In another experiment the surface of lamellae of cultured epitheliocytes was locally sprinkled with a solution containing calcium ionophore A23187 from a micropipette. This agent increased the calcium influx through the plasma membrane into the cytoplasm. Influx of calcium led to local thickening and expansion of lamellae. These experiments do not exclude the possibility that other types of intracellular signals may also be involved in the formation of the pseudopod.

How do the membrane-generated signals cause formation of the actin network at the extension site? The most obvious possibility is that these signals induce local actin polymerization; new filaments then push the membrane to the exterior (Fig. 7.7). This suggestion is supported by experimental data. It was found that activation of platelets is accompanied by a striking increase in the ratio of polymerized to unpolymerized actin. Increase of polymerized actin was also observed in leukocytes placed in the medium with a high concentration of chemotactic peptide inducing formation of pseudopods. The sperm cell of sea cucumber contacting the coat of the egg extends an acrosomal process; this extension is accompanied by rapid polymerization of actin filaments. Thus, at least in these cells, extension of pseudopods is probably the result of actin polymerization. Besides polymerization, various other possible mechanisms of extension have been suggested (Fig. 7.7). For instance, it has been postulated that actin cortex tends to contract and that this contraction exerts general hydrostatic pressure on the more fluid internal cytoplasm; this pressure can squeeze the cytoplasm toward the surface through the "weak" sites of the cortex. Calcium influx can cause local solation of the cortex (see Chapter 1), thus determining the site of cytoplasm flux and of the pseudopod extension.

Another hypothesis is that protrusion of the pseudopod is due to the local calcium-induced swelling of cortical actin gel. Extension caused by cytoplasmic flow or by gel swelling has to be followed by actin polymerization. Otherwise, it would be difficult to explain why the density of the actin network within the pseudopod is often higher than in the nearby zones of the cell.

Thus, the mechanism of pseudopod extension seems to include actin polymerization, possibly supplemented by other local reorganizations of the cortex. When extension and attachment of pseudopods occur repeatedly in the same zone of the cell edge, e.g., in the growth cone of the neuron, this process requires a continuous flow of cytoplasmic material toward that zone. In fact, cytoskeletal proteins slowly move along the axon; this is the so-called slow component of axonal transport. The force driving this transport remains unknown. It is possible that actin and other molecules are actively transported along the preexisting microtubules or filaments. It is also possible that in certain cases, materials for pseudopod extension are supplied by the pressure-induced cytoplasmic flow.

B. Retraction of Pseudopods

Outward extension of pseudopods is usually followed sooner or later by their inward retraction. Retracting pseudopods can disappear rapidly within the cell body. Retracting pseudopods formed at the active edges can also turn upward and then move centripetally at the upper cell surface. In particular, lamellipods bent upward are transformed into wide surface waves, so-called ruffles, moving toward the central part of the cell.

It seems natural to suggest that retraction is due to actin–myosin interactions within the actin network. This suggestion, however, has one difficulty. Immunomorphological studies have shown that the actin cortex of pseudopods contains much smaller amounts of myosin than the cortex of the

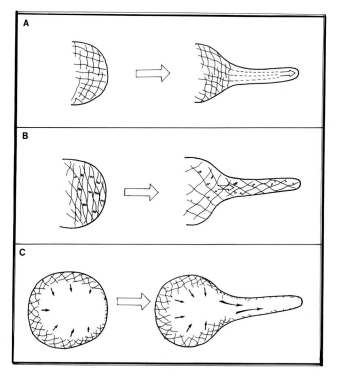

FIG. 7.7. Three hypothetical methods of pseudopod extension. **(A)** Local polymerization of actin. **(B)** Local swelling of actin cortex. **(C)** General contraction of actin cortex leading to local extrusion of internal fluid cytoplasm.

cell body. Exclusion of myosin from the pseudopods was observed in the myxamebae, in the fibroblasts (Fig. 1.7), and in epitheliocytes; in contrast, the pseudopods of leukocytes were reported to contain large amounts of myosin. The apparent absence of myosin is difficult to reconcile with the suggestion that retraction of pseudopods is due to local actin–myosin contraction. Possibly, the pseudopod network is linked with the myosin-containing actin cortex of the cell body and pulled centripetally by this cortex. The centripetally contracted actin network can be disassembled somewhere in the central part of the cell; its material can then be transported back toward the active edge. Thus, as first suggested by Dunn (1980), there can be a continuous circular "actin flow" between the cell edge and the endoplasmic area of locomoting fibroblasts and other spread cells (Fig. 7.8). Actin networks formed at the cell edge can be used for the organization of actin structures in more central parts of the cell. This suggestion is supported by experiments in which reversible disassembly of all types of actin structures of fibroblast was induced by the agents that lowered ATP concentration in the cell (sodium azide and oligomycin). Formation of an actin meshwork at the edge was the first sign of restoration of the cytoskeleton, observed as early as 1–10 min after removal of inhibitors from the medium. Restoration of the actin cortex then progressed toward the central part of the cell. Formation of stress fibers was the last step of recovery, observed only after 60–120 min. Thus, the pseudopods at the edge may be the primary structures from which other parts of the actin cytoskeleton are formed.

As already indicated (see Chapter 5), actin flow can also be involved in the centripetal transport of cross-linked surface receptors. At present, however, evidence for the existence of this flow is still circumstantial.

C. Attachment of Pseudopods

Retraction of the pseudopod can be preceded by its attachment to the substrate or to other cells. Specialized contact structures are formed during this attachment. Generally, formation of cell–cell or of cell–substrate contact seems to involve two types of events: (1) At the outer side of the membrane,

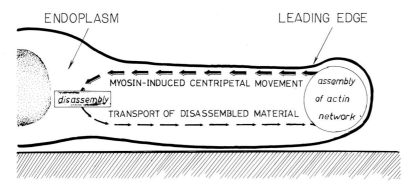

FIG. 7.8. Scheme of hypothetical "actin flow" in the cultured fibroblast. (Based on Dunn, 1980.)

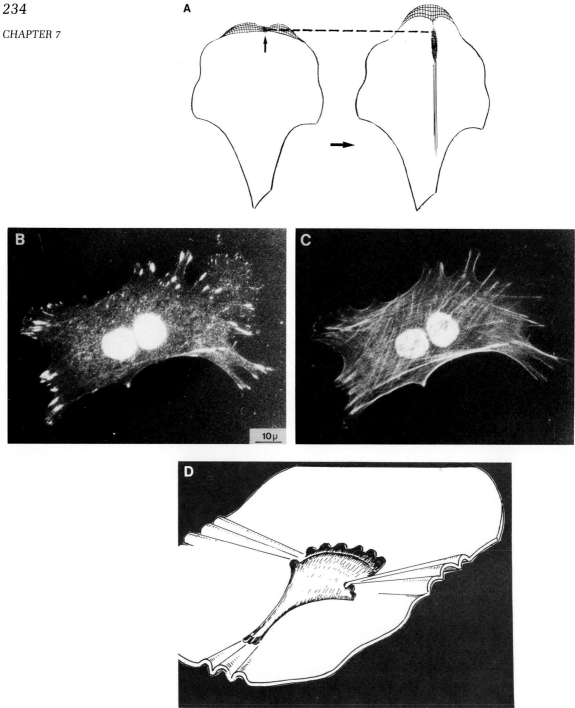

FIG. 7.9. **(A)** Scheme illustrating the two possible stages of the development of focal contact. Left: Formation of initial contact in association with actin network at the leading edge. Right: Maturation of contact and formation of an associated stress fiber. **(B, C)** Initial and mature focal con-

a group of receptor molecules is linked to a group of external ligands bound to the surface of the substrate or the other cell. In particular, membranes of many pseudopod-forming cells contain receptors to various proteins of extracellular matrices, such as collagens of different types, laminin (an important component of the basal laminae), fibronectin, and vitronectin (serum protein easily absorbed at the culture substrate). Attachment of pseudopods can be due to linkage of these receptors to corresponding ligands. Ligands inducing formation of pseudopods are not necessarily identical to those mediating attachment of these same pseudopods. (2) At the inner side of the membrane, the same group of membrane receptors is linked to certain submembranous proteins. These proteins, in turn, become anchored to preexisting cytoskeletal fibrils or nucleate the growth of new fibrils. Possible mechanisms of anchoring of clustered membrane components by the cytoskeleton were discussed in Chapter 5.

New contacts are formed by the surface of pseudopods more readily than by the surface of other cell parts. This conclusion is based on observations of cultured epitheliocytes. As already mentioned, these cells form cell–cell contacts when they touch one another by the lateral surfaces at which pseudopods are continually extended. In contrast, nonactive upper surfaces of the epithelial sheets are nonadhesive. Suspended homologous cells touching these upper surfaces of the sheet do not form any stable contacts with them. Thus, only the pseudopodial surface is able to form cell–cell contacts in this system. Spreading fibroblasts form new cell–substrate contacts, so-called initial focal contacts, only in the zone of the leading edge where new pseudopods are extended. These new contacts are small dots (Fig. 7.9A–C) containing submembranous accumulations of actin and of certain actin-binding proteins, vinculin and α-actinin. Small focal contacts are also formed by the pseudopods of the spreading platelets and of the growth cones of neurons.

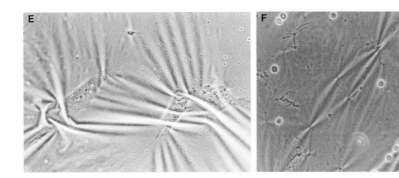

tacts of cultured fibroblast. The same cell doubly stained with antivinculin (B) and with antiactin (C) antibodies. Vinculin is contained in large mature contacts and small initial contacts near the cell edges. The ends of actin bundles are associated only with the ends of mature contacts but not with those of initial contacts. Scale bar: 10 μm. (From Bershadsky *et al.*, 1985.) **(D–F)** The spreading fibroblast wrinkles the elastic substrate. (D) Scheme. (E, F) Microphotographs of wrinkles formed by attached fibroblasts (E) and platelets (F). (E and F from Harris, 1982. Courtesy of A. Harris.)

The reasons why pseudopods are especially prone to form contacts are not clear. Possibly, conditions within the pseudopods favor actin polymerization, and some new filaments are nucleated by the molecules of actin-binding proteins anchored to the receptor clusters or capped by these molecules. Through the contact, attached pseudopods and lamellae exert tension on the substrate. This tension was directly demonstrated in the experiments of Harris (1982), in which the elastic thin films made of silicon rubber were used as the substrate. The spread fibroblasts wrinkled these films (Fig. 7.9D–F). Probably, tension exerted by the attached pseudopods is also responsible for flattening of the cell body during spreading of the fibroblast, epitheliocyte, and platelet. It may also be essential for displacement of the cell body toward the active edge during locomotion of the fibroblast. Tension in the attached pseudopod is probably induced by the same force that causes centripetal retraction of the unattached pseudopod; that is, by the contraction of actin cortex. In fact, wrinkles of elastic substrate, made by the attached fibroblasts, disappeared when cytochalasin was added to the medium.

D. Reorganization of Actin Structures Anchored to the Contacts

Formation of stress fibers associated with focal contacts is the best-known example of this reorganization. Most initial focal contacts formed at the leading edge of fibroblast exist for different periods of time and later disappear. However, some of these contacts undergo further transformation; they considerably increase in size, are elongated, and become associated with the bundles of actin filaments, stress fibers. Dynamic observations of cells injected with fluorescent-labeled actin made by Wang (1984) have shown that new stress fibers are formed by centripetal elongation of the small actin-containing dots present at the cell edges; most likely, these dots correspond to initial focal contacts (Fig. 7.10A–D). Two mechanisms can be involved in the formation of stress fibers: alignment of the existing filaments and polymerization of the new filaments. When some part of the actin network of pseudopods becomes anchored to the initial cell–substrate contact, the centripetal tension will tend to align the attached filaments in parallel to one another. Such an alignment was suggested to be the first step of the formation of stress fiber (Fig. 7.10E). Possibly, mechanically induced alignment promotes some further chemical changes, such as polymerization of new filaments and association of filaments with various actin-binding proteins.

Formation of new stress fibers takes place more often in fibroblasts that do not move on the substrate; rapidly moving cells form these structures more rarely. This observation suggests that development of stress fibers is somehow associated with development of more permanent cell shape.

Thus, local alterations of the actin cytoskeleton during the extension of pseudopods may eventually lead to more general reorganization of the actin cortex; such reorganization is manifested by the formation of highly organized actin systems, such as stress fibers. It is interesting that similar reorganization of the actin cortex seems to be involved in the development of highly specialized actin structures, such as myofibrils of striated muscle. When the tissue of embryonic chick heart is treated with trypsin, isolated

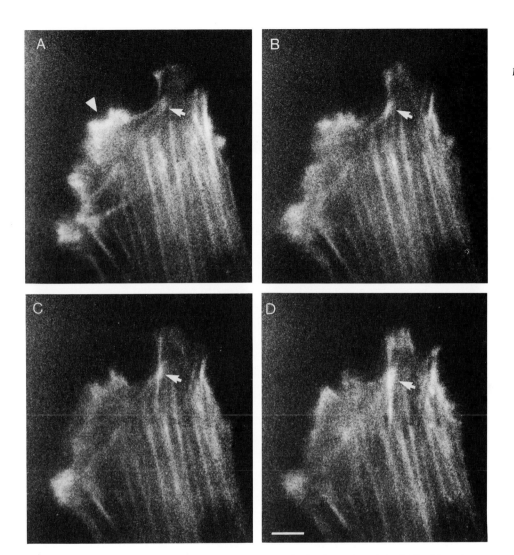

FIG. 7.10. (A–D) Formation of a stress fiber (arrows) at the leading edge of chick fibroblast. Series of time-lapse photographs of a living cell coupled with image-intensified fluorescence microscopy. Photographs were taken at 60 (A), 66 (B), 69 (C), and 75 (D) min after injection of actin molecules fluorescently labeled with tetramethylrhodamine. Scale bar: 5 μm. (From Wang, 1984. Courtesy of Y.-L. Wang.) **(E)** Hypothetical scheme showing possible processes involved in the development of stress fiber. (Top) Initial contact with anchored actin filament network. (Center) Contraction of network leads to alignment of filaments anchored to the contact. (Bottom) Additional actin-binding proteins are incorporated in periodic fashion into the bundle formed by aligned filaments. **(F)** Scheme showing formation of myofibrils in spreading cardiac myocytes. First, stress fibers are formed within the lamellae of the spreading cell; these fibers are continuously stained with actin antibody in immunofluorescence and contain nonmuscle myosin. Later, the stress fibers are reorganized into myofibrils with striated staining of actin and with muscle myosin. (Scheme based on Sanger *et al.*, 1984, and Dlugosz *et al.*, 1984.)

myocytes round up and their myofibrils are disrupted. In culture these rounded cells slowly spread out and after several days reform the myofibrils. Spreading myocytes first develop bundles of filaments morphologically similar to stress fibers. Later these bundles are transformed into myofibrils of the usual structure (Fig. 7.10F).

IV. Factors Controlling the Distribution of Pseudopods: Reorganization of Microtubules

In principle, the cell can extend pseudopods from any site of its surface. However, in fact, the distribution of sites of pseudopodial extensions is highly nonrandom in all the cells. The available data suggest that this nonrandom distribution is primarily determined by external factors and then stabilized by internal reorganization of the cytoskeleton. External factors can modulate the frequency of extension of pseudopods and efficiency of their attachment at various parts of the surface. We have already mentioned several phenomena in which external agents modulate pseudopodial reactions: chemotaxis, contact paralysis, and contact guidance. In all these cases external factors

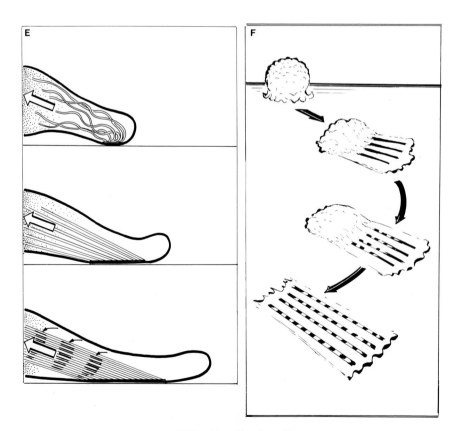

FIG. 7.10. (Continued)

alter either the frequency of extension of pseudopods in certain parts of the surface or the efficiency of their attachment to various parts of the substrate. The primary differences in the pseudopodial activity caused by external factors are further stabilized by intracellular mechanisms. Stabilization can be defined as a process by which the cell continues to extend pseudopods only in directions where previous pseudopods have been attached most efficiently and stops extensions in other, less successful directions.

Reorganization of two types of cytoskeletal elements, actin filaments and microtubules, seems to play the central role in stabilization of pseudopodial activity. It is useful to distinguish microtubule-independent and microtubule-dependent variants of stabilization; by definition, only the second variant is sensitive to colchicine. Microtubule-independent stabilization in "pure" form is observed in small cytoplasmic fragments cut from the fibroblasts (see Chapter 5). Elongated fragments form pseudopods only at the two opposite ends; the lateral edges remain stable. Colchicine does not affect the pseudopodial activity of these fragments. The fragments have also been described to move directionally in the absence of microtubules. It has been suggested that microtubule-independent stabilization is determined by the distribution of actin filaments. New pseudopods are usually extended in directions approximately parallel to the predominant orientation of the actin filaments. When the pseudopod is attached to the substrate, filaments anchored to this attachment may form the bundle, which can orient further extensions of pseudopods from the nearby sites of the surface. As a result, the width of active zone of the edge will increase if the attachment is effective in this zone (Fig. 7.11A). Thus, attachment of pseudopods gradually reorganizes the cortex (see Section III), and this reorganization may stabilize further distribution of pseudopodial activities.

This microtubule-independent stabilization probably acts not only in the small fragments, but also in the large whole cells. However, this mechanism alone cannot stabilize the long edges of polarized fibroblasts and of the neurons. Here the second variant of stabilization, microtubule-dependent stabilization, is necessary to regulate pseudopodial activity. In other words, in these cells the integrity of the microtubular system becomes essential, as shown by experiments with colchicine and similar drugs. Drug-induced depolymerization of microtubules does not stop pseudopodial activity of the polarized fibroblast (Fig. 7.11B). In contrast, the stable lateral zones of the edges disappear; all parts of the outer edges of these cells extend and attach pseudopods. These cells lose elongated shape and acquire an irregular contour. Naturally, these cells cannot move directionally on the substrate; despite their high pseudopodial activity, they remain at the same place or undergo only random displacements.

Depolymerization of microtubules leads to similar changes of the axon-growing neurons. First, activation of pseudopodial extensions at the previous stable lateral sides of axons can be observed. Later, axons can retract completely and the cell acquires an irregular shape.

Destruction of microtubules by colchicine is accompanied by a collapse of vimentin filaments (see Chapter 5). However, when similar collapse was induced by injection of antivimentin serum, which did not affect microtubules, any visible alteration of cell shape and movements was not observed. Thus, the effects of microtubule depolymerization are not mediated by alterations of intermediate filaments.

The exact role of microtubules in regulation of pseudopodial acitivities is not clear. As already mentioned, microtubules grow from the central parts of the cell into the recently formed lamallae of fibroblast or into the processes of neurons. Possibly, these microtubules act on the actin cortex reorganized after the pseudopod attachment. Microtubules may help to stabilize the non-contracted state of this cortex and to maintain the regular distribution of actin filaments. In fact, the stress fibers lose their parallel orientation and are distributed irregularly in colchicine-treated fibroblasts. Elongated, well-oriented focal contacts are also transformed into spots of irregular shape. Microtubules can act mechanically as stiff rods counteracting the contraction of actin cortex; they can also control the transport of cytoplasmic material needed for pseudopodial extension; they can also act in some other unknown way.

The organizing function of microtubules seems to depend on the structural unity of their radiating system. This dependence is suggested by morphological observations showing that the centrosome of the directionally moving fibroblasts and leukocytes is usually located at the anterior side of the nucleus facing the leading lamella. Alterations of the direction of move-

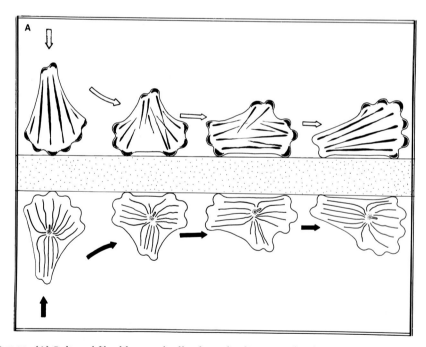

FIG. 7.11. (A) Cultured fibroblast gradually alters the direction of its locomotion on the adhesive substrate and organization of the cytoskeleton after contact of its leading edge with the boundary of nonadhesive substrate (dotted surface). Consecutive stages of movement of the same cell are shown from top to bottom. Two mirror images of this cell show reorganization of the microtubular system and of actin.

ment are accompanied, and possibly even preceded, by movement of the centrosome into an appropriate position. For instance, when the leukocyte is
placed in a chemotactic gradient, its centrosome is reoriented in the direction
of this gradient within 5 min. Centrioles of the endothelial sheets of the aorta
are usually located at the site of the cell nearest to the heart. When this endo-
thelium is wounded, centrioles are reoriented toward the wound. The mech-

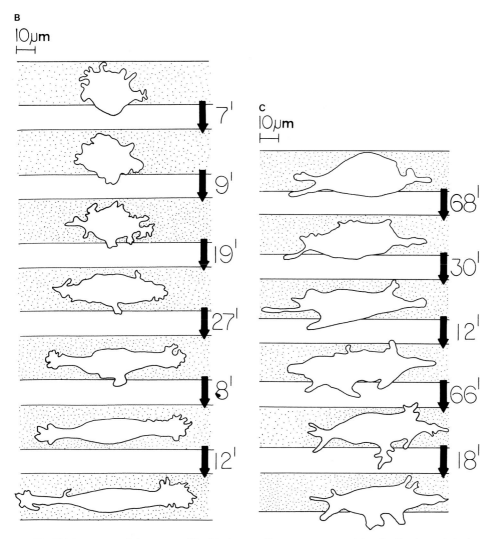

(B, C) Movement of the mouse fibroblast spreading on a narrow strip of adhesive substrate
(glass) bordered by nonadhesive lipid films (edges of films shown by horizontal lines). Tracings
of cell contours from time-lapse films. Numbers next to the arrows show the time between two
consecutive drawings. (B) Control medium. At the beginning of spreading the cell forms pseudopods in all directions, but later cellular edges parallel to those of the strip are stabilized. (C)
Medium containing colcemid (0.1 μg/ml). Stabilization of cellular edges does not take place; the
cell continues to form pseudopods in all directions during the entire observation period. (From
Vasiliev and Gelfand, 1976.)

anism of these intriguing centrosome movements and their exact functions are not clear. As expected (see Chapter 5) reorientation of the microtubular system leads to reorientation of most cellular organelles. For instance, the Golgi apparatus in fibroblasts and macrophages is rapidly repositioned in the direction of subsequent cell migrations in fibroblasts and macrophages that have just received a signal to move.

It has been suggested that this orientation is essential for the insertion of new membrane, via vesicles derived from the Golgi apparatus, into the leading edge. This possibility requires further study.

To summarize, two systems of the cytoskeleton, actin filaments and microtubules, are essential for adaptation of the shape and position of the pseudopod-forming cell to its environment. The actin system plays a leading role in the performance of pseudopodial reactions. Each pseudopod "explores" the environment. If it is successfully attached to the substrate, pseudopodial extension is repeated near the same zone of the cell edge. In contrast, the zones where pseudopodial reactions are less successful can become stable. Each attached cell–substrate contact is linked by anchored filaments with the whole actin cortex. Therefore, each attachment changes somewhat the distribution of filaments in the cortex. The position and orientation of the components of the microtubular system are also gradually changed in association with alterations of cell shape and direction of locomotion. Reciprocally, orientations of filaments and of microtubules determine the site of new pseudopodial extensions. At any given moment, the structure of the cytoskeleton is an integrated result of many previous pseudopodial reactions and a determinant of future reactions.

V. Common and Special Features of Cytoskeletal Reorganization in Various Classes of Pseudopod-Forming Cells

As we have seen, various classes of pseudopod-forming cells have at least two common features of morphogenesis: (1) All use pseudopodial reactions to explore their immediate surroundings and to alter their shape and position. (2) All are able to gradually reorganize all elements of their cytoskeleton in accordance with the statistical results of previous exploratory pseudopodial trials; this reorganization, termed stabilization, directs the sites of future pseudopodial reactions.

Pseudopodial reactions and stabilization processes in the cells of different types have, of course, many special features. In particular, each cell type has its characteristic spectrum of external factors which induce extensions of pseudopods and of the surfaces to which pseudopods can be attached; these differences may depend on the presence of different receptors in the membrane. These differences can determine unequal degrees of spreading of various cells, e.g., fibroblasts and leukocytes, on similar substrates. They may also determine the different ability of epitheliocytes and of fibroblasts to stable cell–cell contacts at their lateral edges.

The shape and size of pseudopods and the degree of their contractility are also very different in the cells of various types.

Cell-specific differences of interrelationships between pseudopodial reactions and stabilization processes are enormous. Stabilization in nervous cells is effective and long lasting, so orientation of the axon, once determined, is almost never changed. In contrast, leukocytes can change their orientation many times in 1 hr. The stability of fibroblast orientation seems to be less than that of neurons, but higher than that of leukocytes. It seems probable that one of the parameters determining these differences is the relationship between actin and tubulin structures. The more developed microtubular system favors stabilization of cell shape, while the more developed actin cortex facilitates extension and retraction of pseudopods; that is, reorganization of cell shape. For instance, actin cortex is less developed and microtubules are more developed in the elongating neuron than in the polarized fibroblast. Accordingly, formation of pseudopods and lamellae takes place only in the small part of the neuron edge and is not followed by forward contraction of the cell body.

The relative development of actin cortex in leukocytes and in amebae seems to be even higher than in fibroblasts. The effects of colchicine on these cells are less striking than on fibroblasts. Accordingly, microtubules may play only a minor role in the short-term stabilization of these cells. Two types of interrelationship between actin and tubulin systems can be observed in platelets. The nonactivated platelets have only a few polymerized actin filaments; they have, however, marginal circular bundles of microtubules that may stabilize their discoid shape. After activation this bundle is disorganized, and a well-developed actin system is formed *de novo*. This actin system is responsible for the spreading of platelets; the course of this spreading seems to be independent of microtubules.

All these considerations are still very general. We have a long way to go before we can perform controlled "engineering" of morphogenetic properties; that is, before we can directionally modify the cytoskeleton and obtain cells with any desired shapes and locomotory behavior.

VI. Integration of Morphogenetic Reactions of Many Cells

In the course of morphogenetic reactions of individual cells, local alterations of cytoskeletal structures lead to integrated reorganization of the whole cytoskeleton.

In the course of morphogenesis of higher eukaryotes, alterations of the cytoskeletons of many cells also became integrated in such a way that organized multicellular structures are formed. Formation of intracellular contacts is one method of this integration. Intermediate filaments and actin filaments anchored to the cell–cell contacts transmit tension within the epithelial sheet, so the whole sheet forms a mechanically integrated system which moves or changes its shape as a whole. For instance, it was suggested that contractions of actin bundles associated with cell–cell contacts lead to alteration of the shape of the sheets (Fig. 7.12). These bundles and contacts are located near the apical parts of the sheet cells. Concerted contraction of the interconnected bundles in some areas of the sheet may cause narrowing of apical parts of the cells and folding of the corresponding part of the sheet. Progressive folding will eventually lead to formation of the epithelial tube.

Formation of epithelial folds and tubes from the sheets is frequently observed during embryonic development of various organs.

Another mechanism of intercellular integration is provided by the interaction of cells with the matrix fibers (Fig. 7.13). Most cell types in the organism are attached to extracellular matrices. Fibrillar structures of various types, especially networks of collagen fibers, are essential components of these matrices. Attached cells exert tension on these fibers. The cells attached simultaneously to several collagen fibers exert tension on these substrates and therefore change their position and orientation. The position of the fibers,

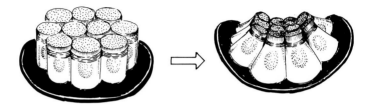

FIG. 7.12. Local contraction of the apical circumferential bundles of actin filaments can alter the form of an epithelial sheet.

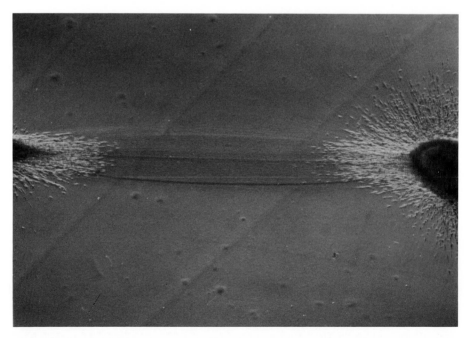

FIG. 7.13. The region between two heart explants grown on a collagen gel for 90 hr viewed with oblique illumination. A dense tract of aligned collagen has formed between the explants which appears discontinuous with the rest of the matrix. (From Stopak and Harris, 1982. Courtesy of A. Harris.)

in turn, determines the orientation of cells that are spread on them. Experiments with fibroblasts, spread on collagen networks in culture, has shown that these reciprocal interactions can produce various types of structures in which the cells and fibers are mutually aligned, e.g., circular aggregates with cells and fibers radiating from them, fusiform groups of cells and fibers. Formation of these cell–matrix complexes models the development of many structures in the organism in which matrix fibers and the cells are mutually oriented, e.g., tendons, ligaments, and many others. Thus, the role of integrated mechanical tensions exerted by cytoskeletons of many cells in various types of morphogenesis can be significant.

VII. Conclusion

Reprogramming of protein synthesis during differentiation determines the type of cytoskeletal organization of individual cells. Modulations of cytoskeleton, discussed in this chapter, usually do not change the cell type, but adapt the shape and position of the cell to its microenvironment. These adaptations of cells within the multicellular organism continue during their entire life-span. Maintenance of the cytoskeleton structure, discussed in Chapter 5, and reorganization, discussed in this chapter, are probably both manifestations of the action of the same integrating mechanisms. These mechanisms bring the actin cortex to the peripheral position and the microtubule-organizing center to the central position in cell fragments and in multinuclear cells. During spreading and locomotion, the same mechanisms bring pseudopodial material to the cell periphery and change the position of the microtubular system. These mechanisms can also regulate the distribution of cellular organelles in accordance with the reorganization of cytoskeleton. Repositioning of cell organelles is regularly observed during locomotion.

Obviously, the cellular mechanisms of regulation of the cytoskeleton discussed in this part of the book are based on the molecular processes described in the first part. In particular, dynamic properties of the cytoskeleton components provide the basis for continuous remodeling of the cytoskeleton organization in the cell. However, at present, there is a wide gap between the understanding of molecular and cellular phenomena related to regulation of the cytoskeleton. To cite but one example, we know at present the effect of changes of calcium concentration on almost all molecules present in the cytoskeleton (see Chapters 1–3). But we still do not know which of these molecular alterations, if any, is responsible for extension and contraction of pseudopods. This gap remains to be filled.

Literature Cited

Bershadsky, A. D., Tint, I. S., Neyfakh, Jr., A. A., and Vasiliev, J. M. (1985) Focal contacts of normal and RSV-transformed quail cells, *Exp. Cell Res.* **158**:433–444.

Dlugosz, A. A., Antin, P. B., Nachmias, V. T., and Holtzer, H. (1984) The relationship between stress fiber-like structures and nascent myofibrils in cultured cardiac myocytes, *J. Cell Biol.* **99**:2268–2278.

Dunn, G. A. (1980) Mechanisms of fibroblast locomotion, in *Cell Adhesion and Motility* (A. S. C. Curtis and J. D. Pitts, eds.), Cambridge University Press, Cambridge, United Kingdom, pp. 409–423.

Harris, A. (1982) Traction and its relations to contraction in tissue cell locomotion, in *Cell Behavior* (R. Bellairs, A. Curtis, and G. Dunn, eds.), Cambridge University Press, Cambridge, United Kingdom, pp. 109–134.

Lewis, J. C. (1984) Cytoskeleton in platelet function, in *Cell and Muscle Motility*, Vol. 5, *The Cytoskeleton* (J. W. Shay, ed.), Plenum Press, New York, pp. 341–377.

Sanger, J. W., Mittal, B., and Sanger, J. M. (1984) Formation of myofibrils in spreading chick cardiac myocytes, *Cell Motil.* **4**:405–416.

Stopak, D., and Harris, A. K. (1982) Connective tissue morphogenesis by fibroblast traction. I. Tissue culture observations, *Dev. Biol.* **90**:383–398.

Vasiliev, J. M. (1985) Spreading of non-transformed and transformed cells, *Biochim. Biophys. Acta* **780**:21–65.

Vasiliev, J. M., and Gelfand, I. M. (1976) Effects of colcemid on morphogenetic processes and locomotion of fibroblasts, in *Cell Motility* (R. Goldman, T. Pollard, and J. Rosenbaum, eds.), Cold Spring Harbor Laboratory, Cold Spring Harbor, New York, pp. 279–304.

Vasiliev, J. M., and Gelfand, I. M. (1981) *Neoplastic and Normal Cells in Culture*, Cambridge University Press, Cambridge, United Kingdom.

Wang, Y-L. (1984) Reorganization of actin filament bundles in living fibroblasts, *J. Cell Biol.* **99**:1478–1485.

Yumura, S., and Fukui, Y. (1985) Reversible cyclic AMP-dependent change in distribution of myosin thick filaments in *Dictyostelium*, *Nature* **314**:194–196.

Additional Readings

General

Albrecht-Buehler, G. (1985) Is cytoplasm intelligent too? in *Cell and Muscle Motility*, Vol. 6 (J. W. Shay, ed.), Plenum Press, New York, pp. 1–21.

Bellairs, D., Curtis, A., and Dunn, G., eds. (1982) *Cell Behaviour. A Tribute to Michael Abercrombie*, Cambridge University Press, Cambridge, London, New York.

Bereiter-Hahn, J. (1985) Architecture of tissue cells. The structural basis which determines shape and locomotion of cells, *Acta Biotheoret.* **34**:139–148.

Bretscher, M. S. (1984) Endocytosis: Relation to capping and cell locomotion, *Science* **224**:681–686.

Buckley, I. K. (1981) Fine-structural and related aspects of nonmuscle-cell motility, in *Cell and Muscle Motility* (R. M. Dowben and J. W. Shay, eds.), Plenum Press, New York, pp. 135–202.

Fleischer, M., and Wohlfarth-Bottermann, K. E. (1975) Correlation between tension, force generation, fibrillogenesis and ultrastructure of cytoplasmic actomyosin during isometric and isotonic contractions of protoplasmic strands, *Cytobiologie* **10**:339–365.

Hay, E. D. (1983) Interaction of embryonic cell surface and cytoskeleton with extracellular matrix, *Am. J. Anat.* **165**:1–12.

Lackie, J. M. (1986) *Cell Movement and Cell Behaviour*, Allen and Unwin, London.

Middleton, C. A., and Sharp, J. A. (1984) *Cell Locomotion* in Vitro, Croom Helm, London, Canberra.

Oster, G. F. (1984) On the crawling of cells, *J. Embryol. Exp. Morph.* **83**(suppl):329–364.

Oster, G., Murray, J. D., and Harris, A. K. (1983) Mechanical aspects of mesenchymal morphogenesis, *J. Embryol. Exp. Morph.* **78**:83–125.

Trinkaus, J. (1984) *Cells into Organs: Forces That Shape the Embryo*, 2nd ed., Prentice-Hall, Englewood Cliffs, NJ.

Trinkaus, J. P. (1985) Protrusive activity of the cell surface and the initiation of cell movement during morphogenesis, *Exp. Biol. Med.* **10**:130–173.

Vasiliev, J. M. (1982) Spreading and locomotion of tissue cells: Factors controlling the distribution of pseudopodia, *Phil. Trans. R. Soc. Lond.* **299**:159–167.

Abercrombie, M. (1980) The crawling movement of metazoan cells, *Proc. R. Soc. Lond. (Biol.)* **207**:129–147.

Albrecht-Buehler, G. (1976) Filopodia of spreading 3T3 cells. Do they have a substrate exploring function? *J. Cell Biol.* **69**:275–284.

Buckley, I. K., and Porter, K. R. (1967) Cytoplasmic fibrils in living cultured cells: A light and electron microscope study, *Protoplasma* **64**:349–380.

Chen, W.-T. (1979) Induction of spreading during fibroblast movement, *J. Cell Biol.* **81**:684–691.

Couchman, J. R., Badley, R. A., and Rees, D. A. (1983) Redistribution of microfilament-associated proteins during the formation of focal contacts and adhesions in chick fibroblasts, *J. Muscle Res. Cell Motil.* **4**:647–661.

Geiger, B., Avnur, Z., Kreis, T. E., and Schlessinger, J. (1984) The dynamics of cytoskeletal organization in areas of cell contact, in *Cell and Muscle Motility*, Vol. 5 (J. W. Shay, ed.), Plenum Press, New York, pp. 195–234.

Gotlieb, A. I., Heggeness, M. H., Ash, J. F., and Singer, S. J. (1979) Mechanochemical proteins, cell motility and cell–cell contacts: The localization of mechanochemical proteins inside cultured cells at the edge of an *in vitro* "wound," *J. Cell Physiol.* **100**:563–578.

Gotlieb, A. I., May, L. M., Subrahmanyan, L., and Kalnins, V. I. (1981) Distribution of microtubule organizing centers in migrating sheets of endothelial cells, *J. Cell Biol.* **91**:589–594.

Grinnell, F., and Geiger, B. (1986) Interaction of fibronectin-coated beads with attached and spread fibroblasts, *Exp. Cell Res.* **162**:449–461.

Harris, A. K., Wild, P., and Stopak, D. (1980) Silicone rubber substrata: A new wrinkle in the study of cell locomotion, *Science* **208**:177–179.

Harris, A. K., Stopak, D., and Wild, P. (1981) Fibroblast traction as a mechanism for collagen morphogenesis, *Nature* **290**:249–251.

Heath, J. P. (1983) Behavior and structure of the leading lamella in moving fibroblasts. I. Occurrence and centripetal movement of arc-shaped microfilament bundles beneath the dorsal cell surface, *J. Cell Sci.* **60**:331–354.

Heath, J. P., and Dunn, G. A. (1978) Cell to substratum contacts of chick fibroblasts and their relation to the microfilament system. A correlated interference–reflexion and high-voltage electron-microscopy study, *J. Cell Sci.* **29**:197–212.

Herman, I. M., Grisona, N. J., and Pollard, T. D. (1981) Relation between cell activity and the distribution of cytoplasmic actin and myosin, *J. Cell Biol.* **90**:84–91.

Hynes, R. O., Destree, A. T., and Wagner, D. D. (1982) Relationships between microfilaments, cell–substratum adhesion, and fibronectin, *Cold Spring Harbor Symp. Quant. Biol.* **46**:659–669.

Ingram, V. M. (1969) A side view of moving fibroblasts, *Nature* **222**:641–644.

Isenberg, G., Rathke, P. C., Hülsmann, N., Franke, W. W., and Wohlfarth-Bottermann, K-E. (1976) Cytoplasmic actomyosin fibrils in tissue culture cells. Direct proof of contractility by visualization of ATP-induced contraction in fibrils isolated by laser micro-beam dissection, *Cell Tiss. Res.* **166**:427–443.

Izzard, C. S., and Lochner, L. R. (1976) Cell-to-substrate contacts in living fibroblasts: An interference reflexion study with an evaluation of the technique, *J. Cell Sci.* **21**:129–159.

Izzard, C. S., and Lochner, L. R. (1980) Formation of cell-to-substrate contacts during fibroblast motility: An interference–reflexion study, *J. Cell Sci.* **42**:81–116.

Kreis, T. E., and Birchmeier, W. (1980) Stress fiber sarcomeres of fibroblasts are contractile, *Cell* **22**:555–561.

Kupfer, A., Louvard, D., and Singer, S. J. (1982) The polarization of the Golgi apparatus and microtubule-organizing center in cultured fibroblasts at the edge of an experimental wound, *Proc. Natl. Acad. Sci. USA* **79**:2603–2607.

Maher, P. A., Pasquale, E. B., Wang, J. V. J., and Singer, S. J. (1985) Phosphotyrosine-containing proteins are concentrated in focal adhesions and intercellular junctions in normal cells, *Proc. Natl. Acad. Sci. USA* **82**:6576–6580.

McAbee, D. D., and Grinnell, F. (1983) Fibronectin-mediated binding and phagocytosis of polystyrene latex beads by baby hamster kidney cell, *J. Cell Biol.* **97**:1515–1523.

Mittal, A. K., and Bereiter-Hahn, J. (1985) Ionic control of locomotion and shape of epithelial cells: I. Role of calcium influx, *Cell Motil.* **5**:123–136.

Opas, M., and Kalnins, V. I. (1985) Spatial distribution of cortical proteins in cells of epithelial sheets, *Cell Tiss. Res.* **239**:451–454.

Owaribe, K., Kodama, R., and Egchi, G. (1981) Demonstration of contractility of circumferential actin bundles and its morphogenetic significance in pigmented epithelium *in vitro* and *in vivo*, *J. Cell Biol.* **90**:507–514.

Rees, D. A., Couchman, J. R., Smith, C. C., Woods, A., and Wilson, G. (1982) Cell–substratum interactions in the adhesion and locomotion of fibroblasts, *Phil. Trans. R. Soc. Lond. B* **299**:169–176.

Sanger, J. W., Sanger, J. M., and Jockusch, B. M. (1983) Differences in the stress fibers between fibroblasts and epithelial cells, *J. Cell Biol.* **96**:961–969.

Schlessinger, J., and Geiger, B. (1983) The dynamic interrelationships of actin and vinculin in cultured cells, *Cell Motil.* **3**:399–403.

Small, J. V. (1981) Organization of actin in the leading edge of cultured cells: Influence of osmium tetroxide and dehydration on the ultrastructure of actin meshworks, *J. Cell Biol.* **91**:695–705.

Small, J. V., and Rinnerthaler, G. (1985) Cytostructural dynamics of contact formation during fibroblast locomotion *in vitro*, *Exp. Biol. Medi.* **10**:54–68.

Small, J. V., Isenberg, G., and Celis, J. E. (1978) Polarity of actin at the leading edge of cultured cells, *Nature* **272**:638–639.

Svitkina, T. M., Neyfakh, A. A., Jr., and Bershadsky, A. D. (1986) Actin cytoskeleton of spread fibroblasts appears to assemble at the cell edges, *J. Cell Sci.* **82**:235–248.

Turksen, K., Opas, M., Aubin, J. E., and Kalnius, V. I. (1983) Microtubules, microfilaments and adhesion patterns in differentiating chick retinal pigmental epithelium (RPE) cells *in vitro*, *Exp. Cell Res.* **147**:379–391.

Wehland, J., Osborn, M., and Weber, K. (1979) Cell-to-substratum contacts in living cells: A direct correlation between interference–reflexion and indirect immunofluorescence microscopy using antibodies against actin and α-actinin, *J. Cell Sci.* **37**:257–273.

Ameboid cells

Amato, P. A., Unanue, E. R., and Taylor, L. (1983) Distribution of actin in spreading macrophages: A comparative study on living and fixed cells, *J. Cell Biol.* **96**:750–761.

Fechheimer, M., and Zigmond, S. H. (1983) Changes in cytoskeletal proteins of polymorphonuclear leukocytes induced by chemotactic peptides, *Cell Motil.* **3**:349–361.

Howard, T. H., and Oresajo, C. O. (1985) The kinetics of chemotactic peptide-induced change in F-actin content, F-actin distribution and the shape of neutrophils, *J. Cell Biol.* **101**:1078–1085.

Lehto, V. P., Hovi, T., Vartio, T., Badley, R. A., and Virtanen, I. (1982) Reorganization of cytoskeleton and contractile elements during transition of human monocytes into adherent macrophages, *Lab. Invest.* **47**:391–399.

Nemere, I., Kupfer, A., and Singer, S. J. (1985) Reorientation of the Golgi apparatus and the microtubule-organizing center inside macrophages subjected to a chemotactic gradient, *Cell Motil.* **5**:17–29.

Preston, T. M. (1985) A prominent microtubule cytoskeleton in acanthamoeba, *Cell Biol. Int. Rep.* **9**:307–315.

Rubino, S., Fighetti, M., Unger, E., and Cappuccinelli, P. (1984) Location of actin, myosin, and microtubular structures during directed locomotion of *Dictyostelium amebae*, *J. Cell Biol.* **98**:382–390.

Taylor, D. L., and Condeelis, J. S. (1979) Cytoplasmic structure and contractility in amoeboid cells, *Int. Rev. Cytol.* **56**:57–144.

Taylor, D. L., Blinks, J. R., and Reynolds, G. (1980) Contractile basis of ameboid movement. VIII. Aequorin luminescence during ameboid movement, endocytosis, and capping, *J. Cell Biol.* **86**:599–607.

Valerius, N. H., Stendahl, O., Hartwig, J. H., and Stossel, T. P. (1981) Distribution of actin-binding protein and myosin in polymorphonuclear leukocytes during locomotion and phagocytosis, *Cell* **24**:195–202.

Wallace, P. J., Wersto, R. P., Packman, C. H., and Lichtman, M. A. (1984) Chemotactic peptide-induced changes in neutrophilactin conformation, *J. Cell Biol.* **99**:1060–1065.

Wehland, J., Weber, K., Gawlitta, W., and Stockem, W. (1979) Effects of the actin-binding protein DNAase I on cytoplasmic streaming and ultrastructure of *Amoeba proteus*. An attempt to explain amoeboid movement, *Cell Tiss. Res.* **199**:353–372.

Yumura, S., Mori, H., and Fukui, Y. (1984) Localization of actin and myosin for the study of ameboid movement in *Dictyostelium* using improved immunofluorescence, *J. Cell Biol.* **99**:894–899.

Axonal growth

Bray, D. (1979) Mechanical tension produced by nerve cells in tissue culture, *J. Cell Sci.* **37**:391–410.

Bray, D., and Chapman, K. (1985) Analysis of microspike movements on the neuronal growth cone, *J. Neurosci.* **5**:3204–3213.

Bray, D., and Gilbert, D. (1981) Cytoskeletal elements in neurons, *Ann. Rev. Neurosci.* **4**:505–523.

Harrison, R. G. (1910) The outgrowth of the nerve fiber as a mode of protoplasmic movement, *J. Exp. Zool.* **9**:787–846.

Katz, M. J., George, E. B., and Gilbert, L. J. (1984) Axonal elongation as a stochastic walk, *Cell Motil.* **4**:351–370.

Letourneau, P. C. (1981) Immunocytochemical evidence for co-localization in nerve growth cones of actin and myosin and their relationship to cell–substratum adhesions, *Dev. Biol.* **85**:113–122.

Letourneau, P. C. (1983) Axonal growth and guidance, *Trends Neur. Sci. (TINS)* **6**:451–455.

Marsh, L., and Letourneau, P. C. (1984) Growth of neurites without filopodial or lamellipodial activity in the presence of cytochalasin B., *J. Cell Biol.* **99**:2041–2047.

Shaw, G., and Bray, D. (1977) Movement and extension of isolated growth cones, *Exp. Cell Res.* **104**:55–62.

Solomon, F. (1981) Guiding growth cones, *Cell* **24**:279–280.

Solomon, F., and Magendantz, M. (1981) Cytochalasin separates microtubule disassembly from loss of asymmetric morphology, *J. Cell Biol.* **89**:157–161.

Yamada, K. M., Spooner, B. S., and Wessells, N. K. (1971) Ultrastructure and function of growth cones and axons of cultured nerve cells, *J. Cell Biol.* **49**:614–635.

Other systems

Allen, R. D., Zacharski, L. R., Widirstky, S. T., Rosenstein, R., Zaitlin, L. M., and Burgess, D. R. (1979) Transformation and motility of human platelets. Details of the shape change and release reaction observed by optical and electron microscopy, *J. Cell Biol.* **83**:126–142.

Edds, K. T. (1984) Differential distribution and function of microtubules and microfilaments in sea urchin coelomocytes, *Cell Motil.* **4**:269–281.

Fay, F. S., Fujiwara, K., Rees, D. D., and Fogarty, K. E. (1983) Distribution of α-actinin in single isolated smooth muscle cells, *J. Cell Biol.* **96**:783–795.

Lehtonen, E., Lehto, V-P., Badley, R. A., and Virtanen, I. (1983) Formation of vinculin plaques precedes other cytoskeletal changes during retinoic acid-induced teratocarcinoma cell differentiation, *Exp. Cell Res.* **144**:191–197.

Lewis, J. C., White, M. S., Prater, T., Porter, K. R., and Steele, R. J. (1983) Three-dimensional organization of the platelet cytoskeleton during adhesion *in vitro*: Observations on human and nonhuman primate cells, *Cell Motil.* **3**:589–608.

Marchisio, P. C., Cirillo, D., Naldini, L., Primavera, M. V., Teti, A., and Zambonin-Zallone, A. (1984) Cell–substratum interaction of cultured avian osteoclasts is mediated by specific adhesion structures, *J. Cell Biol.* **99**:1696–1705.

Naib-Majani, W., Stockem, W., Wohlfarth-Bottermann, K-E., Osborn, M., and Weber, K. (1982) Immunocytochemistry of the acellular slime mold *Physarum polycephalum*. II. Spatial organization of cytoplasmic actin, *J. Cell Biol.* **28**:103–114.

Small, J. V. (1985) Geometry of actin–membrane attachments in the smooth muscle cell: The localizations of vinculin and α-actinin, *EMBO J.* **4**:45–49.

Tilney, L. G., and Inoue, S. (1985) Acrosomal reaction of the *Thyone* sperm. III. The relationship between actin assembly and water influx during the extension of the acrosomal process, *J. Cell Biol.* **100:**1273–1283.

White, G. E., Gimbrone, M. A., Jr., and Fujiwara K. (1983) Factors influencing the expression of stress fibers in vascular endothelial cells *in situ,* J. Cell Biol. **97:**416–424.

Wong, A. J., Pollard, T. D., and Herman, I. M. (1983) Actin filament stress fibers in vascular endothelial cells *in vivo, Science* **219:**867–869.

8

Reorganization of Cytoskeleton
Cell Division

I. Introduction

Cell division is a multistage reorganization of morphology in which alterations of cytoskeleton play the central role. Division, like reorganization of
interphase cells, described in Chapter 7, is based on alteration of the assembly
and distribution of cytoskeletal structures. There are, however, important differences between these two types of reorganization. Morphogenetic reorganization adapts the morphology of the interphase cell to its environment; the
course of this reorganization is directed by external factors. In contrast, the
course of cell division is directed by a strict internal program; external factors
can only disturb this course.

Cell division includes two sequences of events: nuclear division (mitosis)
and cytoplasmic division (cytokinesis). These events are described in all general biology courses. Here we will only review the main alterations of the
three types of cytoskeletal fibrils which take place during mitosis and
cytokinesis.

The microtubular system undergoes the most complex reorganization
during mitosis (Fig. 8.1A–E). These reorganizations include

1. Formation of two polar microtubule-organizing centers (MTOCs). In
   many cells duplicated centrioles separate and migrate to the positions
   of future spindle poles; these centrioles become surrounded by halos
   of pericentriolar material. In some cells polar MTOCs without centrioles are formed *de novo* (see Chapter 2). The polar MTOCs of many
   lower eukaryotes develop from specialized regions of the nuclear
   envelope.
2. Destruction of the interphase microtubular array and growth of two
   half-spindle arrays from the two polar centers.
3. Development of kinetochores around the centromeric regions of chromosomes and association of these kinetochores with the spindle.
4. Association of the two half-spindles into an integrated spindle.

251

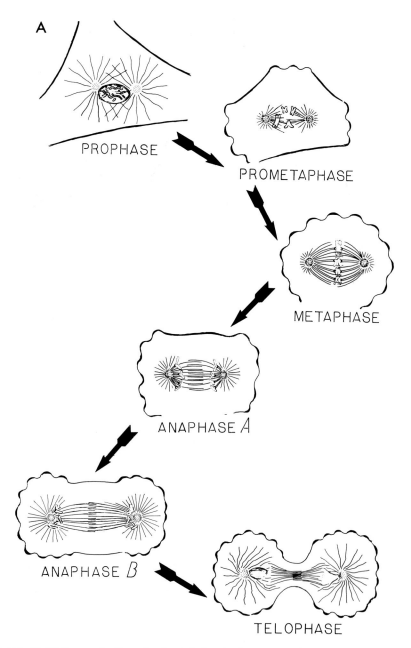

FIG. 8.1. (A–E) Reorganization of microtubules in the course of mitosis and cytokinesis. (A) Scheme. (B) Tubulin immunofluorescence in cultured 3T3 cell. (a) Interphase; (b) prophase; (c) prometaphase; (d) metaphase; (e) telophase; (f) telophase showing midbody. (From Cox *et al.*, 1979. Courtesy of B. R. Brinkley.) (C) Anaphase in cultured epithelial cell (PE line). Electron micrograph of the section. Centrioles at the poles (arrows). (From Vorobjev and Chentsov, 1982. Courtesy of I. A. Vorobjev.) (D) Metaphase in cultured cell (PTK line). Tubulin immunofluorescence. (Courtesy of B. R. Brinkley.) (E) Electron micrograph of midbody of the cultured cell (HeLa line). ×30,000, reproduced at 37%. (Courtesy of B. R. Brinkley.)

(F) Scheme of redistribution of actin during division of cultured animal cell. Stress fibers disappear in the prophase and the cell is rounded. It is not clear whether the spindle contains actin. A contractile ring of actin filaments is formed during cytokinesis. Cytokinesis is followed by respreading of daughter cells and formation of new stress fibers.

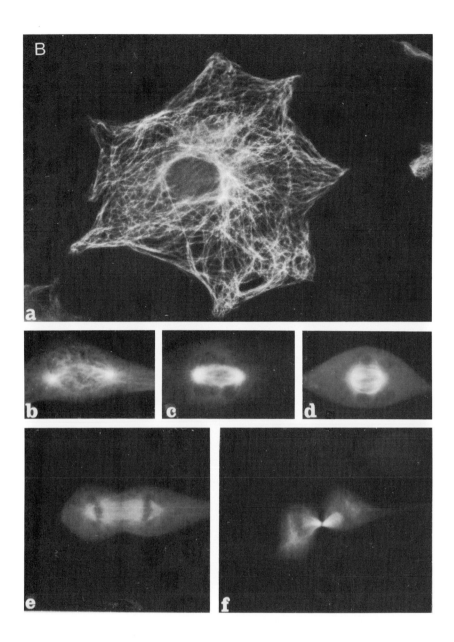

(G) Scheme of reorganization of intermediate filaments during mitosis. Formation of the "cage" of filaments around the spindle (black arrow) and of the globular aggregates of IF-protein (white arrow).

(H) Reorganization of actin and tubulin structure during division of meristematic plant cells. Root tip cells of *Allium cepa* were double-stained for tubulin (indirect immunofluorescence using fluorescein-labeled antibodies) and for F-actin (rhodamine-labeled phalloidin). The same cells were photographed twice using fluorescence microscopy with two sets of filters revealing only the fluorescence of fluorescein or of rhodamine. Left: Actin; right: tubulin within the same cell. First row: Interphase cell with actin cables and with cortical array of microtubules arranged at right angle to actin cables and to long axis of the cell. Second row: Preprophase cell with prepro-phase bundle of microtubules but no equivalent distribution of actin. Third row: Metaphase. Fourth row: Anaphase. Fifth row: Telophase cell. Codistribution of microtubules and of actin in the phragmoplast. Scale bar: 10 μm. (From Clayton and Lloyd, 1985. Courtesy of L. Clayton.)

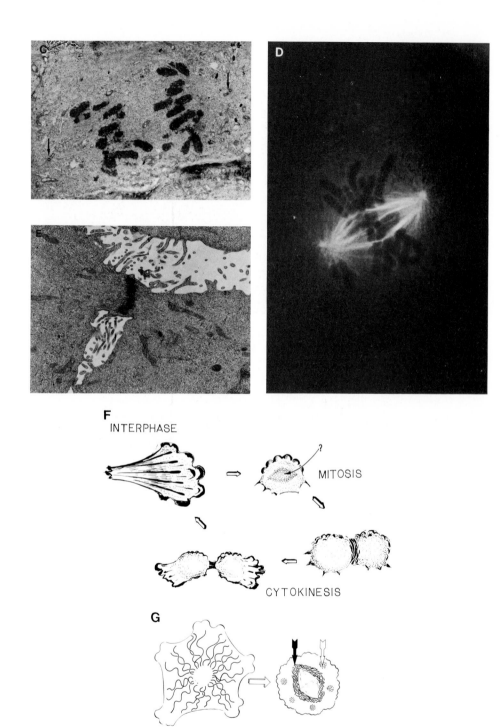

FIG. 8.1. (Continued)

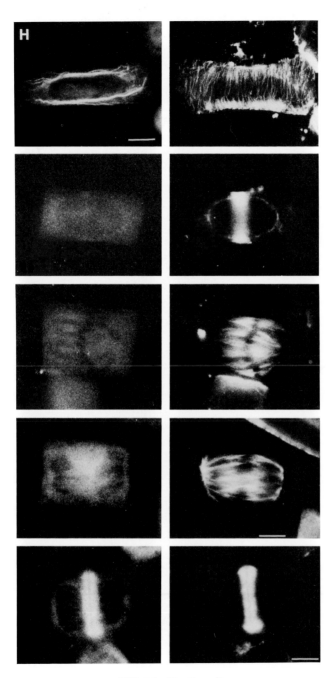

FIG. 8.1. (Continued)

5. Movement of kinetochores with sister chromatids toward the opposite poles (anaphase A).
6. Mutual sliding of two half-spindles away from one another (anaphase B).
7. Disassembly of the remaining spindle microtubules and reassembly of interphase arrays of two daughter cells: disassembly of kinetochores.

These microtubular reorganizations are observed during division of all types of eukaryotic cells. An additional type of alteration is observed in plant cells. Here the loosely distributed cortical microtubules of the interphase cells (see Chapter 2) are replaced by a dense circular microtubular bundle. This preprophase band disappears before the beginning of mitosis (Fig. 8.1H). The plane of division of the two daughter cells usually corresponds to the previous position of the preprophase band. Actin reorganizations during division of animal cells (Fig. 8.1F) include

1. Disappearance of well-organized actin structures characteristic of interphase cells, e.g., of the stress fibers of fibroblasts or of myofibrils of myocytes; this disappearance is often accompanied by cell rounding.
2. Formation of a contractile ring; that is, of a circular dense bundle of actin filaments at the cell periphery.
3. Progressive contraction of this ring associated with the formation and growth of division furrow.
4. Reformation of interphase actin structures and restoration of interphase shapes of daughter cells.

Intermediate filaments in many cells are concentrated at the periphery of the mitotic apparatus (Fig. 8.1G). As described in Chapter 3, in some cells intermediate filaments disappear during division and their material is collected into globular aggregates. There is no evidence that intermediate filaments play an active role at any stage of cell division.

In the following sections we shall briefly discuss the possible mechanisms of several selected types of cytoskeletal reorganizations occurring during cell division. Besides the intact cells, two experimental systems have been widely used in recent studies of these mechanisms: (1) models of dividing cells permeabilized by nonionic detergents and (2) isolated spindles obtained by lysis of mitotic cells in microtubule-stabilizing buffers and further centrifugation. Permeabilized cells and even isolated spindles are able to continue the anaphasic movements of chromosomes in the presence of ATP. Permeabilized cells incubated with ATP are also able to continue cytokinesis.

II. Formation of the Spindle

Formation of the spindle includes the three main types of cytoskeletal reorganization: radial growth of microtubules from the poles, formation of kinetochore-associated microtubules, and association of the two half-spindles. Radial growth of microtubules from the poles is the first obvious manifestation of formation of the spindle. This growth is observed at the end of the prophase. New spindle microtubules are not added to the interphase array, but replace it. Interphase microtubules are broken down and destroyed at the beginning of the prophase, while spindle microtubules start to grow *de novo* from the same centers. When the MTOC of the interphase cell is switched over to spindle formation, its nucleating ability is considerably increased; the number of microtubules growing from the poles at the end of the prophase can be about 5–10 times greater than that in the interphase. Centrosomes isolated from mitotic cells nucleate more microtubules *in vitro* than those isolated from interphase cells.

Nucleation of the spindle microtubules seems to be regulated by the state of the surrounding cytoplasm. Centrosomes isolated from mammalian cells and injected into the cytoplasm of fertilized frog eggs induced the formation of large astral microtubular arrays. In contrast, centrosomes injected into unfertilized eggs did not nucleate the spindle. When the unfertilized egg was activated by pricking, its cytoplasm aquired the ability to form a spindle around the injected centrosome. Further studies of the changes caused by activation may elucidate the specific factors controlling the nucleating properties of centrosomes. Besides centrosomes, mitotic cells contain a second type of potential MTOC, kinetochores. As discussed in Chapter 2, kinetochores can nucleate microtubules *in vitro*, although less efficiently than centrosomes. The polarity of microtubules nucleated by kinetochores can be opposite to that of microtubules nucleated by centrosomes. Nucleation of microtubules by kinetochores was also observed *in vivo* during restoration of the spindle after its destruction by microtubule-specific drugs. In these conditions the mitotic cell contains an unusually high concentration of nonpolymerized tubulin, which can promote nucleation even from kinetochores. It is not clear whether the kinetochore nucleates microtubules during the normal course of mitosis.

Besides nucleation, or instead of it, kinetochores can perform another function during mitosis: they can capture the microtubules growing from the poles. These kinetochore microtubules become attached to the MTOCs at both ends and therefore have increased stability (see Chapter 2). Association of the two kinetochores of each chromosome with the ends of spindle microtubules growing from the opposite poles is a long process that may involve much trial and error. Some chromosomes become attached to the polar microtubules near the pole and then move toward the equatorial region. Formation of stable bipolar attachments in this region is often preceded by rotation and oscillatory movements of chromosomes to and from the poles. Recent experiments of Mitchison and Kirschner (1985) seem to reproduce *in vitro* some of these complicated prometaphase movements. They had shown that kinetochore-capped microtubules placed in tubulin solution continue to elongate not only at the free ends, but also at the kinetochore-attached ends (so-called proximal assembly). Simultaneously with this assembly, the kinetochore moves to the end of the microtubule, so that this end remains capped. This translocation, but not the proximal assembly, is ATP-dependent.

The exact mechanisms of complex exploratory prometaphase movements remain obscure. Theoretically, several types of processes can be responsible for these movements.

It is possible that kinetochores slide along the microtubules attached to the poles. The force for this sliding can be provided by some ATP-driven translocator molecules attached either directly to the kinetochores or to the kinetochore-nucleated microtubules. Immunomorphological studies of sea urchin eggs have shown that mitotic spindle contains proteins reacting with antibodies against kinesin (see Chapter 5) and against dynein. Both these ATPases can be involved in prometaphase movements; however, at present their function is not yet clear.

Still another possibility is that chromosomes are pulled by special fibrils attached to them. Electron-dense material, termed "collar," extending from the poles to the chromosomes was in fact found in the mitotic cells of certain diatoms; its composition and function are not yet clear. Still another possibility is that growing polar microtubules push the chromosomes around until the stable kinetochore–pole connection is formed (Fig. 8.2A). Experiments with photobleaching indicate that the spindle is a highly dynamic structure,

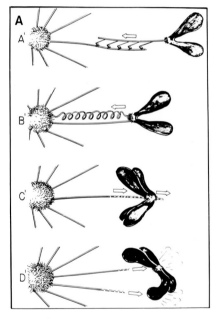
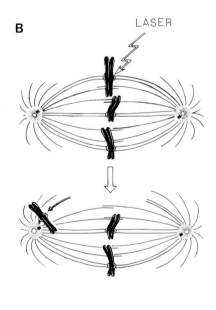

FIG. 8.2. (A) Several hypothetical mechanisms of prometaphase chromosome movements: (A) Dynein-mediated sliding of microtubules. (B) Pulling action of special contractile matrix. (C) Proximal assembly of kinetochore microtubule and active chromosome translocation along the microtubule toward the plus end. (D) Pushing action of growing polar microtubules. The first two mechanisms may also be involved in the anaphase A movements. **(B)** Inactivation of one kinetochore by laser microbeam causes movement of the metaphase chromosome toward the pole connected with the other kinetochore. (Based on McNeil and Berns, 1981.)

where microtubules can disappear and regrow every few seconds (see Chapter 2). Obviously, the chromosome may make many transitory contacts with these oscillating microtubules before a stable kinetochore–microtubule attachment is established.

The real mechanisms of spindle–chromosome interactions are yet to be found; possibly, various mechanisms act in various cells, e.g., in diatoms and in higher eukaryotes. It is also possible that different mechanisms are responsible for various components of prometaphase movements, e.g., chromosome migrations to and from the pole.

Association of two kinetochores of each bivalent chromosome with the microtubules of two half-spindles obviously binds these half-spindles to one another. Another factor essential for this binding is interaction of polar microtubules; that is, of spindle microtubules with free distal ends not captured by kinetochores. These microtubules, growing from two opposite poles, interdigitate in the central part of the spindle. Analysis of the distribution of these antiparallel microtubules in the center of the diatom spindle has shown that they form a regular structure with definite spacing between the microtubules. Lateral links were seen between the interdigitating microtubules. Possibly, these links stabilize the association of two half-spindles.

III. Reorganization of the Spindle Leading to Chromatid Separation

A. Dynamic Equilibrium of the Metaphase Spindle

The fully formed metaphase spindle consists of two interconnected half-spindles, which contain immobilized chromosomes at the equator. This metaphase position of bivalent chromosomes is a result of equilibrium between the opposite traction forces acting on their two kinetochores and directed toward the corresponding poles of the spindle.

Several factors strongly suggest the existence of these traction forces. First, abnormal chromosomes with unequal numbers of kinetochores in their sister chromatids do not occupy the central position in the spindle during the metaphase, but are shifted closer to the pole toward which the greater number of kinetochores are pulling. Second, when one of a pair of kinetochores with bivalent chromosomes is inactivated by laser irradiation, the whole double chromosome immediately starts to move toward the pole to which the second unirradiated kinetochore is attached (Fig. 8.2B).

Third, any cytoplasmic particle that happens to be present within the metaphase spindle moves toward the nearer pole. Thus, poleward traction forces seem to act not only on the kinetochores, but also on the particles of diverse nature.

The equilibrated structure of the spindle undergoes two types of change during the anaphase. Chromosomes move toward the poles (anaphase A), and the poles move away from one another—the spindle elongates (anaphase B).

B. Poleward Movement of Kinetochores

Anaphase begins by separation of sister chromatids; these chromatids then begin to move toward the pole. Probably, the forces moving these chromatids are identical to the forces that at the previous stage pulled the bivalent chromosomes toward the opposite poles. In other words, separation of chromatids may be necessary and sufficient for initiation of their poleward movement. The presence of spindle microtubules is essential for these movements, as their depolymerization by specific drugs stops chromosome migration in permeabilized anaphase cells.

The possible nature of forces pulling chromosomes is still discussed in the literature (Fig. 8.2A). The two most obvious possibilities are that chromosome movements are due either to actin–myosin interactions or to dynein–microtubule interactions. Cytochalasins, phalloidin, and other agents specifically acting on actin–myosin structures did not affect chromosome movements in permeabilized cell models. These data argue against a significant role of actin–myosin structures in these movements. The role of dynein-based mechanisms seems to be more likely, but is also not yet proven. The role of other microtubule-binding translocators also remains to be proven. It is also possible that poleward migration of chromosomes is due to the action of some special, unusual structures. For instance, the collar material of mitotic diatoms may participate not only in prometaphase movements (see section II), but also in anaphase migrations. Evidence for the presence of similar structures in the mitotic cells of higher eukaryotes is still lacking. One cannot exclude the role of such unconventional mechanisms as spasmin-based motility (see Chapter 4).

Poleward movements of chromosomes are somewhat similar to the radial movements of organelles in interphase cells (see Chapter 5). In particular, both types of movements are strictly dependent on the presence of intact microtubules. However, poleward chromosome migrations also have a very special feature: they are associated with progressive shortening of the microtubules to which the kinetochores are attached. This shortening has not been observed in any other system. The shortening can be due to some local depolymerization of the microtubule; for instance, the moving kinetochore can induce this depolymerization by some unknown mechanism. Another possibility is that dynamic kinetochore microtubules repeatedly shrink, regrow from the pole, and reattach themselves to the kinetochore. As the kinetochore moves poleward, each subsequent reattachment will occur at a shorter distance from the pole; that is, the length of attached microtubule will be progressively decreased. Microtubule shortening seems to be essential for the migrations of kinetochore, as these migrations can be stopped by the microtubule-stabilizing drug taxol. It has been suggested that kinetochore microtubules act as sticks limiting the rate of poleward displacement of the chromosome.

Thus, kinetochore-to-pole movements during anaphase seem to be caused by the traction forces of unknown nature which develop within the fully formed metaphase spindle and set kinetochores in motion after the separation of sister chromatids. These movements are accompanied by progressive shortening of the kinetochore-attached microtubules and may be dependent on this shortening.

C. Elongation of the Spindle

Elongation of the spindle manifested by an increase in interpolar distance is the second component of anaphase movements. This elongation can be very large; e.g., the length of the spindle of certain amebae increases three- or fourfold from metaphase to anaphase. The elongation is probably caused by mutual sliding of the antiparallel polar microtubules, which interdigitate in the central part of the spindle. Morphological examination of the mitotic diatoms has confirmed that the length of this central zone of interdigitating microtubules decreases during the anaphase. The average length of the polar microtubules can increase during the anaphase. Thus, sliding of these microtubules can be accompanied by their growth; both these processes will push the spindle poles apart. Elongation of the isolated spindles of diatoms is induced by addition of ATP; this elongation leads to full separation of the two half-spindles.

There are three possible versions of the mechanisms of spindle sliding: actin, dynein, or unknown molecules. Sliding of the half-spindles in permeabilized cell models and in isolated spindles is not inhibited by cytochalasins on phalloidin, but is inhibited by vanadate. These data suggest that sliding is based not on the actin–myosin mechanism, but on the action of dyneinlike molecules. However, elongation of the spindle is different from classical dynein-based flagellar bending in one important aspect. The sliding microtubules in the flagellum have similar polarities, while the microtubules of two half-spindles sliding along one another have opposite polarities. Thus, even if the sliding of the spindle is based on dyneinlike mechanisms, the characteristics of this mechanism would be significantly different from that acting in the flagellum. In other words, at present the most plausible version is that sliding is driven by some unknown molecules, possibly related to dynein or kinesin.

IV. Cytokinesis

Formation and constriction of the furrow followed by separation of the daughter cells are the main processes occurring in the telophase of animal cell division. High sensitivity to cytochalasins is a characteristic feature of these processes distinguishing them from metaphase and anaphase movements; even low concentrations of cytochalasins block cytokinesis, so that mitosis is terminated by the formation of binucleated cells. Cleavage of permeabilized mitotic cells is blocked not only by cytochalasins, but also by two other agents interacting with actin, phalloidin and N-ethylmaleimide-modified subfragment-I of myosin. Thus, cytokinesis is the only part of animal cell division that is obviously based on actin–myosin interactions.

Formation of a ring of actin filaments in the cortex precedes furrowing.

In addition to actin, immunofluorescence reveals myosin (Fig. 8.3), α-actinin, and filamin in this ring. Furrowing is accompanied by contraction of the ring. Probably, the filaments of the ring are somehow anchored to the cell membrane, so constriction of the ring causes invagination of the membrane. It is not yet known in what form and by what mechanism material for the ring is transported to the cortex. Spindle microtubules probably control this transport, because the position of the ring is determined by the localization of the equator of the spindle. If the orientation of the spindle is altered by micromanipulation at an early stage of mitosis, the furrow is formed at the site corresponding to the new position of the spindle equator. The process of furrow formation seems to be similar in certain respects to the processes of pseudopod formation occurring at the active edges of fibroblasts (see Chapter 7). In both situations cytoplasmic transport of materials is directed toward certain sites of the cell surface located near the plus ends of microtubules. This transport is followed by formation at these sites of a contractile filament-containing network coated by the external membrane. As a result of these processes, either exvagination (pseudopod formation) or invagination (furrowing) of the surface take place.

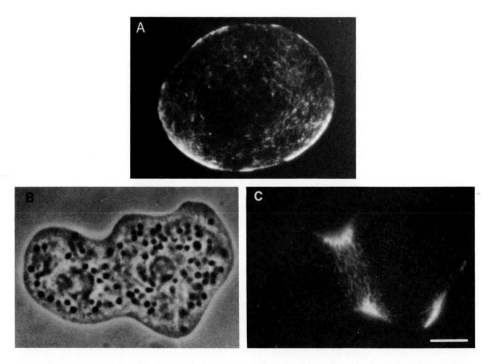

FIG. 8.3. Concentration of myosin in the contractile ring of dividing *Dictyostelium discoideum* myxamebae. Immunofluorescence using antimyosin antibody (A,C). Phase contrast (B). **(A)** Interphase cell containing numerous myosin rods within the cytoplasm. **(B,C)** Cell performing cytokinesis; myosin in the contractile ring. Scale bar: 5 μm. (From Yumura and Fukui, 1985. Courtesy of S. Yumura.)

Contraction eventually leads to formation of an intercellular bridge. The central part of this bridge, called the midbody, contains closely packed parallel microtubules and dense matrix material. Probably, formation of the midbody is the result of constriction of the central part of the spindle; the matrix may consist of concentrated material gluing together the two half-spindles.

Furrow contraction leading to formation of the bridge is not always immediately followed by complete separation of the sister cells. Sometimes, the cells can remain connected by the bridge for several hours, so-called post-telophasic delay of separation. After this delay, the bridge begins to elongate and is eventually broken. Possibly, formation of the furrow and final break of the intercellular bridge can have somewhat different mechanisms. The morphology of the cytokinesis of plant cells is very different from that of animal cells. The telophasic plant cell forms a cylindrical array of parallel microtubules, phragmoplast, in the equatorial zone. The membrane-coated vesicles are then transported along these microtubules to the equatorial plane; fusion of these vesicles leads to formation of the cell plate, eventually separating the sister cells. Recently it was found that actin filaments are concentrated in the zone of the equatorial plane before the formation of the cell plate (Fig. 8.1F); the function of this actin is not clear. Despite all the obvious differences between the cytokinesis of plant and animal cells, these processes apparently have some similarities. In both systems microtubular arrays perpendicular to the plane of division are formed: the phragmoplast in plant cells and the midbody in animal cells. In both systems actin filaments are accumulated in the plane of division. Possibly, the transport of membrane material toward the plane of division takes place not only in the plant cells, but also in the animal cells. This material in animal cells may be incorporated into the constricting furrow; in the plant cell, into the cell plate.

V. Induction and Coordination of Mitotic Events

We have discussed possible mechanisms of reorganization occurring at different stages of cell division. In addition to the specific problems arising with regard to each of these events, there is also another more general problem, coordination of the various events. How are all the processes, having different molecular mechanisms, coordinated into a highly ordered sequence? What signals at the end of each stage induce the beginning of the next stage? Fusion of a mitotic and an interphase cell induces prophasic changes (envelope breakdown and chromatin condensation) in the interphase nucleus. This fact suggests that mitotic events can be triggered by some factors present in the cytoplasm. Discussion of the nature of concrete signal changes triggering mitosis and regulating its further progress is usually centered around the two familiar possibilities, calcium alteration and protein phosphorylation. The importance of calcium in regulation of mitosis is suggested by several facts. Calcium-binding protein, calmodulin, was shown immunomorphologically to be accumulated in the spindle, especially in its central parts. Electron microscopy revealed the presence of numerous membranous vesicles within the spindle. These vesicles are formed at the beginning of mitosis from elements of the endoplasmic reticulum and of the Golgi apparatus. The vesicles present in the spindle were shown to contain calcium; possibly, they can take up this ion from the surrounding cytoplasm and release it. In this way the local concentration of calcium in the medium around the spindle microtubules can be regulated. Calcium, in conjunction with calmodulin, can control the state of the spindle and, especially, the degree of polymerization of microtubules. As yet, there are, however, no direct data about the changes of calcium concentration at various stages of mitosis and about their critical regulatory roles.

The role of protein phosphorylation in regulation of mitotic events is suggested by the recent observations of Vandre et al. (1984) in which a monoclonal antibody recognizing various phosphoproteins present in the mitotic cultured cells was used. Immunomorphological examination of cultured cells has shown that material reacting with this antibody is present in small amounts in the nucleus and in the centrosome during the interphase. In the prophase the amount of this material increases, and it appears in the cytoplasm, where it is present throughout mitosis. These data suggest that the degree of phosphorylation of cellular proteins may increase during mitosis, but they give no indication of the identity of these proteins, as the antibody used in these experiments reacts with a large number of various phosphorylated molecules. Increased phosphorylation of vimentin was observed during mitosis in another investigation. Data on phosphorylation of other cytoskeletal proteins are still lacking.

Control of the timing of mitotic events depends not only on the local environment around the cytoskeletal structures, but also on the structural integrity of these structures. In the experiments of Sluder and Begg (1983), the spindle was depolymerized by colcemid and then restored after removal of the drug; as a result the last stage of division was considerably delayed. In another experiment a similar delay was observed after mechanical separation of the spindle into two half-spindles. These results suggest that the onset of the following stages of cell division can be postponed until termination of previous stages. Obviously, this regulation can safeguard the exact chromosome distribution between the daughter cells against reversible alterations of normal spindle structure.

Thus, the nature of mitotic events and of their coordination remains a major challenge for cell biologists. Watching the time-lapse movies of mitosis, e.g., the classic films of Bajer, one cannot help admire the visual beauty, order, and complexity of this process. But we still do not understand how this beauty and order are achieved.

Literature Cited

Clayton, L., and Lloyd, C. W. (1985) Actin organization during the cell cycle in meristematic plant cells. Actin is present in the cytokinetic phragmoplast, *Exp. Cell Res.* **156**:231–238.

Cox, S. M., Rao, P. N., and Brinkley, B. P. (1979) Action of nitrous oxide and griseofulvin on microtubules and chromosome movement in dividing cells, in *Effects of Drugs on Cell Nucleus* (H. Busch, S. T. Crooke, and Y. Dascal, eds.), Academic Press, New York, pp. 521–549.

McNeil, P. A., and Berns, M. W. (1981) Chromosome behavior after laser microirradiation of a single kinetochore in mitotic PtK₂ cells, *J. Cell Biol.* **88**:543–553.

Mitchison, T., and Kirschner, M. W. (1985) Properties of the kinetochore *in vitro*. II. Microtubule capture and ATP-dependent translocation, *J. Cell Biol.* **101**:766–777.

Sluder, G., and Begg, D. A. (1983) Control mechanisms of the cell cycle: Role of the spatial arrangement of spindle components in the timing of mitotic events, *J. Cell Biol.* **97**:877–886.

Vandre, D. D., Kronebusch, P., and Borisy, G. G. (1984) Interphase–mitosis transition: Microtubule rearrangements in cultured cells and sea urchin eggs, in *Molecular Biology of the Cytoskeleton* (G. G. Borisy, D. W. Cleveland, and D. B. Murphy, eds.), Cold Spring Harbor Laboratory, Cold Spring Harbor, New York, pp. 3–16.

Vorobjev, I. A., and Chentsov, Y. S. (1982) Centrioles in the cell cycle. I. Epithelial cells, *J. Cell Biol.* **98**:938–949.

Yumura, S., and Fukui, Y. (1985) Reversible cyclic AMP-dependent change in distribution of myosin thick filaments in *Dictyostelium*, *Nature* **314**:194–196.

Additional Readings

Aubin, J. H., Osborn, M., and Weber, K. (1981) Inhibition of cytokinesis and altered contractile ring morphology induced by cytochalasins in synchronized PtK₂ cells, *Exp. Cell Res.* **136**:63–79.

Bajer, A. S. (1982) Functional autonomy of monopolar spindle and evidence for oscillatory movement in mitosis, *J. Cell Biol.* **93**:33–48.

Bajer, A., and Mole-Bajer, J. (1956) Cine-micrographic studies on mitosis in endosperm, *Chromosoma* **7**:558–607.

Brinkley, B. R., and Brenner, S. L. (1982) Chromosome movement: A mini-review, in *Axoplasmic Transport* (D. G. Weiss, ed.), Springer-Verlag, Berlin, Heidelberg, pp. 27–38.

Cande, W. Z. (1982) Nucleotide requirements for anaphase chromosome movement in permeabilized mitotic cells: Anaphase B but not anaphase A requires ATP, *Cell* **28**:15–22.

Cande, W. Z., McDonald, K., and Meeusen, R. L. (1981) A permeabilized cell model for studying cell division: A comparison of anaphase chromosome movement and cleavage furrow constriction in lysed PtK₁ cells, *J. Cell Biol.* **88**:618–629.

Cande, W. Z., McDonald, K., Wordeman, L., and Coltrin, D. (1985) *In vitro* anaphase spindle elongation using isolated diatom spindles, *Cell Motil.* **5**:169–170.

De Brabander, M., Geuens, G., De Mey, J., and Jonaiau, M. (1979) Light microscopic and ultrastructural distribution of immunoreactive tubulin in mitotic mammalian cells, *Biol. Cell.* **34**:213–226.

Euteneuer, U., Ris, H., and Borisy, G. G. (1983) Polarity of kinetochore microtubules in chinese hamster ovary cells after recovery from a colcemide block, *J. Cell Biol.* **97**:202–208.

Evans, R. M., and Fink, L. M. (1982) An alteration in the phosphorylation of vimentin-type intermediate filaments is associated with mitosis in cultural mammalian cells, *Cell* **29**:43–52.

Hays, T. S., and Salmon, E. D. (1985) Poleward traction force on a kinetochore at metaphase is a product of the number of kinetochore microtubules and fiber length, *J. Cell Biol.* **101** (5,pt2):6a (abstr. 20).

Hirokawa, N., Takemura, R., and Hisanaga, S. J. (1985) Cytoskeletal architecture of isolated mitotic spindle with special reference to microtubule-associated proteins and cytoplasmic dynein, *J. Cell Biol.* **101**:1858–1870.

Hollenbeck, P. J., and Cande, W. Z. (1985) Microtubule distribution and reorganization in the first cell cycle of fertilized eggs of *Lytechinus pictus*, *Eur. J. Cell Biol.* **37**:140–148.

Hollenbeck, P. J., Suprynowicz, F., and Cande, F. Z. (1984) Cytoplasmic dynein-like ATPase cross-links microtubules in an ATP-sensitive manner, *J. Cell Biol.* **99**:1251–1258.

Kiehart, D. P., Mabuchi, I., and Inoue, S. (1982) Evidence that myosin does not contribute to force production in chromosome movement, *J. Cell Biol.* **94**:165–178.

McIntosh, J. R. (1981) Microtubule polarity and interaction in mitotic spindle function, in *International Cell Biology, 1980–1981* (H. G. Schweiger, ed.), Springer-Verlag, Berlin, pp. 359–368.

Nicklas, R. B. (1983) Measurements of the force produced by the mitotic spindle in anaphase, *J. Cell Biol.* **97**:542–548.

Nicklas, R. B., and Gordon, G. W. (1985) The total length of spindle microtubules depends on the number of chromosomes present, *J. Cell Biol.* **100**:1–7.

Paweletz, N., Mazia, D., and Finze, E-M. (1984) The centrosome cycle of sea urchin eggs, *Exp. Cell Res.* **152**:47–65.

Pickett-Heaps, J. D., and Spurck, T. P. (1982) Studies on kinetochore function in mitosis. I. The effects of colchicine and cytochalasin on mitosis in the *Hantzschia amphioxys, Eur. J. Cell Biol.* **28**:72–82.

Pickett-Heaps, J. D., and Spurck, T. P. (1982) Studies on kinetochore function in mitosis. II. The effect of metabolic inhibitors on mitosis and cytokinesis in the diatom *Hantzschia amphioxis, Eur. J. Cell Biol.* **28**:83–91.

Pickett-Heaps, J. D., Tippit, D. H., and Porter, K. R. (1982) Rethinking mitosis, *Cell* **29**:729–743.

Pickett-Heaps, J., Spurck, T. P., and Tippit, D. (1984) Chromosome motion and the spindle matrix, *J. Cell Biol.* **99**(I, pt 2):1375–1435.

Piperno, G. (1984) Monoclonal antibodies to dynein subunits reveal the existence of cytoplasmic antigens in sea urchin eggs, *J. Cell Biol.* **98**:1842–1850.

Pratt, M. (1984) ATPases in mitotic spindles, *Int. Rev. Cytol.* **87**:83–105.

Ring, D., Hubble, R., and Kirschner, M. (1982) Mitosis in a cell with multiple centrioles, *J. Cell Biol.* **9**:549–556.

Scholey, J. M., Porter, M. E., Grissom, P. M., and McIntosh, J. R. (1985) Identification of kinesin in sea urchin eggs, and evidence for its localization in mitotic spindle, *Nature* **318**:483–486.

White, J. I. (1985) The astral relaxation theory of cytokinesis revisited, *Bio Essays* **2**:267–272.

Wolniak, S. N., Hepler, P. K., and Jackson, W. T. (1983) Ionic changes in the mitotic apparatus at the metaphase/anaphase transition, *J. Cell Biol.* **96**:598–605.

9

Neoplastic Transformations
Possible Unexplored Functions of Cytoskeleton

I. Introduction

There are at least two reasons why neoplastic transformations of one particular cell type, cultured fibroblasts, deserve special discussion in a book on the cytoskeleton. First, these transformations are the best-studied genetic changes of cells of higher eukaryotes, manifested by alterations of morphogenesis. Therefore, analysis of these changes can help further our understanding of the mechanisms of regulation of morphogenesis. Second, neoplastic transformations are manifested not only by morphological changes, but also by alterations of growth regulation. Therefore, analysis of the expression of neoplastic transformations can help to reveal the possible relationship of the cytoskeleton to this regulation and, more generally, to the control of biochemical reorganizations of the cell.

II. Deficient Morphogenetic Reactions of Transformed Fibroblasts

A. General Characteristics of Transformations

Most studies of neoplastic transformations *in vitro* have been performed with two types of fibroblastic cultures: primary cultures, obtained directly from embryo tissues, or permanent minimally changed lines, selected from primary cultures. Genetically stable neoplastic transformations can occur in these cultures spontaneously with low frequency; a higher rate of these transformations can be induced by oncogenic viruses, carcinogenic chemicals, or ionizing radiation. Several consecutive transformations can occur within the same population; new cell clones with additional altered properties emerge after each of these transformations.

Each transformation results from the action of a special protein, oncoprotein, which is coded by a gene, oncogene, in the cell genome. Sequential transformations usually involve the appearance of several different oncogenes. At present about 40 different types of oncogenes and oncoproteins are known. In most cases these oncogenes are altered variants of normal cell genes, called protooncogenes. Oncogene can be formed from protooncogene by mutation in the coding part; as a result, abnormal protein is synthesized in the cell. Another possible method of oncogene formation is a genetic change leading to altered regulation of the expression of protooncogene; as a result, abnormal quantities of normal protein are synthesized. Relative roles of qualitative and quantitative changes of the products of protooncogenes in various transformations are not yet completely clear.

Transformations caused by various oncogenes can have nonidentical manifestations and, possibly, nonidentical mechanisms of action. According to the localization of the gene products, nuclear and cytoplasmic oncogenes can be distinguished. We shall describe here only the "typical" transformations caused by cytoplasmic oncogenes (genes called *src*, *sis*, *erbB*, *ras*, *abl*, and others) in permanent lines and, sometimes, in primary cultures of fibroblastic cells. These transformations are manifested by simultaneous alterations of many cell characters (Fig. 9.1). One of them is tumorigenicity; transformed cells become able to form malignant tumors when they are implanted into immunologically compatible animals. Two other groups of alterations can be directly detected in culture: alterations of cell morphology and abnormal growth regulation.

B. Morphological Alterations of Transformed Cells

In Chapter 7 we described the morphological reorganizations of normal fibroblasts and, in particular, their spreading on the substrate. This spreading

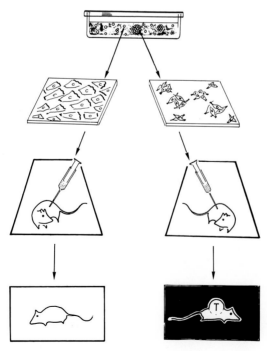

FIG. 9.1. Some altered properties of transformed cells. Neoplastically transformed cell suspended in agar forms colony; nontransformed cells in the same suspension do not multiply. Transformed cells on the substrate form the colonies of badly spread cells. When these cells are injected into immunologically compatible animals, they form tumors.

becomes deficient in transformed cultures. Transformed cells spread on the substrate to smaller areas than their normal progenitors; they form smaller lamellae. Instead of having a flattened, well-spread shape, these cells are often fusiform or almost spherical (Fig. 9.2A,B). The degree of the shape alterations may be different in various transformed lines. In some cases the character of the alterations depends directly on the nature of the inducing oncogenes. For instance, chick fibroblasts transformed by the virus carrying the complete *src* gene are round; the same cells transformed by the virus with certain deletion mutants of *src* gene have fusiform morphology. Like other transformed characters, morphological alterations can increase after each consecutive transformation.

Actin cytoskeleton organization, characteristic of the normal spread fibroblast, is also altered by transformations. The most conspicuous change is

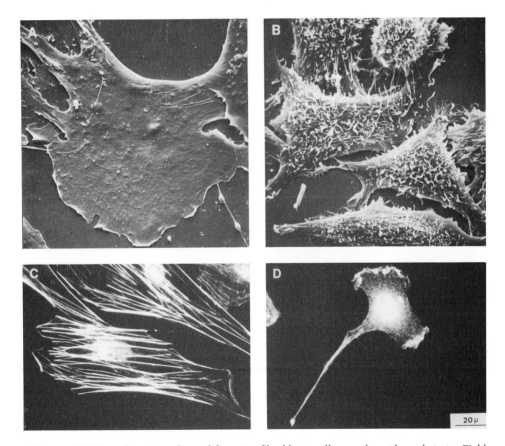

FIG. 9.2. (A) Normal nontransformed hamster fibroblast well spread on the substrate. Field width: 140 μm. **(B)** Badly spread hamster fibroblasts transformed by SV40 virus (HEK line). Scanning electron micrographs. Field width: 70 μm. (Courtesy of Y. A. Rovensky.) **(C,D)** Normal rat fibroblast (C) and spontaneously transformed neoplastic rat fibroblast of the SAM-LEW line (D). Indirect immunofluorescence using antiactin antibodies. Notice disappearance of bundle in transformed cell; actin is accumulated within the pseudopods of the leading edge of this cell. (From Bershadsky *et al.*, 1982.)

deficiency of the filament bundles; these bundles become thinner and less numerous or disappear (Fig. 9.2C,D).

Focal contacts also decrease in size and number. Certain types of transformed cell form only the small "initial" contacts located near the active edges; larger "mature" contacts associated with the ends of actin bundles are

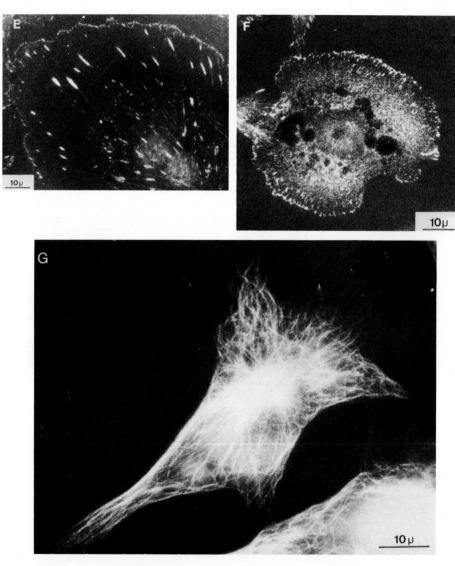

Fig. 9.2. *(Continued)* **(E,F)** Focal contacts in nontransformed quail fibroblast (E) and in quail fibroblast transformed by Rous sarcoma virus (F). Immunofluorescence microscopy using antivinculin antibodies. The nontransformed cell has dotlike initial contacts at the edge and elongated dashlike mature contacts in more central parts of the cell. The transformed cell has only initial contacts at the edge; mature contacts are absent; the thick central part of the badly spread cell is diffusely stained. (From Bershadsky *et al.*, 1985.) **(G)** Well-developed radiating microtubular system in the transformed mouse fibroblast (L line). Indirect immunofluorescence using antitubulin antibody. (Courtesy of I. S. Tint.)

not formed (Fig. 9.2E,F). Deficiency of the bundles and contacts is associated
with decreased tension exerted by the spread cell on the substrate; transformed fibroblast wrinkles the elastic substrate (see Chapter 7) much less
than the normal parent cell. Transformation also decreases formation of specialized cell–cell contacts; even in dense cultures, these contacts become
scarce or disappear completely.

Alterations of intermediate filaments and of microtubules are much less
conspicuous than those of actin filament bundles. Certain lines of transformed cells have a partially collapsed network of vimentin filaments, which
disappear from the cell periphery and are concentrated around the nucleus.
Decreased numbers of microtubules were reported in several types of transformed cells. However, many other transformed cultures show no obvious
morphological changes of these two groups of cytoskeletal fibrils.

Thus, the main morphogenetic defect of transformed fibroblasts can be
described as decreased efficiency of attachment of pseudopods (see Chapter
7) associated with deficient reorganization of the actin cortex and deficient
formation of contact structures.

Pseudopods are used by normal fibroblasts to test their local environment. Deficient pseudopodial attachment decreases the ability of transformed
cells to adapt their position to the environment. Although many transformed
calls remain elongated, they orient themselves poorly along the substrates
and do not form regular arrays of mutually parallel cells in colonies. Data on
the chemical alterations of cytoskeletal components accompanying these
morphological changes are still scarce. A decreased total amount of tropomyosin was found in several transformed lines. More detailed analysis has
shown that many lines of transformed rat cells had similar alterations of the
ratios of five molecular isoforms of tropomyosin. One or two of three major
isoforms decreased after transformation, while one or both minor isoforms
increased (Fig. 9.3). Increased concentration of proteins with actin-capping
activity was also observed in several transformed lines. As described in Chapter 2, tropomyosin can stabilize actin filaments, while capping proteins can
control their length. Possibly, alterations of these actin-binding proteins contribute to changes of the organization of actin cytoskeleton in transformed
cells.

It was also reported that mouse and rat transformed fibroblasts synthesize
only β- and γ-actins, while the synthesis of α-actin was greatly inhibited by
transformation. This modulation of the synthesis of one electrophoretic form
of actin can possibly lead to some alterations of the actin cortex structure.
However, it is not clear whether any of these alterations is common for all
types of transformed cells. Deficient morphogenesis is characteristic not only
of transformed fibroblasts in culture, but of neoplastic cells of many types in
the organism. For instance, the cells of many malignant tumors poorly form
specialized contacts with other cells and with extracellular matrix. Many features of pathological behavior of malignant cells in the organism, such as
abnormal ability to form tissue structure, invasion, and metastasis, may be
due to their altered morphogenetic properties. It was found, for instance, that
certain sublines of transformed cells, selected for high frequency of metastasis *in vivo*, show more deficient spreading *in vitro* than parent lines, giving
low frequency of metastases.

C. Mechanisms of Abnormal Regulation of the Cytoskeletal Organization in Transformed Cells

Morphological changes of certain types of transformed cells can be due to alterations of the genes directly coding cytoskeletal proteins (see Section III). However, in most well-analyzed cases these changes are due to genetic alterations of another type, disturbing the regulation of cytoskeletal organization by membrane-generated signals.

This conclusion is based on studies of primary structures of a number of oncogenes. It was found that proteins coded by these oncogenes, oncoproteins, are homologous to various proteins of a normal regulatory pathway by which certain external ligands, growth factors, induce a sequence of intracellular alterations. We shall briefly describe this pathway before returning to the action of oncoproteins.

Growth factors, by definition, stimulate the sensitive cells to synthesize DNA and to divide. However, in this section we shall discuss not the final result of their action, but the initial effects of these molecules on the plasma

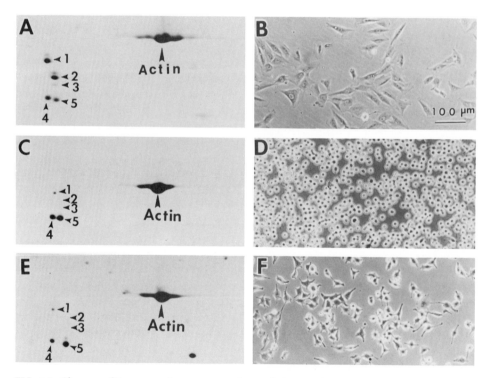

FIG. 9.3. Changes of tropomyosin patterns and morphologies of cells (NRK line) after transformation with two different RNA viruses. **(A,B)** Nontransformed NRK cells. **(C,D)** NRK cells transformed by Kirsten virus. **(E,F)** NRK cells transformed by Rous virus. (B,D,F) Phase contrast micrographs showing that transformed cells are less flattened due to decreased spreading on the substrate. (A,C,E) Parts of two-dimensional gel electrophoresis of isolated microfilaments containing different isoforms of tropomyosin (1–5) and actin. The level of tropomyosins 1, 2, and 3 is decreased and the level of tropomyosin 5 is increased after transformations by both viruses. (From Matsumura *et al.*, 1983. Courtesy of F. Matsumura.)

membrane; these effects can be closely related to alterations of the cytoskeleton.

The two best-known growth factors, active in various fibroblastic and epithelial cultures, are (1) platelet-derived growth factor (PDGF), which has two molecular variants with mol. wt. 31 kD and 28 kD; PDGF is stored in the nonactivated platelets and released into the medium after activation; and (2) epidermal growth factors (EGF) with mol. wt. about 5 kD; EGF is usually isolated from mouse salivary gland extracts.

Each type of growth factor molecule binds to specific receptors present on the surface of sensitive cells. The cells of each type have a characteristic spectrum of receptors to the molecules of various growth factors. Binding of these molecules to corresponding receptors induces a complex of immediate membrane changes, which occur within the first 10 min of exposure. Many of these changes are common for various growth factors bound to different receptors. One of these changes is activation of tyrosine-specific protein kinase. This enzyme catalyzes the attachment of phosphate groups to the tyrosine residues of polypeptide chains. Membrane receptors to several types of growth factors have built-in tyrosine-specific kinase activity localized in the submembranous domains of their molecules. This activity is stimulated by binding of the ligand to the external part of the receptor molecule.

Certain growth factors, e.g., PDGF, also induce activation of specific membrane-bound phosphodiesterase (phospholipase C), which catalyzes the hydrolysis of one of the membrane lipids, phosphatidylinositol 4,5-biphosphate, into diacylglycerol and inositol triphosphate.

Inositol triphosphate then mobilizes the release of calcium ions into the cytoplasm from intracellular stores, especially endoplasmic reticulum. Diacylglycerol, in conjunction with calcium, activates another membrane-bound enzyme, protein kinase C; this kinase is serin-specific. Protein kinase C can phosphorylate many substrates. Among other targets, this kinase stimulates the Na^+–H^+ exchanger; that is, the system that is responsible for the exchange of sodium ions for protons across the plasma membrane. This effect can lead to alkalinization of cytoplasm.

Certain growth factors may also modify the activity of membrane-bound adenylate cyclase and thus alter the production of cAMP.

These manifold membrane alterations act as "second signals" inducing many types of changes inside the cells, including formation of pseudopods. PDGF, EGF, and other growth factors were observed to induce extension of numerous ruffles on the upper surfaces of sensitive cells within the first few minutes of exposure. Fibroblasts have a chemotactic response to the gradient of PDGF in the medium. Thus, growth factors belong to a large group of external ligands which induce extension of pseudopods (see Chapter 7). Large concentrations of growth factors were also observed to induce disruption of the stress fibers, disappearance of vinculin from the focal contacts, and even the rounding of certain sensitive cells; the role of these alterations in growth activation is not clear.

Analysis of the primary structures of several oncoproteins has shown that they are closely related to normal proteins participating in growth factor–membrane interactions. Some oncoproteins are homologous to the growth factor molecules; others, to the growth factor receptors; still others, to the membrane proteins participating in signal generation. For instance, oncoprotein coded by the oncogene *sis* is homologous to PDGF. The transformed cells carrying this oncogene in their genome release the corresponding oncoprotein into the medium. This oncoprotein probably acts as endogenous growth factor; that is, the *sis* protein can be bound to the receptors of the same cell that produced it. There are data indicating that binding of the *sis* to the receptors can even take place within the cell, without previous release of the oncoprotein into the medium. In this way the cell transformed by *sis* oncogene can continuously stimulate its own membrane. Protein *erbB* is homologous to a part of the normal membrane receptor for another exogenous growth factor, EGF. *ErbB* protein, like the normal receptor, is located in the plasma membrane and has a submembranous domain with tyrosine-specific protein kinase activity. Unlike the normal receptor, *erbB* protein does not bind EGF. Oncoprotein *src* is also localized in the plasma membrane and has tyrosine-specific protein kinase activity. These oncoproteins can act as abnormal stimuli continuously causing a "false alarm." Without binding any ligands, these molecules can continuously activate formation of "second signals" sent by the membrane into the cell. In other words, all these oncoproteins are likely to cause pathological hyperstimulation of the normal regulatory pathway. During normal morphogenetic reactions membrane-generated signals can cause local reorganization of the actin cytoskeleton, leading to formation of pseudopods (see Chapter 7). Similar signals continuously generated within the membrane of the transformed cell can cause general reorganization of its actin cytoskeleton. Thus, the transformed cell continuously activated by pathological "endogenous" stimuli pays less attention to normal external stimuli, e.g., the substrate. Therefore, its spreading becomes deficient.

Here, as in many normal situations, the exact sequence of processes connecting membrane alterations and cytoskeleton changes is not yet known. After transformation, oncoprotein coded by the oncogene *src* is accumulated submembranously in the areas of focal contact (Fig. 9.4). Injection of this oncoprotein, designated pp60^{src} into the normal cell caused rapid disorganization of the stress fibers. These results suggest that oncoproteins of this type can directly act on some components of the actin cytoskeleton. One possible specific method of this action is suggested by the finding that vinculin and talin present in the focal contact can be phosphorylated by pp60^{src} which possesses an inherent tyrosine-specific protein kinase activity. However, this oncoprotein also phosphorylates many other cytoplasmatic proteins; in particular, a submembranous actin-binding protein of about 36 kD was found to be a major substrate for tyrosine phosphorylation.

Second messengers induced by oncoproteins or normal growth factors caused other modifications of cytoskeletal components. For instance, activation of protein kinase C leads to phosphorylation of serine residues of many substrates including several cytoskeletal proteins, such as myosin light chain, troponin, vinculin, and others. Phosphatidylinositol 4,5-biphosphate was found to dissociate the complexes of profilin with actin; under appropriate conditions this dissociation promoted polymerization of actin *in vitro*. Diacylglycerol was observed to promote the formation of complexes between α-actinin and actin. Numerous effects of increased Ca^{2+} concentration on the cytoskeleton have been discussed in many chapters of this book. Obviously, pH alteration caused by membrane activation can also affect the state of the cytoskeleton. The main problem at present is that we cannot choose between these many possibilities to determine which particular known or yet unknown protein modification leads to reorganization of the actin system in transformed cells.

Thus, studies of transformed cells suggest that the state of the actin cytoskeleton can be regulated by the integral level of signals generated at various sites of the membrane. These general regulatory mechanisms can modulate the reactions of the cytoskeleton to specific external stimuli; e.g., they can modulate the degree of spreading induced by contact with the substrate. Further studies of pathological overstimulation of these regulations in transformed cells can help to elucidate their normal mechanisms.

III. Possible Role of Cytoskeleton in the Regulation of Proliferation

A. Abnormal Regulation of Proliferation in Transformed Cells

Normal fibroblastic cells, attached to the substrate, proliferate only when appropriate growth factors are present in the medium. In the absence of growth factors these fibroblasts do not begin the new cell cycle after division, but are arrested for an indefinite time in the G0 or resting state.

Activation of proliferation by growth factors is a long, multistage process. First, growth factor molecules bound to the receptors of sensitive cells induce immediate membrane responses, described in Section II. Then the stage of prereplicative changes begins; this stage can take from 6 to 20 hr or more.

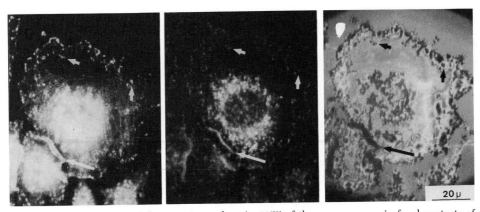

FIG. 9.4. Concentration of the protein product (pp60^{src}) of the oncogene *src* in focal contacts of rat cell (NRK line) transformed by the Rous sarcoma virus. Fluorescence microscopy of the same cell double-stained with antibodies to α-actinin (left) and to pp60^{src} (middle). (Right) The same field seen in the interference reflection microscope revealing focal contacts as dark spots and lines. Arrows point to several prominent contacts. (From Shriver and Rohrschneider, 1981. Courtesy of L. Rohrschneider.)

During this stage sequential activation of the synthesis of various types of RNAs and of many proteins and other metabolic changes takes place. The activated cell then enters the replication stage, which includes the phase of synthesis of DNA (S-phase), postsynthetic G2 phase, and mitosis. The requirement of transformed cells for growth factors is decreased in comparison with nontransformed parent cells. In the course of consecutive transformations, minimal concentrations of growth factors needed for activation of proliferation become lower and lower. Fully transformed cells can continuously proliferate in medium that does not contain any exogenous growth factors at all. Homology between oncoproteins and proteins of the normal pathway, discussed in Section II, explains the decreased requirement of the transformed cell for growth factors. Normal stimulation of proliferation is an end result of a sequence of changes initiated by membrane binding of growth factors. Endogenous stimulation of membrane-generated changes by oncoproteins can make the proliferation of transformed cells independent from exogenous stimuli.

B. Is Reorganization of Cytoskeleton Essential for Growth Activation by Oncoproteins?

Transformation by a single oncogene can simultaneously induce morphological alterations and activation of growth. It is possible that membrane-induced reorganization of the cytoskeleton in transformed cells is essential only for the development of morphological changes and is not related to growth activation. An alternative possibility is that cytoskeletal reorganization is an essential intermediate step in the multistage program of activation of proliferation (Fig. 9.5).

The data available at present are not sufficient to choose between these two alternatives. To prove the second hypothesis it would be necessary to find the genetic changes of cytoskeletal components leading to neoplastic transformation. In other words, it would be necessary to find oncoproteins that would be incorporated directly into the cytoskeleton and would alter both morphology and growth. Two pieces of evidence suggest that oncoproteins of this type do exist. In the first group of experiments it was found that point mutation of the β-actin gene is characteristic of a certain line of human fibroblasts transformed by chemical carcinogens (Fig. 9.6). A subclone of this line was then selected; this subclone had more severe morphological alterations, higher oncogenicity, and more altered growth regulation than the parent line. This clone was found to contain β-actin with a more altered structure than the mutated parent line actin. This clone had additional mutations of the β-actin gene that probably led to increased expression of transformed properties.

Another group of investigators deciphered the structure of the *fgr* oncogene and found that a part of its sequence is highly homologous to that of the γ-actin gene. The localization of *fgr* oncoprotein in cells transformed by the *fgr* oncogene is not yet clear. It seems probable that it can incorporate itself into the actin cortex and alter the properties of this cortex. A human oncogene called Onc-D or *trk* formed by the fusion of truncated tropomyosin gene and protein tyrosine kinase sequences has also been described.

Thus, the evidence indicating that certain oncogenes may directly alter the cytoskeleton seems to be growing.

C. Dependence of Proliferation of Normal Cells on Their Shape

The importance of the cytoskeleton in regulation of proliferation is also supported by other data obtained in experiments with normal cultured fibroblasts.

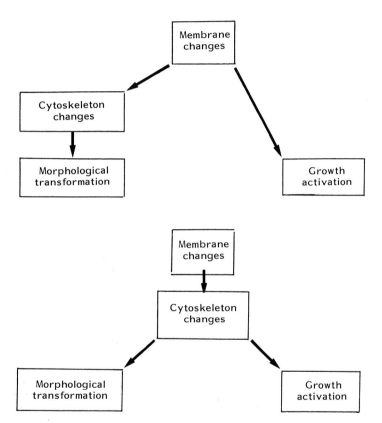

FIG. 9.5. Two possible types of interrelationships between cytoskeletal reorganization and growth activation in transformed cells.

	240							250		
Normal human cytoplasmic β-actin · · · · · · · · · · –	Glu –	Leu –	Pro –	Asp –	Gly	– Gln –	Val –	Ile –	Gly –	· · ·
β-Actin from transformed HuT-14 cells · · · –	Glu –	Leu –	Pro –	Asp –	Asp	– Gln –	Val –	Ile –	Gly –	· · ·

FIG. 9.6. Normal and mutant β-actins of human fibroblast have only one different amino acid residue; parts of the sequences with this residue are shown. (Based on Vandekerckhove *et al.*, 1980.)

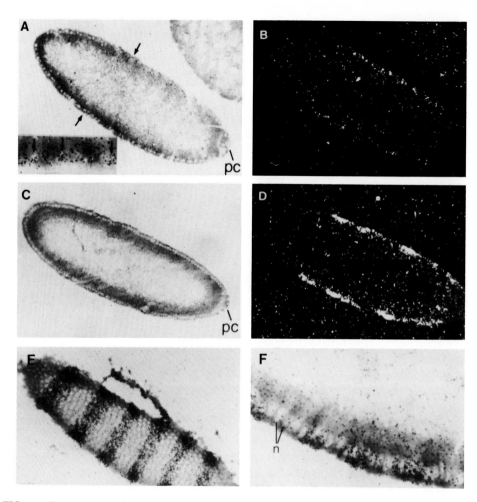

FIG. 9.7. Transcription of a gene is determined by the position of the nucleus within the cortical cytoplasm of the *Drosophila* embryo. After fertilization, the zygote nucleus undergoes 13 synchronous divisions without the formation of cell membranes. The first divisions take place in the central yolk, but after the seventh nuclear division, the majority of nuclei migrate to the periphery. Later the nuclei located in the cortex become enclosed by the cell membrane. At 3.5 hr of development the embryo consists of a single-layer epithelium, the cellular blastoderm. Hafen *et al.* (1984) studied the distribution of transcripts of the gene *ftz+* at these stages of development. This gene is involved in the control of segmentation of embryos; at late stages of development embryos homozygous for mutations of this gene have only half the number of segments of wild-type embryos. The recombinant plasmid containing the cloned DNA segment that is homologous to most of the coding region of the *ftz+* gene was isolated and labeled with tritium. The labeled plasmid was hybridized *in situ* to *ftz+* RNA contained in tissue sections of *Drosophila* embryos; the sections were then autoradiographed. Photographs show these sections in bright and dark-field illumination. Only the silver grains showing the distribution of label are seen in the dark-field photographs. The anterior end of the section always points to the left and the posterior end to the right.

(A,B) Longitudinal section of the embryo after the eleventh nuclear division. The peripheral portion of the section exhibits a higher accumulation of silver grains in the zone between 15% and 65% egg length. Inset is an enlargement of a portion of this zone. The arrows indicate nuclei that are labeled with the *ftz+* probe and therefore most likely to contain newly synthesized *ftz+* transcripts. pc, Polar cells.

It was found that proliferation is efficiently stimulated by growth factors only when the cells are spread on the substrate. In usual culture, cells suspended in fluid medium rapidly settle down on the bottom of the culture flask. However, when a substance forming a semisolid gel, e.g., agar or methylcellulose, is added to the medium, the cells remain suspended in this gel. These unspread cells do not start replication in medium containing standard amounts of growth factors. In control experiments, small pieces of glass were added to a semisolid medium with suspended cells; the cells, which managed to attach themselves to these pieces, started to proliferate and formed colonies.

When the cells are attached to the substrate, their rate of proliferation is proportional to the degree of their spreading. This was shown in experiments in which the cells were placed on a series of substrates with diminished adhesiveness. The rate of DNA synthesis in culture decreased proportionally with the decrease in average area occupied by the spread cell on the substrate. Spreading affects the synthesis of many types of macromolecules in sensitive cells. Detachment of normal spread cells from the substrate and their reversal to a round shape was observed not only to switch off the DNA synthesis and division, but also to decrease drastically the synthesis of various types of RNAs and of proteins.

Neoplastic transformation can make cell proliferation independent not only of growth factors, but also of spreading. In contrast to their normal progenitors, transformed cells suspended in agar or methylcellulose synthesize DNA, divide, and form colonies. This phenomenon is known as loss of anchorage dependence of growth. Detailed studies of Penman and collaborators (1982) have shown that the degree of loss of anchorage dependence of various macromolecular syntheses may be different in various transformed lines. For instance, suspension of 3T3 cells inhibited synthesis of both heterogeneous nuclear RNA and mRNA. Synthesis of both these groups of RNA became nonsensitive to suspension in the HDP3T6 line; however, DNA synthesis was inhibited by detachment from the substratum. Metabolism of the fully transformed and anchorage-independent SvPy3T3 line was totally unaffected by suspension.

The mechanisms of the effect of spreading on proliferation remain obscure. Possibly, reorganization of the cytoskeleton during spreading pro-

---

(C,D) Longitudinal section of the embryo after the twelfth nuclear division. Regional differences are noted in the signal intensity within the labeled domain.

(E) Three-hour embryo. Cellular blastoderm stage. Superficial section. Bands of labeled and unlabeled cells alternate in the blastoderm.

(F) Transverse section. Alternating groups of labeled and unlabeled cortical cells. n, Nuclei appearing white in the preparation. Scale bar: 0.1 mm.

Thus, the ftz+ gene is expressed in a segmental manner at the blastoderm stage. This pattern is gradually established at previous stages, when the peripheral nuclei are not yet separated from one another by the cell boundaries. It is suggested that some position-dependent regional differences of cortical cytoplasm surrounding the nuclei determine the degree of ftz+ gene expression in these nuclei; this cytoplasm contains numerous cytoskeletal elements. (From Hafen et al., 1984. Micrographs courtesy of W. J. Gehring.)

motes the induction of proliferation. This profound reorganization can facilitate additional reorganization induced by bound growth-factor molecules. In this way, spreading can increase cell sensitivity to the growth factors.

The important role of cytoskeletal reorganizations in regulation of proliferation is also suggested by experiments with cytoskeleton-specific drugs. These drugs modulate the proliferation of normal substrate-spread fibroblasts. In particular, colchicine and other microtubule-destroying drugs were observed to activate DNA synthesis in certain types of culture, e.g., in primary cultures of mouse embryo fibroblasts; however, this effect was not observed in all fibroblastic lines. Cytochalasin B and D reversibly inhibited proliferation in mouse fibroblastic cultures; these drugs were effective even at low doses, which did not drastically decrease cell spreading. An opposite effect, stimulation of cell response to growth factors by cytochalasin, was also reported. Further study of these opposite effects may further our understanding of the role of various components of the cytoskeleton in the regulation of proliferation.

IV. Conclusion: Possible Role of Cytoskeleton in the Reprogramming of Cellular Syntheses

Initiation of proliferation is an example of an integrated alteration of the program of cellular biochemical activities; it includes multistage changes of expression of many genes, of synthesis of many proteins, and of rates of almost all metabolic reactions. Analysis of the neoplastic transformations shows that initiation of proliferation is correlated with reorganization of the cytoskeleton. A similar correlation is indicated by the dependence of proliferation on spreading of the cell. These correlations suggest (but do not prove) that the cytoskeleton plays an essential role in regulation of synthetic activities of the cells. In the regulatory pathways, alterations of the cytoskeleton can be important intermediary stages between membrane changes induced by external factors and changes of gene activities. Reprogramming of cell syntheses occurs, of course, not only during the initiation of proliferation, but also during various differentiation processes. It can be suggested that cytoskeleton organization is not only altered during differentiation, but in some cases regulates the course of this differentiation. As described in Chapter 6, spreading, induced by cell–substrate contacts, can alter the spectrum of intermediate filament proteins synthesized by epitheliocytes. This phenomenon may be an example of regulation of cell-specific syntheses by cytoskeletal reorganization. Embryological experiments showed many years ago that the direction of differentiation of various cells formed during cleavage can be determined by the polar organization of the egg, in particular of the egg cortex (see Introduction). Possibly, regional differences in the organization of the egg cortex cytoskeleton somehow determine the patterns of gene expression in cells formed from various parts of the egg (Fig. 9.7). These intriguing phenomena are yet to be analyzed. How can the cytoskeleton control the expression of genes? Theoretically, there are many possible methods of this control. For instance, the cytoskeleton can regulate the generation of signals in the membrane by altering the distribution of membrane molecules. The cytoskeleton can also regulate the transmission of these signals from the membrane through the cytoplasm into the nucleus by altering the movements of cytoplasmic molecules. One cannot exclude the possibility that cytoskeletal fibrils attached to the nuclear membrane somehow regulate the organization of the nucleus. Additional hypotheses of this type can be suggested on the basis of data presented in Chapters 5–7. It remains to be determined which of these hypothetical regulatory mechanisms, if any, really functions in the cell.

To summarize, in this chapter we have described several enigmatic phenomena that suggest that the cytoskeleton has still unexplored functions in the cell. It is possible that the cytoskeleton controls not only the morphological organization of cell components, but also the organization of biochemical processes. It is possible that the cytoskeleton, in interaction with other structures, especially the plasma membrane and the nucleus, regulates reprogramming of cell synthesis during normal development and during pathological perversions of this development; that is, neoplastic transformations. Further work in this field may bring exciting discoveries.

Literature Cited

Bershadsky, A. D., Urbancova, H., and Vesely, P. (1982) Cytoskeleton of rat neoplastic cells of spontaneous origin and their derivatives after supertransformation with avian sarcoma virus B77 *in vitro*, *Folia Biol.* **28**:177–184.

Bershadsky, A. D., Tint, I. S., Neyfakh, A. A., Jr., and Vasiliev, J. M. (1985) Focal contacts of normal and RSV-transformed quail cells. Hypothesis of the transformation-induced deficient maturation of focal contacts, *Exp. Cell Res.* **158**:433–444.

Hafen, E., Kuroiwa, A., and Gehring, W. (1984) Spatial distribution of transcripts from the segmentation gene fushi tarazu during *Drosophila* embryonic development, *Cell* **37**:833–841.

Martin-Zanca, D., Hughes, S. H., and Barbacid, M. (1986) A human oncogene formed by the fusion of truncated tropomyosin and protein tyrosine kinase sequences, *Nature* **319**:743–748.

Matsumura, F., Lin, J. J-C., Yamashiro-Matsumura, S., Thomas, G. P., and Topp, W. C. (1983) Differential expression of tropomyosin forms in the microfilaments isolated from normal and transformed rat cultured cells, *J. Biol. Chem.* **258**:13954–13964.

Pasquale, E. B., Maher, P. A., and Singer, S. J. (1986) Talin is phosphorylated on tyrosine in chicken embryo fibroblasts transformed by Rous sarcoma virus, *Proc. Natl. Acad. Sci. USA* **83**:5507–5511.

Penman, S., Fulton, A., Capco, D., Ben Ze'ev, A., Wittelsberger, S., and Tse, C. F. (1982) Cytoplasmic and nuclear architecture in cells and tissue: Form, functions and mode of assembly, *Cold Spring Harbor Symp. Quant. Biol.* **46**(Part 2):1013–1027.

Shriver, K., and Rohrschneider, L. (1981) Organization of pp60[src] and selected cytoskeletal proteins within adhesion plaques and junctions of Rous sarcoma virus-transformed rat cells, *J. Cell Biol.* **89**:526–535.

Vandekerckhove, J., Leavitt, J., Kakunaga, T., and Weber, K. (1980) Coexpression of a mutant beta-actin and the two normal beta- and gamma-cytoplasmic actins in a stably transformed human cell line, *Cell* **22**:893–899.

Additional Readings

Berridge, M. J., and Irvine, R. F. (1984) Inositol triphosphate, a novel second messenger in cellular signal transduction, *Nature* **312**:315–321.

Boschek, C. B., Jockusch, B. M., Friis, R. R., Grundmann, E., and Bauer, H. (1981) Early changes in the distribution and organization of microfilament proteins during cell transformation, *Cell* **24**:175–184.

Burn, P., Rotman, A., Meyer, R. K., and Burger, M. M. (1985) Diacylglycerol in a large α-actinin/actin conplexes and in the cytoskeleton of activated platelets, *Nature* **314**:469–472.

Cooper, G. M., and Lane, M-A. (1984) Cellular transforming genes and oncogenesis, *Biochim. Biophys. Acta* **738**:9–20.

Crossin, K. L., and Carney, D. H. (1981) Evidence that microtubule depolymerization early in the cell cycle is sufficient to initiate DNA synthesis, *Cell* **23**:61–71.

Foe, V. E., and Alberts, B. M. (1983) Studies of nuclear and cytoplasmic behaviour during the five mitotic cycles that precede gastrulation in *Drosophila* embryogenesis, *J. Cell Sci.* **61**:31–70.

Folkman, J., and Moscona, A. (1978) Role of cell shape in growth control, *Nature* **273**:345–347.

Friedman, E., Verderame, M., Winawer, S., and Pollack, R., (1984) Actin cytoskeletal organization loss in the benign-to-malignant tumor transition in cultured human colonic epithelial cells, *Cancer Res.* **44**:3040–3050.

Herman, B., and Pledger, W. J. (1985) Platelet-derived growth factor-induced alterations in vinculin and actin distribution in BALB/c-3T3 cells, *J. Cell Biol.* **100**:1031–1041.

Iwashita, S., Kitamura, N., and Yoshida, M. (1983) Molecular events leading to fusiform morphological transformation by partial *src* deletion mutant of Rous sarcoma virus, *Virology* **125**:419–431.

Kakunaga, T., Leavitt, J., and Hamada, H. (1984) A mutation in actin associated with neoplastic transformation, *Fed. Proc.* **43**:2275–2279.

Lassing, I., and Lindberg, U. (1985) Specific interaction between phosphatidylinositol 4,5-biphosphate and profilactin, *Nature* **314**:472–474.

Leavitt, I., Gunning, P., Kedes, L., and Iariwalla, R. (1985) Smooth muscle α-actin is a transformation-sensitive marker for mouse NIH 3T3 and rat-2 cells, *Nature* **316**:840–842.

Lin, J.J.-C., Helfman, D. M., Hughes, S. H., and Chou, C. S. (1985) Tropomyosin isoforms in chicken embryo fibroblasts: Purification, characterization and changes in Rous sarcoma virus–transformed cells, *J. Cell Biol.* **100**:692–703.

Magargal, W. M., and Lin, S. (1986) Transformation-dependent increase in endogenous cytochalasin like activity in chicken embryo fibroblasts infected by Rous sarcoma virus, *Proc. Natl. Acad. Sci. USA* **83**:8201–8205.

Maness, P. F., and Levy, B. T. (1983) Highly purified pp60[src] induces the actin transformation in microinjected cells and phosphorylates selected cytoskeletal proteins *in vitro*, *Mol. Cell Biol.* **3**:102–112.

Naharro, G,. Robbins, K. C., and Reddy, E. P. (1984) Gene product of v-fgr onc: Hybrid protein containing a portion of actin and tyrosine-specific protein kinase, *Science* **223**:63–66.

Osborn, M., and Weber, K. (1977) The display of microtubules in transformed cells, *Cell* **12**:561–571.

Pegrum, S. M. (1982) Contact relationship between chick embryo cells growing in monolayer culture after infection with Rous sarcoma virus, *Exp. Cell Res.* **138**:147–157.

Pollack, R., Osborn, M., and Weber, K. (1975) Patterns of organization of actin and myosin in normal and transformed cultured cells, *Proc. Natl. Acad. Sci. USA* **72**:994–998.

Radke, K., Carter, V. C., Moss, P., Dehazya, P., Schliwa, M., and Martin, S. (1983) Membrane association of a 36000-dalton substrate for tyrosine phosphorylation in chicken embryo fibroblasts transformed by avian sarcoma viruses, *J. Cell Biol.* **97**:1601–1611.

Raz, A., and Geiger, B. (1982) Altered organization of cell–substrate contacts and membrane-associated cytoskeleton in tumor cell variants exhibiting different metastatic capabilities, *Cancer Res.* **42**:5180–5190.

Rohrschneider, L. R. (1979) Immuno-fluorescence on avian sarcoma virus transformed cells: Localization of the *src* gene product, *Cell* **16**:11–24.

Sefton, B. M., Hunter, T., Ball, E. H., and Singer, S. J. (1981) Vinculin: A cytoskeletal target of the transforming protein of Rous sarcoma virus, *Cell* **24**:165–174.

Steinberg, B. M., Smith, K., Colozzo, M., and Pollack, R. (1980) Establishment and transformation diminish the ability of fibroblasts to contract a native collagen gel, *J. Cell Biol.* **87**:304–308.

Takai, Y., Kaibuchi, K., Tsuda, T., and Hoshijima, M. (1985) Role of protein kinase C in transmembrane signaling, *J. Cell Biochem.* **29**:143–155.

Tucker, R. W., Pardee, A. B., and Fujiwara, K. (1979) Centriole ciliation is related to quiescence and DNA synthesis is 3T3 cells, *Cell* **17**:527–535.

Vasiliev, J. M. (1985) Spreading of non-transformed and transformed cells, *Biochim. Biophys. Acta* **780**:21–65.

Vasiliev, J. M., and Gelfand, I. M. (1981) *Neoplastic and Normal Cells in Culture*, Cambridge University Press, Cambridge.

Weinberg, R. A. (1985) The action of oncogenes in the nucleus and cytoplasm, *Science* **230**:770–776.

Index